Biomathematics

Volume 3

Edited by

K. Krickeberg · R. C. Lewontin · J. Neyman
M. Schreiber

M. Iosifescu · P. Tăutu

ACADEMY OF THE SOCIALIST REPUBLIC OF ROMANIA
Centre of Mathematical Statistics, Bucharest

Stochastic processes and applications in biology and medicine

I

THEORY

Editura Academiei
București

Springer-Verlag
Berlin · Heidelberg · New York

1973

This work is based on the book
"Procese stohastice şi aplicaţii în medicină şi biologie"

Editura Academiei, 1968
str. Gutenberg 3 bis, Bucureşti

AMS Subject Classifications (1970)
Primary 60J10, 60J15, 60J80, 60J70, 92A05, 92A10, 92A15
Secondary 60G05, 60J40, 60K05, 60K25, 92A25

ISBN-13: 978-3-642-80752-7 e-ISBN-13: 978-3-642-80750-3
DOI: 10.1007/ 978-3-642-80750-3

Softcover reprint of the hardcover 1st edition 1973

Preface

This volume is a revised and enlarged version of Chapters 1 and 2 of a book with the same title, published in Romanian in 1968. The revision resulted in a new book which has been divided into two parts because of the large amount of new material. The present part is intended to introduce mathematicians and biologists with a strong mathematical and probabilistic background to the study of stochastic processes.

We hope some readers will be able to discover by themselves the new features of our treatment such as the inclusion of some unusual topics, the special attention paid to some usual topics, and the grouping of the material.

We draw the reader's attention to the numbering, because there are structural differences between the two parts. In Part I there are Chapters, Sections, Subsections, Paragraphs and Subparagraphs. Thus the numbering a.b.c.d.e refers to Subparagraph e of Paragraph d of Subsection c of Section b of Chapter a. Definitions, theorems lemmas and propositions are numbered a.b.n, n = 1,2, ..., where a indicates the chapter and b the section. In Part II there are Sections, Subsections, Paragraphs, and Subparagraphs. Thus the numbering a.b.c.d refers to Subparagraph d of Paragraph c of Subsection b of Section a. Theorems and lemmas are numbered a.n, n = 1, 2,..., where a indicates the section. In the present part references to Part II are given by II. Thus II; a.b.c refers to Paragraph c of Subsection b of Section a of Part II.

We also wish to call the reader's attention to the bibliography: it exceeds the number of works quoted in the course of the volume in order not to overburden the text. Thus the bibliography should be consulted with the purpose of discovering historical sources, parallel researches and, in a general way, starting points for new investigations.

The authors desire to express their thanks to Dr. Klaus Peters of Springer-Verlag for his interest and patience in editing this new book, to Mrs. Iulia Bidianu, Mrs. Sorana Gorjan of the Publishing House of the Romanian Academy, and to Mr. K. Wickwire of Springer-Verlag for their careful checking of the language.

March 1972

Marius Iosifescu
Petre Tăutu

Contents

CHAPTER 1

Discrete parameter stochastic processes

CHAPTER 2

Continuous parameter stochastic processes

CHAPTER 1

DISCRETE PARAMETER STOCHASTIC PROCESSES

The object of the theory of stochastic (random) processes is the study of families of random variables defined on a given probability space. The simplest case is that in which the parameter of the family varies over the set of natural numbers, the negative integers or all the integers. In this case the family is called a *chain*, a *random sequence* or, more sophisticatedly, a *discrete parameter stochastic process*. The theory of discrete parameter stochastic processes, as opposed to that of continuous parameter stochastic processes, will be closely related to the classical chapters of probability theory.

The simplest example of a chain is any sequence of independent random variables. The study of sequences of independent random variables originally led to the consideration of chains with more complicated properties. If $(\xi_n)_{n \in N^*}$ is a sequence of independent real-valued random variables, the sequence of its partial sums

$$S_n = \xi_1 + \ldots + \xi_n, \quad n \in N^*,$$

is called a *chain with independent increments* (the random variables $S_1, S_2 - S_1, \ldots, S_{n+1} - S_n$ are independent for every $n \in N^*$). It follows immediately that every chain with independent increments is the sequence of partial sums of a sequence of independent random variables. We shall see that the notion "independent increments" takes on added emphasis in the case of continuous parameter stochastic processes.

The above sequence $(S_n)_{n \in N^*}$ enjoys another property, whose study, as will be seen in the sequel, has proved very fruitful. This is the *Markov property* [1)] which is expressed by the relation

$$\mathsf{P}(S_{n+1} < x \mid S_i, \ 1 \leqslant i \leqslant n) = \mathsf{P}(S_{n+1} < x \mid S_n)$$

for all $x \in R$, $n \in N^*$.

[1)] The independence of random variables ξ_n, $n \in N^*$, is essential for this property to hold.

To conclude these preliminary considerations, whose aim is to emphasize that the origins of the theory were sequences of independent random variables, we note that in analogy to sequences of independent identically distributed random variables, stationary chains have been studied. A chain $(\xi_n)_{n \in N^*}$ is said to be (strictly) *stationary* if the distribution of the random vector $(\xi_{n+h}, \ldots \xi_{n+m+h})$ does not depend on $h \in N$ for all $n \in N^*$, $m \in N$. We shall not consider stationary chains and stationary continuous parameter processes here since the most important problems of stationarity belong to functional analysis. A complete treatment of this aspect may be found in ROZANOV (1967). For purely probabilistic aspects of stationarity the reader is referred to CRAMÉR and LEADBETTER (1967).

1.1. Denumerable Markov chains

1.1.1. Preliminaries

The notion of a Markov chain was introduced for a special case by A. A. MARKOV in 1906. It proved to possess extraordinary fecundity in both theoretical developments and applications to the most varied fields. In the first period of development (toward the end of the forties) important research devoted to the finite case was carried out by B. HOSTINSKY, M. FRÉCHET, G. MIHOC, O. ONICESCU and V. I. ROMANOVSKI. The next period, devoted to the denumerable case, was inaugurated by A. N. KOLMOGOROV and continued by K. L. CHUNG, W. DOEBLIN, W. FELLER.

1.1.1.1. Definition of a Markov chain

1.1.1.1.1. Let I be at most a denumerable set. Consider a sequence of *stochastic* matrices ${}^n\mathbf{P} = ({}^n p_{ij})_{i,j \in I}$, $n \in N$, that is matrices whose elements satisfy

$$ {}^n p_{ij} \geqslant 0, \sum_{j \in I} {}^n p_{ij} = 1, \qquad i \in I. $$

Definition 1.1.1. An I-valued sequence $(\xi_n)_{n \in N}$ of random variables on the probability space $(\Omega, \mathcal{K}, \mathsf{P})$ is said to be a *(denumerable nonhomogeneous) Markov chain* with *state space I* and *transition matrices* ${}^n\mathbf{P}$, $n \in N$, if

$$ \mathsf{P}(\xi_{n+1} = i_{n+1} \mid \xi_\nu = i_\nu,\ 0 \leqslant \nu \leqslant n) = \mathsf{P}(\xi_{n+1} = i_{n+1} \mid \xi_n = i_n) = {}^n p_{i_n i_{n+1}} $$

for all $n \in N$, $i_\nu \in I$, $0 \leqslant \nu \leqslant n$, whenever the left member is defined.

Definition 1.1.2. A Markov chain for which the transition matrices ${}^n\mathbf{P}$, $n \in N$, do not depend on n, that is for which

$$ {}^n\mathbf{P} = \mathbf{P} = (p_{ij})_{i,\, j \in I}, $$

is said to be *homogeneous* (or to have *stationary transition probabilities*).

Denoting by $\mathcal{K}_n$ the σ-algebra generated by the random variables ξ_ν, $0 \leqslant \nu \leqslant n$, the equalities in Definition 1.1.1 may be written

$$ \mathsf{P}(\xi_{n+1} = i \mid \mathcal{K}_n) = \mathsf{P}(\xi_{n+1} = i \mid \xi_n) = {}^n p_{\xi_n i}, \quad \mathsf{P} - \text{a.s.} $$

for all $n \in N$, $i \in I$. [2]

It is also easily seen that the equalities

$$ \mathsf{P}(\xi_{n+1} = i_{n+1} \mid \xi_\nu = i_\nu,\ 0 \leqslant \nu \leqslant n) = \mathsf{P}(\xi_{n+1} = i_{n+1} \mid \xi_n = i_n), $$
$$ n \in N,\ i_\nu \in I,\ 0 \leqslant \nu \leqslant n, $$

are equivalent to the equalities

$$ \mathsf{P}(\xi_{t_{n+1}} = i_{n+1} \mid \xi_{t_\nu} = i_\nu,\ 0 \leqslant \nu \leqslant n) = \mathsf{F}(\xi_{t_{n+1}} = i_{n+1} \mid \xi_{t_n} = i_n), $$
$$ n \in N,\ t_\nu \in N,\ t_\nu < t_{\nu+1},\ i_\nu \in I,\ 0 \leqslant \nu \leqslant n, \tag{1.1.1} $$

which seem more general.

1.1.1.1.2. A very important consequence of (1.1.1) is the relation

$$ \mathsf{P}(\xi_{t_\lambda} = i_\lambda,\ n+1 \leqslant \lambda \leqslant n+m \mid \xi_{t_\nu} = i_\nu,\ 0 \leqslant \nu \leqslant n) = $$
$$ \mathsf{P}(\xi_{t_\lambda} = i_\lambda,\ n+1 \leqslant \lambda \leqslant n+m \mid \xi_{t_n} = i_n) $$

for all $n \in N$, $m \in N^*$, $t_0 < \ldots < t_{m+n} \in N$ and $i_\mu \in I$, $0 \leqslant \mu \leqslant m+n$. The proof by induction with respect to m is immediate.

Using the last relation we deduce

$$ \mathsf{P}(\xi_\nu = i_\nu,\ \xi_\lambda = i_\lambda,\ 0 \leqslant \nu \leqslant n,\ n+2 \leqslant \lambda \leqslant n+m+1 \mid \xi_{n+1} = i) = $$
$$ \mathsf{P}(\xi_\nu = i_\nu,\ 0 \leqslant \nu \leqslant n \mid \xi_{n+1} = i) \times $$
$$ \times \mathsf{P}(\xi_\lambda = i_\lambda,\ n+2 \leqslant \lambda \leqslant n+m+1 \mid \xi_{n+1} = i), $$

---

[2] The reader will be able to deduce that if $A \in \mathcal{K}_{n-1}$, $n \in N^*$, then

$$ \mathsf{P}(\xi_{n+1} = i \mid \xi_n = j,\ A) = \mathsf{P}(\xi_{n+1} = i \mid \xi_n = j). $$

that is, for Markov chains, the future and the past are conditionally independent, given the present. It is easy to see that this property is equivalent to the Markov property as stated in Definition 1.1.1.

Finally, let us note the *time-reversibility* of the Markov property, namely

$$\mathsf{P}(\xi_n = i \mid \xi_\lambda = i_\lambda,\ n+1 \leqslant \lambda \leqslant n+m) = \mathsf{P}(\xi_n = i \mid \xi_{n+1} = i_{n+1})$$

for all $n \in N$, $m \in N^*$, $i, i_\lambda \in I$, $n+1 \leqslant \lambda \leqslant n+m$. In other words, the chain $(\eta_h)_{h \in -N}$, $\eta_h = \xi_{-h}$ with reversed parameter-scale is again a Markov chain. The proof by induction with respect to m is again immediate.

1.1.1.2. The existence theorem

The existence of Markov chains with given transition matrices follows from

Theorem 1.1.3. *Given the set I and the sequence of stochastic matrices $({}^n\mathbf{P})_{n \in N}$, there exist a probability space $(\Omega, \mathcal{K}, \mathsf{P})$ and a Markov chain $(\xi_n)_{n \in N}$ defined on that probability space with state space I and transition matrices ${}^n\mathbf{P}$, $n \in N$.*

Proof. We shall take

$$(\Omega,\ \mathcal{K}) = (I^N,\ \{\mathcal{P}(I)\}^N),$$

that is, the space of sequences $(i_n)_{n \in N}$ of elements from I with the minimal σ-algebra containing all cylinders

$$C(i_0, \ldots, i_l) = \{i_0\} \times \ \ldots \ \times \{i_l\} \times I \times I \times I \times \ldots$$

with $i_\nu \in I$, $0 \leqslant \nu \leqslant l$, $l \in N$.

Let $\mathbf{p} = (p_i)_{i \in I}$, $p_i \geqslant 0$, $\sum_{i \in I} p_i = 1$ be an arbitrary probability distribution on I. If we define the measure P on the class of all cylinders by the equalities

$$\mathsf{P}\{C(i_0)\} = p_{i_0},$$

$$\mathsf{P}\{C(i_0, \ldots, i_l)\} = p_{i_0} \prod_{n=0}^{l-1} {}^n p_{i_n i_{n+1}}, \quad l \geqslant 1,$$

then following Ionescu Tulcea's theorem (see Loève (1963, p. 137)) this measure may be extended to the whole $\mathcal{K}$.

Now let us set

$$\xi_n(\omega) = i_n$$

if

$$\omega = (i_n)_{n \in N}.$$

We shall prove that the sequence $(\xi_n)_{n\in N}$ satisfies the theorem. If $\mathsf{P}(\xi_n = i_n) > 0$, then

$$\mathsf{P}(\xi_{n+1} = i_{n+1} \mid \xi_n = i_n) = \frac{\mathsf{P}(\xi_{n+1} = i_{n+1},\ \xi_n = i_n)}{\mathsf{P}(\xi_n = i_n)} =$$

$$\frac{\mathsf{P}\left\{\bigcup_{i_0,\ldots,i_{n-1}\in I} C(i_0, \ldots, i_{n-1}, i_n, i_{n+1})\right\}}{\mathsf{P}\left\{\bigcup_{i_0,\ldots,i_{n-1}\in I} C(i_0, \ldots, i_{n-1}, i_n)\right\}} =$$

$$= \frac{\sum_{i_0,\ldots,i_{n-1}\in I} p_{i_0} \prod_{t=0}^{n} {}^t p_{i_t i_{t+1}}}{\sum_{i_0,\ldots,i_{n-1}\in I} p_{i_0} \prod_{t=0}^{n-1} {}^t p_{i_t i_{t+1}}} = {}^n p_{i_n i_{n+1}}.$$

If $\mathsf{P}(\xi_\nu = i_\nu,\ 0 \leqslant \nu \leqslant n) > 0$, we have also

$$\mathsf{P}(\xi_{n+1} = i_{n+1} \mid \xi_\nu = i_\nu,\ 0 \leqslant \nu \leqslant n) =$$

$$\frac{\mathsf{P}\{C(i_0, \ldots, i_{n+1})\}}{\mathsf{P}\{C(i_0, \ldots, i_n)\}} = \frac{p_{i_0} \prod_{t=0}^{n} {}^t p_{i_t i_{t+1}}}{p_{i_0} \prod_{t=0}^{n-1} {}^t p_{i_t i_{t+1}}} = {}^n p_{i_n i_{n+1}}. \diamondsuit$$

It is obvious that

$$\mathsf{P}(\xi_0 = i) = p_i,\ i \in I.$$

The distribution $\mathbf{p}$ is called the *initial distribution* of the Markov chain. Occasionally, to emphasize the dependence on $\mathbf{p}$ of the probability P, we shall write $\mathsf{P}_{\mathbf{p}}$. In particular, if $\mathbf{p}$ is concentrated at $i \in I$, the corresponding $\mathsf{P}_{\mathbf{p}}$ will be denoted by P_i, and the corresponding Markov chain will be said to start in i.

1.1.1.3. The n-step transition probabilities

Let us set for $i, j \in I$, $m, n \in N$

$${}^m p_{ij}^{(n)} = \begin{cases} \delta_{ij} & \text{if } n = 0 \\ {}^m p_{ij} & \text{if } n = 1, \end{cases}$$

$${}^m p_{ij}^{(n)} = \sum_{k\in I} {}^m p_{ik}^{(n-1)}\, {}^{m+n-1} p_{kj} \quad \text{if } n > 1.$$

Then we may write

$$({}^m p_{ij}^{(0)})_{i,\,j\in I} = \mathbf{I} \text{ (the unit matrix)}$$

$$({}^m p_{ij}^{(n)})_{i.\,j\in I} = \prod_{s=m}^{m+n-1} {}^s\mathbf{P}, \quad n \geqslant 1. \tag{1.1.2}$$

It follows that the matrices $({}^m p_{ij}^{(n)})_{i.\,j\in I}$ are also stochastic matrices for all $m,\ n \in N$.

It is easily seen that

$${}^m p_{ij}^{(n)} = \mathsf{P}(\xi_{m+n} = j \mid \xi_m = i).$$

We call ${}^m p_{ij}^{(n)}$ the *n-step transition probability* from state i at time m to state j.

Using (1.1.2) we get

$${}^m p_{ij}^{(n+n')} = \sum_{k\in I} {}^m p_{ik}^{(n)}\, {}^{m+n} p_{kj}^{(n')}, \quad i, j \in I,\ m,\ n,\ n' \in N.$$

This is the famous *Chapman-Kolmogorov equation.*

In the homogeneous case the probabilities ${}^m p_{ij}^{(n)}$ do not depend on m and we shall denote them by $p_{ij}^{(n)}$. We have

$$(p_{ij}^{(n)})_{i,\,j\in I} = \mathbf{P}^n.$$

Since in this case

$$p_{ij}^{(n)} = \mathsf{P}(\xi_{m+n} = j \mid \xi_m = i)$$

for all $i, j \in I,\ m, n \in N,\ p_{ij}^{(n)}$ is called the n-step transition probability from state i to state j. The Chapman-Kolmogorov equation becomes

$$p_{ij}^{(n+n')} = \sum_{k\in I} p_{ik}^{(n)} p_{kj}^{(n')}, \quad i, j \in I, \quad n, n' \in N.$$

1.1.1.4. Strong Markov property

To close this section we shall mention a more powerful variant of the Markov property, the so-called *strong Markov property* in which all Markov chains join (differing from continuous parameter Markov processes where the strong Markov property is verified only by some classes of such processes). An N-valued random variable τ on $(\Omega, \mathcal{K}, \mathsf{P})$ is called a (finite) *Markov time* of $(\xi_n)_{n\in N}$ if $(\tau = n) \in \mathcal{K}_n$ for all $n \in N$. In other words, the event $(\tau = n)$ is defined by what happened up to and at time n inclusive for every $n \in N$. (We do not consider the general case in which τ is not necessarily finite to avoid some complications which can obscure the ideas). A typical example of a Markov time of $(\xi_n)_{n\in N}$ is the least non-negative entire number τ such that ξ_τ belong to a given set of states $A \subset I$. Let us consider the random variables $\eta_n = \xi_{\tau+n}$, $n \in N$. We shall prove that $(\eta_n)_{n\in N}$ is still a Markov chain. Set

$$A = (\eta_\nu = i_\nu,\ 0 \leqslant \nu \leqslant n)$$

$$A_m = A \cap (\tau = m).$$

We have $A_m \in \mathcal{K}_{m+n}$, $m, n \in N$, and then

$$\mathsf{P}\{A_m \cap (\eta_{n+1} = i_{n+1})\} = \mathsf{P}\{A_m \cap (\xi_{n+m+1} = i_{n+1})\} =$$

$$= \int_{A_m} {}^{n+m}p_{i_n i_{n+1}} \mathsf{P}(d\omega) = \int_{A_m} {}^{n+\tau}p_{i_n i_{n+1}} \mathsf{P}(d\omega).$$

Summing over $m \in N$ we get

$$\mathsf{P}(\eta_{n+1} = i_{n+1}, \eta_\nu = i_\nu,\ 0 \leqslant \nu \leqslant n) = \int_A {}^{n+\tau}p_{i_n i_{n+1}} \mathsf{P}(d\omega).$$

If we denote by $\mathcal{K}_n(\tau)$ the σ-algebra generated by the random variables η_ν, $0 \leqslant \nu \leqslant n$, it follows that

$$\mathsf{P}(\eta_{n+1} = i_{n+1} \mid \mathcal{K}_n(\tau)) = {}^{n+\tau}p_{\eta_n i_{n+1}}$$

P — almost surely. Replacing the set A by the set $B = (\eta_n = i_n)$, we deduce analogously

$$\mathsf{P}(\eta_{n+1} = i_{n+1} \mid \eta_n) = {}^{n+\tau}p_{\eta_n i_{n+1}}$$

P — almost surely. As we have already noted, these relations imply that $(\eta_n)_{n\in N}$ is a Markov chain. This is the strong Markov property. It is an obvious generalization of the classical Markov property (where $\tau \equiv 0$).

We note that in the homogeneous case, $(\eta_n)_{n\in N}$ is a homogeneous Markov chain with the same transition probabilities as those of $(\xi_n)_{n\in N}$.

1.1.2. Classification of states

In the following (up to Section 1.3) we will consider only homogeneous Markov chains. To use an intuitive language we shall say that the Markov chain $(\xi_n)_{n\in N}$ is in state i at the time n (or at the n-th step), to mean the event $(\xi_n = i)$. Similar expressions like starting, passing, reaching, entering, returning, avoiding, moving, etc., will be interpreted in the same spirit.

The states of a Markov chain may be classified according to whether or not the chain returns to the state in question or whether or not other states are reachable from a given state. We shall study both possibilities, as well as the relationship between the corresponding classifications.

1.1.2.1. Return states and recurrent states

1.1.2.1.1. Let us consider the probability that a Markov chain starting in state i reaches state j in exactly n steps, that is

$$\mathsf{P}(\xi_{m+n} = j, \xi_{m+\nu} \neq j, 0 < \nu < n \mid \xi_m = i).$$

This probability does not depend on m, being

$$p_{ij} \quad \text{if} \quad n = 1$$

and

$$\sum_{i_\nu \neq j (0<\nu<n)} p_{ii_1} p_{i_1 i_2} \cdots p_{i_{n-1} j} \quad \text{if} \quad n > 1.$$

Let us denote it by $f_{ij}^{(n)}$, and set

$$f_{ij}^* = \sum_{n \in N^*} f_{ij}^{(n)}.$$

Obviously, f_{ij}^* is the probability that the Markov chain, starting in state i, reaches state j at least once, that is

$$f_{ij}^* = \mathsf{P}\Big(\bigcup_{u \in N^*} (\xi_{m+n} = j) \mid \xi_m = i\Big)$$

for every $m \in N^*$.

Now, we can determine the probability g_{ij} that the chain, starting in state i, reaches state j infinitely often (i.o.). We have

$$g_{ij} = \mathsf{P}\Big(\bigcap_{s \in N^*} \bigcup_{n=s}^{\infty} (\xi_{m+n} = j) \mid \xi_m = i\Big) = \lim_{s \to \infty} \sum_{k \in I} p_{ik}^{(s)} f_{kj}^*, \qquad (1.1.3)$$

and we note that the sum $\sum_{k \in I} p_{ik}^{(s)} f_{kj}^*$ does not increase when s increases.

To obtain a more useful result put

$$g_{ij}(s) = \mathsf{P}(\xi_{m+n} = j \text{ for at least } s \text{ values of } n \in N^* \mid \xi_m = i).$$

We have [3]

$$g_{ij}(s+1) =$$

$$= \mathsf{P}(\xi_{r+m} = j \text{ for at least } s+1 \text{ values of } r \in N^* \mid \xi_m = i) =$$

$$= \sum_{n \in N^*} \mathsf{P}(\xi_\nu \neq j,\ m < \nu < m+n,\ \xi_{m+n} = j;\ \xi_{m+n+r} = j$$

$$\text{for at least } s \text{ values of } r \in N^* \mid \xi_m = i) =$$

$$= \sum_{n \in N^*} \mathsf{P}(\xi_\nu \neq i,\ m < \nu < m+n,\ \xi_{m+n} = j \mid \xi_m = i) \times$$

$$\times \mathsf{P}(\xi_{m+n+r} = j \text{ for at least } s \text{ values of } r \in N^* \mid \xi_{m+n} = j).$$

Thus

$$g_{ij}(s+1) = f_{ij}^* g_{jj}(s),$$

whence, on account of the fact that $g_{ij}(1) = f_{ij}^*$, we have

$$g_{ij}(s+1) = f_{ij}^*(f_{jj}^*)^s.$$

Since

$$g_{ij} = \lim_{s\to\infty} g_{ij}(s),$$

it follows that

$$g_{ij} = \begin{cases} 0 \text{ if } f_{jj}^* < 1 \\ f_{ij}^* \text{ if } f_{jj}^* = 1 \end{cases} \tag{1.1.4}$$

for all $i, j \in I$. Relation (1.1.4) implies

$$g_{ij} = f_{ij}^* g_{jj} \tag{1.1.5}$$

for all $i,\ j \in I$.

Definition 1.1.4. A state $i \in I$ is said to be *recurrent* or *nonrecurrent* [4] according as $f_{ii}^* = 1$ or $f_{ii}^* < 1$. It is said to be a *return* state or a *non-return* state according as $f_{ii}^* > 0$ or $f_{ii}^* = 0$.

---

[3] Here we use the so-called *method of first entrance* which amounts to the decomposition of an event into disjoint events associated with the time of the first entrance into state j.

[4] Sometimes for recurrent and nonrecurrent one uses the names *persistent* and *transient*.

Recurrent states are boundary cases of return states while nonreturn states are boundary cases of nonrecurrent states.

It follows from (1.1.3) that with probability one, the Markov chain returns to a recurrent state infinitely often, while with probability one it returns to a nonrecurrent state only a finite number of times.

1.1.2.1.2. Now we give criteria for return and recurrence.

Theorem 1.1.5. *A state $i \in I$ is a return state or a nonreturn state according as $p_{ii}^{(n)} > 0$ for at least an $n \in N^*$ or $p_{ii}^{(n)} = 0$ for all $n \in N^*$.*

Proof. The proof follows from the obvious inequalities

$$\sup_{n \in N^*} p_{ii}^{(n)} \leqslant f_{ii}^* \leqslant \sum_{n \in N^*} p_{ii}^{(n)}. \ \diamond$$

Theorem 1.1.6. *A state $i \in I$ is recurrent or nonrecurrent according as the series $\sum_{n \in N^*} p_{ii}^{(n)}$ diverges or converges.*

Proof. We have the relation

$$p_{jk}^{(n)} = \sum_{m=1}^{n} f_{jk}^{(m)} p_{kk}^{(n-m)}, \quad n \in N^*, \ j, \ k \in I, \tag{1.1.6}$$

whose verification is immediate by applying the method of first entrance into state k.

Summing up (1.1.6) over n, $1 \leqslant n \leqslant s$, we get

$$\sum_{n=1}^{s} p_{jk}^{(n)} = \sum_{n=1}^{s} \sum_{m=1}^{n} f_{jk}^{(m)} p_{kk}^{(n-m)} = \sum_{m=1}^{s} \left(f_{jk}^{(m)} \sum_{n=m}^{s} p_{kk}^{(n-m)} \right),$$

whence

$$\left(1 + \sum_{n=1}^{s} p_{kk}^{(n)}\right) \sum_{m=1}^{s} f_{jk}^{(m)} \geqslant \sum_{n=1}^{s} p_{jk}^{(n)} \geqslant \left(1 + \sum_{n=1}^{s-s'} p_{kk}^{(n)}\right) \sum_{m=1}^{s'} f_{jk}^{(m)} \tag{1.1.6'}$$

for all $s' < s$. Dividing by $1 + \sum_{n=1}^{s} p_{kk}^{(n)}$ and letting first $s \to \infty$ then $s' \to \infty$, we obtain the *Doeblin formula*

$$\sum_{m \in N^*} f_{jk}^{(m)} = \lim_{s \to \infty} \frac{\sum_{n=1}^{s} p_{jk}^{(n)}}{1 + \sum_{n=1}^{s} p_{kk}^{(n)}}. \tag{1.1.7}$$

In particular for $j = k = i$, we have

$$1 - f_{ii}^{*} = \lim_{s\to\infty} \frac{1}{1 + \sum_{n=1}^{s} p_{ii}^{(s)}} \tag{1.1.8}$$

what concludes the proof. ◇

Theorem 1.1.6 may be generalized as follows

Theorem 1.1.7. *The series*

$$\sum_{n\in N^*} p_{ij}^{(n)} \tag{1.1.9}$$

converges or diverges according as $g_{ij} = 0$ *or* $g_{ij} > 0$.

Proof. If $g_{ij} = 0$, relation (1.1.5) implies $f_{ij}^{*} g_{ij} = 0$, that is, either $f_{ij}^{*} = 0$ or $g_{jj} = 0$. For $f_{ij}^{*} = 0$, all terms of the series (1.1.9) are zero. For $g_{jj} = 0$, state j is nonrecurrent and thus according to Theorem 1.1.6 the series $\sum_{n\in N^*} p_{jj}^{(n)}$ converges; then (1.1.7) implies that the series (1.1.9) converges.

If $g_{ij} > 0$, relation (1.1.4) implies that $f_{ij}^{*} > 0$ and $g_{jj} > 0$. It follows that state j is recurrent and thus according to Theorem 1.1.6 the series $\sum_{n\in N^*} p_{jj}^{(n)}$ diverges; then relation (1.1.8) implies that the series (1.1.9) diverges. ◇

Corollary. *If the state j is nonrecurrent, then the series* (1.1.9) *converges. If the state j is recurrent, then the series* (1.1.9) *either diverges or all its terms are zero.*

Note that $\sum_{n\in N} p_{ij}^{(n)}$ represents the mean number of visits to state j starting in state i. The function G defined on $I \times I$ by $G(i,j) = \sum_{n\in N} p_{ij}^{(n)}$ is called the *Green function* and is related to the potential theory of Markov chains. For a comprehensive work on this subject see Kemeny, Snell, and Knapp (1966).

1.1.2.2. Regenerative phenomena

1.1.2.2.1. The successive entrances of a Markov chain into a given state constitute one of the most important examples of a so-called *regenerative phenomenon.*

Let $(\Omega, \mathcal{K}, \mathsf{P})$ be a probability space and $\mathcal{E} = (E_n)_{n\in N}$ a sequence of events (i.e. elements of the σ-algebra $\mathcal{K}$) on it.

Definition 1.1.8. [5] The sequence $\mathcal{E}$ is said to be a *regenerative phenomenon* iff $E_0 = \Omega$ and

$$\mathsf{P}\left(\bigcap_{r=1}^{s+1} E_{n_r}\right) = \mathsf{P}(E_{n_1}) \prod_{r=1}^{s} \mathsf{P}(E_{n_{r+1}-n_r}) \tag{1.1.10}$$

for all $s \in N^*$ and all increasing sequences $n_1 < n_2 < \ldots < n_{s+1}$.

If $(\xi_n)_{n \in N}$ is a Markov chain, then by the Markov property,

$$\mathsf{P}_i\left(\bigcap_{r=1}^{s+1} (\xi_{n_r} = i)\right) = p_{ii}^{(n_1)} \prod_{r=1}^{s} p_{ii}^{(n_{r+1}-n_r)} = \mathsf{P}_i(\xi_{n_1} = i) \prod_{r=1}^{s} \mathsf{P}_i(\xi_{n_{r+1}-n_r} = i)$$

for every state $i \in I$. Thus $((\xi_n = i))_{n \in N}$ is a regenerative phenomenon on $(\Omega, \mathcal{K}, \mathsf{P}_i)$.

A regenerative phenomenon $\mathcal{E} = (E_n)_{n \in N}$ is said to "occur" at just those times n for which the sample-point ω lies in the measurable subset E_n of the sample-space Ω. The requirement (1.1.10) can be interpreted by saying that the mechanism responsible for the occurrence of the phenomenon $\mathcal{E}$ suffers a complete loss of memory each time the phenomenon occurs.

Now write $\mathcal{K}_{(m,\, n)}$ for the σ-algebra generated by the sets E_s, $m < s < n$. Relation (1.1.10) appears to be rather weak, since it only involves intersections of sets E_s, and not the more general $\mathcal{K}_{(0,\, \infty)}$-sets. But it is easy to verify that for any $s \in N^*$ and any sets $A \in \mathcal{K}_{(0,\, s)}$, $B \in \mathcal{K}_{(s,\, \infty)}$ we have

$$\mathsf{P}(AE_sB)\, \mathsf{P}(E_s) = \mathsf{P}(AE_s)\, \mathsf{P}(E_sB) \tag{1.1.11}$$

Let us set

$$p_n = \mathsf{P}(E_n), \quad n \in N,$$

$$f_1 = \mathsf{P}(E_{n+1} \mid E_n),$$

$$f_r = \mathsf{P}(E_{n+r}E_{n+r-1}^c \cdots E_{n+1}^c \mid E_n), \quad r \geqslant 2,$$

(the independence of $n \in N$ of the f_r follows easily from (1.1.10)). Thus $(f_r)_{r \in N^*}$ may be thought of as the *interarrival time distribution* of $\mathcal{E}$.[6]

---

[5] This definition due to Kingman (1963) is a formalization of the idea of a *recurrent event* in W. Feller's sense. Following Kendall (1967) we prefer to speak of a regenerative phenomenon rather than of a recurrent event. The adjective "recurrent" has acquired a conventional meaning in the theory of Markov chains and to speak of $\mathcal{E}$ as if it were an element of $\mathcal{K}$ could be confusing.

[6] If $\mathcal{E}$ occurs at the instants $\tau_0 = 0$, $\tau_0 + \tau_1$, $\tau_0 + \tau_1 + \tau_2$, $\cdots$, then $\tau_1, \tau_2, \cdots$ are identically distributed random variables with $\mathsf{P}(\tau_1 = r) = f_r$, $r \in N^*$. Moreover, it is easy to see that the τ_n are independent.

The greatest common divisor of the natural numbers r for which $f_r > 0$ will be called the *period* of $\mathcal{E}$.

We have $p_0 = 1$, $0 \leqslant p_n \leqslant 1$, $f_r \geqslant 0$, $\sum_{r \in N^*} f_r \leqslant 1$, and, by making use of (1.1.11),

$$p_n = \sum_{r=1}^{n} f_r p_{n-r}, \qquad n \in N^*. \tag{1.1.12}$$

Conversely, if the sequence $(f_r)_{r \in N^*}$ is given and satisfies the conditions $f_r \geqslant 0$, $\sum_{r \in N^*} f_r \leqslant 1$ and if $(p_n)_{n \in N}$ is defined by (1.1.12) and by $p_0 = 1$ then we can construct a regenerative phenomenon in such a way that for it the f_r and the p_n have the meanings given above. Moreover, this regenerative phenomenon may be chosen to be the successive entrances into a given state of a Markov chain. The reader will be able to verify that for the Markov chain with state space $\{0, 1, \ldots, l\}$ where l is the smallest integer such that $\sum_{r=1}^{l} f_r = 1$ or ∞ and with the transition probabilities

$$p_{ij} = \begin{cases} f_{i+1} \left(1 - \sum_{r=1}^{i} f_r\right)^{-1} & \text{if } j = 0 \\ \left(1 - \sum_{r=1}^{i+1} f_r\right)\left(1 - \sum_{r=1}^{i} f_r\right)^{-1} & \text{if } j = i+1 \\ 0 \quad \text{otherwise} & \end{cases}$$

we have $f_{00}^{(r)} = f_r$, $p_{00}^{(n)} = p_n$.

The recurrence relation (1.1.12) is, of course, equivalent to the power series identity

$$P(z) = \frac{1}{1 - F(z)}, \quad |z| < 1, \tag{1.1.13}$$

where

$$P(z) = \sum_{n \in N} p_n z^n, \quad F(z) = \sum_{r \in N^*} f_r z^r,$$

and it is obviously a particular instance of the renewal equation (2.3.81). This is the reason for which sequences $(p_n)_{n \in N}$ associated with regenerative phenomena are sometimes called *renewal sequences* (see e.g. KENDALL (1967)).

1.1.2.2.2. The asymptotic behaviour of p_n as $n \to \infty$ is clarified by

Theorem 1.1.9. *Let the period of $\mathcal{E}$ be d. Then, we have*

$$\lim_{n\to\infty} p_{nd} = \frac{d}{\sum_{r\in N^*} rf_r + \infty\left(1 - \sum_{r\in N^*} f_r\right)}$$ [7)]

and $p_n = 0$ if $n \not\equiv 0$ (mod d).

Proof. The fact that $p_n = 0$ if $n \not\equiv 0$(mod 0) is a consequence of the identity (1.1.13). Note also that if $\sum_{r\in N^*} f_r < 1$, then the same identity implies the convergence of the series $\sum_{n\in N} p_n$, i.e. $\lim_{n\to\infty} p_{nd} = 0$ in accordance with the claim. Suppose that $\sum_{r\in N^*} f_r = 1$. Following ERDÖS, FELLER and POLLARD (1949) let us set

$$a_n = \sum_{r=n+1}^{\infty} f_r, \quad n \in N.$$

Then $a_0 = 1$ and

$$\sum_{n\in N} a_n = \sum_{r\in N^*} rf_r$$

which is verified by summation by parts. Further, $f_r = a_{r-1} - a_r$, consequently (1.1.12) yields

$$p_n = \sum_{r=1}^{n} (a_{r-1} - a_r)\, p_{n-r}$$

or

$$a_0 p_n + a_1 p_{n-1} + \dots + a_n p_0 = a_0 p_{n-1} + a_1 p_{n-2} + \dots + a_{n-1} p_0.$$

This means that the sums do not depend on the value of n; since $a_0 p_0 = 1$ we have

$$\sum_{m=0}^{n} a_m p_{n-m} = 1, \quad n \in N. \tag{1.1.14}$$

---

7) The symbol ∞ will be subjected to the rules; $1/\infty = 0$, $\infty + c = \infty$ and $\infty \times c = \infty$ or 0 according as $c > 0$ or $c = 0$.

Set $\lambda = \overline{\lim_{n\to\infty}}\, p_{nd}$ and let $\{n_s\} \subset N^*$ be a sequence for which $\lim_{s\to\infty} p_{n_s d} = \lambda$. If $r \in N^*$ is such that $f_r > 0$ (implying d divides r), we have

$$\lambda = \lim_{s\to\infty} p_{n_s d} = \lim_{s\to\infty} \left(f_r\, p_{n_s d - r} + \sum_{\substack{m=1\\ m\neq r}}^{n_s d} f_m p_{n_s d - m} \right)$$

$$\leqslant f_r \underline{\lim}_{s\to\infty} p_{n_s d - r} + \sum_{\substack{m=1\\ m\neq r}}^{\infty} f_m \overline{\lim_{s\to\infty}}\, p_{n_s d - m}$$

$$\leqslant f_r \underline{\lim}_{s\to\infty} p_{n_s d - r} + (1 - f_r)\, \lambda.$$

Thus $\underline{\lim}_{s\to\infty} p_{n_s d - r} \geqslant \lambda$. Consequently, by the very definition of λ, we have $\lim_{s\to\infty} p_{n_s d - r} = \lambda$. This result is true for all $r \in N^*$ for which $f_r > 0$ and every sequence $\{n_s\} \subset N^*$ for which $\lim_{s\to\infty} p_{n_s d} = \lambda$. By applying it a finite number of times we conclude that $\lim_{s\to\infty} p_{n_s d - t} = \lambda$ for any t of the form $\sum_{j=1}^{l} c_j r_j, l, c_j, r_j \in N^*$, with $f_{r_j} > 0, 1 \leqslant j \leqslant l$. Let us choose l and the r_j such that their greatest common divisor equals d. Elementary number theory implies the existence of natural numbers c_j, $1 \leqslant j \leqslant l$, such that $rd = \sum_{j=1}^{l} c_j r_j$ for sufficiently large $r (r \geqslant r_0)$. This allows us to assert that

$$\lim_{s\to\infty} p_{(n_s - r)d} = \lambda, \quad r \geqslant r_0.$$

Putting $n = (n_s - r_0)\, d$ in (1.1.14) and taking into account the fact that $p_n = 0$ for $n \not\equiv 0 \pmod{d}$, we get

$$\sum_{m=0}^{n_s - r} a_{md}\, p_{(n_s - r_0 - m) d} = 1.$$

Hence, letting $s \to \infty$, we deduce

$$\lambda \sum_{m \in N} a_{md} = 1$$

provided that the series converges; otherwise λ must vanish so that in either case

$$\lambda = \frac{1}{\sum_{m \in N} a_{md}}.$$

But, since $f_r = 0$ for $r \not\equiv 0 \pmod{d}$, we have

$$a_{md} = \frac{1}{d} \sum_{n=md}^{md+d-1} a_n,$$

so that

$$\lambda = \frac{d}{\sum_{m \in N} a_m} = \frac{d}{\sum_{r \in N^*} r f_r}.$$

Reasoning as in case of $\overline{\lim}$ we obtain the same value for $\underline{\lim}$.

Theorem 1.1.9 is therefore proved. ◇

For further research concerning the asymptotic behaviour of the ratio p_{n+1}/p_n when $\lim_{n \to \infty} p_n = 0$ see GARSIA (1963) and GARSIA, OREY, and RODEMICH (1962).

1.1.2.2.3. Now, following FELLER (1966 a), we shall deduce a representation of the probabilities $p_n = \mathsf{P}(E_n)$ associated with a regenerative phenomenon $\mathcal{E} = (E_n)_{n \in N}$ as Fourier coefficients of a probability distribution on the unit circle. We use a method devised by CHUNG and FUCHS (1951).

Without any loss of generality we may suppose $\mathcal{E}$ to be aperiodic (that is of period one). The generating function $P(z) = (1 - F(z))^{-1}$ is continuous except, perhaps, at $z = 1$ in the closed unit disk $|z| \leqslant 1$. Letting $F = a + ib$ one has for $|z| < 1$

$$2\mathrm{Re}\, P(z) = \frac{2(1-a)}{(1-a)^2 + b^2} = 1 + \frac{1 - a^2 - b^2}{(1-a)^2 + b^2} > 1 \quad (1.1.15)$$

Writing $z = \rho\, \mathrm{e}^{\mathrm{i}\theta}$ it follows for fixed $\rho < 1$ and $n \in N$,

$$\rho^n p_n = \frac{1}{2\pi} \int_{-\pi}^{\pi} \mathrm{e}^{-\mathrm{i}n\theta}\, P(\rho \mathrm{e}^{\mathrm{i}\theta})\, d\theta,$$

whereas for $n < 0$ the integral equals 0. Adding these relations for $\pm n$ one gets

$$\rho^n p_n = \frac{1}{2\pi} \int_{-\pi}^{\pi} \mathrm{e}^{-\mathrm{i}n\theta}\, [2\mathrm{Re}\, P(\rho \mathrm{e}^{\mathrm{i}\theta}) - 1]\, d\theta$$

for $n \in N$. On account of (1.1.15), the expression within brackets is positive. For $n = 0$ we have

$$1 = \frac{1}{2\pi} \int_{-\pi}^{\pi} [2\mathrm{Re}\, P(\rho e^{i\theta}) - 1]\, d\theta.$$

It follows that $\rho^n p_n$, $n \in N$, are the Fourier coefficients of a probability distribution μ_ρ on the unit circle. Their convergence (as $\rho \to 1$) implies the existence of a probability distribution μ_1 which is the weak limit of μ_ρ as $\rho \to 1$. Outside a neighborhood of $\theta = 0$ the density of μ_ρ converges boundedly to a limit, which is therefore the density of μ_1 — except for a possible atom of weight say α at the origin. Thus

$$p_n = \alpha + \frac{1}{2\pi} \int_{-\pi}^{\pi} e^{in\theta} \frac{1 - a^2 (e^{i\theta}) - b^2 (e^{i\theta})}{[1 - a (e^{i\theta})]^2 + b^2 (e^{i\theta})}\, d\theta \qquad (1.1.16)$$

By the Riemann-Lebesgue theorem (see e.g. HARDY and ROGOSINSKI (1956, p. 23)) $p_n \to \alpha$ as $n \to \infty$. The value of α is given by Theorem 1.1.9.

1.1.2.2.4. If in Definition 1.1.8 we require that $E_0 = \emptyset$ instead of $E_0 = \Omega$ we have the definition of a so-called *delayed regenerative phenomenon*. In this case the distribution of the first occurrence of $\mathcal{E}$ differs from the interarrival time distribution. A typical example of a delayed regenerative phenomenon on $(\Omega, \mathcal{K}, \mathsf{P}_i)$ is $\mathcal{E} = ((\xi_n = j))_{n \in N}$, $i \neq j$ where $(\xi_n)_{n \in N}$ is a Markov chain. For details and applications the reader is referred to FELLER (1968, p. 316).

1.1.2.3. Positive states and limit theorems

1.1.2.3.1. Turning again to Markov chains we shall introduce two types of states characterized by the expected return time. Denote by τ_{ij}, $i, j \in I$, a discrete random variable taking on the values $m \in N^*$ and ∞ with the probabilities $f_{ij}^{(m)}$ and $1 - f_{ij}^*$. The variable τ_{ij} may be interpreted as the *(first) passage time* (in terms of the number of steps required) from state i to state j. The mean value of τ_{ij}

$$\sum_{m \in N^*} m f_{ij}^{(m)} + \infty \times (1 - f_{ij}^*)$$

will be called the *expected passage time* from state i to state j (the *expected return time* to state i if $j = i$). We denote this mean by $1/\nu_{ij}$ so that ν_{ij} may be considered as the *mean frequency of passage* from state i to state j (the *mean frequency of return* to state i if $j = i$).

Definition 1.1.10. A state $i \in I$ is said to be *positive* or *null* according as $\nu_{ii} > 0$ or $\nu_{ii} = 0$.

It is obvious that a nonrecurrent state is null while a positive one is recurrent.

1.1.2.3.2. In order to establish a criterion for this new dichotomy we introduce the concept of period of a state.

Definition 1.1.11. The *period* d_i of the return state $i \in I$ is the greatest common divisor of all natural numbers n for which $p_{ii}^{(n)} > 0$. The period of a nonreturn state is ∞. A state $i \in I$ is said to be *periodic* or *aperiodic* according as $d_i > 1$ or $d_i = 1$.

One can prove that, whatever the return state i, after a certain lapse of time, the Markov chain returns to i with positive probability after every interval of time d_i. We have indeed

Proposition 1.1.12. *If $i \in I$ is a return state, then $p_{ii}^{(md_i)} > 0$ for all sufficiently large $m \in N^*$.*

Proof. Let n_j, $1 \leqslant j \leqslant l$, be positive integers such that $p_{ii}^{(n_j)} > 0$ and whose greatest common divisor equals d_i. By the elementary number theory, for m sufficiently large, md_i can be written as $\sum_{j=1}^{l} c_j n_j$ for suitable positive integers c_j. Thus

$$p_{ii}^{(md_i)} \geqslant \prod_{j=1}^{l} p_{ii}^{(c_j n_j)} > 0. \ \diamond$$

The proposition below shows that in Definition 1.1.11 we may replace $p_{ii}^{(n)}$ by $f_{ii}^{(n)}$.

Proposition 1.1.13. *If $i \in I$ is a return state, then the greatest common divisor of all numbers n for which $f_{ii}^{(n)} > 0$ coincides with d_i.*

Proof. Consider the generating functions

$$P_{ii}(z) = \sum_{n \in N} p_{ii}^{(n)} z^n, \qquad F_{ii}(z) = \sum_{n \in N^*} f_{ii}^{(n)} z^n, \qquad |z| < 1.$$

Relation (1.1.6) yields (for $j = k = i$)

$$P_{ii}(z) = 1 + F_{ii}(z)\, P_{ii}(z)$$

whence

$$P_{ii}(z) = \frac{1}{1 - F_{ii}(z)}, \quad F_{ii}(z) = 1 - \frac{1}{P_{ii}(z)} \tag{1.1.17}$$

These equations lead immediately to the desired result. $\diamond$

1.1.2.3.3. The following theorem is the basic limit theorem of Markov chains.

Theorem 1.1.14 (A. N. KOLMOGOROV). *If $i \in I$ is null, then*

$$\lim_{n\to\infty} p_{ii}^{(n)} = 0;$$

if $i \in I$ is positive, then

$$\lim_{n\to\infty} p_{ii}^{(nd_i)} = d_i \nu_{ii}.$$

In either case

$$p_{ii}^{(n)} = 0, \quad n \not\equiv 0 \pmod{d_i}.$$

Proof. As we have already noted, $\mathcal{E} = ((\xi_n = i))_{n\in N}$ is a regenerative phenomenon on $(\Omega, \mathcal{K}, \mathsf{P}_i)$ for which $p_n = p_{ii}^{(n)}$, $f_r = f_{ii}^{(r)}$. We need only apply Theorem 1.1.9 taking into account Proposition 1.1.13. ◇

Corollary. *A state $i \in I$ is null or positive according as $\overline{\lim_{n\to\infty}} p_{ii}^{(n)} = 0$ or > 0.*

By virtue of Theorem 1.1.14, the aperiodic positive states are sometimes called *ergodic*.

1.1.2.3.4. To obtain the limiting behaviour of "non-diagonal" probabilities $p_{ij}^{(n)}$ we need the quantities

$$f_{ij}^*(r) = \sum_{m\in N} f_{ij}^{(md_j+r)}, \quad 1 \leqslant r \leqslant d_j.$$

We have

$$f_{ij}^*(r) = \mathsf{P}_i(\xi_n = j \text{ for some } n \equiv r \pmod{d_j}))$$

and

$$\sum_{r=1}^{d_j} f_{ij}^*(r) = f_{ij}^*, \qquad i, j \in I.$$

Theorem 1.1.15. *If $j \in I$ is a null state, then*

$$\lim_{n\to\infty} p_{ij}^{(n)} = 0$$

for any $i \in I$; if $j \in I$ is a positive state, then

$$\lim_{n\to\infty} p_{ij}^{(nd_j+r)} = f_{ij}^*(r)\, d_j \nu_{jj}, \quad 1 \leqslant r \leqslant d_j,\ i \in I.$$

Proof. If $j \in I$ is a null state, then on account of Theorem 1.1.14,

$$\lim_{n\to\infty} p_{jj}^{(n)} = 0.$$

By making use of (1.1.6) we get

$$p_{ij}^{(n)} \leqslant \sum_{m=1}^{n'} f_{ij}^{(m)} p_{jj}^{(n-m)} + \sum_{m=n'+1}^{n} f_{ij}^{(m)}, \quad i \in I.$$

for $n > n'$. Letting first $n \to \infty$ then $n' \to \infty$ yields

$$\lim_{n\to\infty} p_{ij}^{(n)} = 0, \quad i \in I.$$

If $j \in I$ is a positive state, then Theorem 1.1.14 implies that

$$\lim_{n\to\infty} p_{jj}^{(nd_j)} = d_j \nu_{jj},$$

$$p_{jj}^{(n)} = 0, \ n \not\equiv 0 \ (\mathrm{mod}\ d_j).$$

Using again (1.1.6) we obtain

$$0 \leqslant p_{ij}^{(nd_j+r)} - \sum_{m=0}^{n'} f_{ij}^{(md_j+r)} p_{jj}^{([n-m]d_j)} \leqslant \sum_{m=n'+1}^{n} f_{ij}^{(md_j+r)}$$

for $n > n'$. Letting first $n \to \infty$ then $n' \to \infty$ yields

$$\lim_{n\to\infty} p_{ij}^{(nd_j+r)} = f_{ij}^{*}(r)\, d_j \nu_{jj}, \quad i \in I. \ \diamond$$

Corollary. *The Cesaro limit*

$$\lim_{n\to\infty} \frac{1}{n} \sum_{m=1}^{n} p_{ij}^{(m)} = \pi_{ij}$$

exists and equals $f_{ij}^{*} \nu_{jj}$, $i, j \in I$.

The matrix $\mathbf{\Pi} = (\pi_{ij})_{i,j\in I}$ is *substochastic*, that is

$$0 \leqslant \pi_{ij} \leqslant 1, \quad \sum_{j\in I} \pi_{ij} \leqslant 1, \quad i, j \in I.$$

It is not difficult to prove the equalities

$$\mathbf{\Pi P} = \mathbf{P\Pi} = \mathbf{\Pi}^2 = \mathbf{\Pi}.$$

As concerns the limits π_{ij} (sometimes called *final probabilities*), there are three possibilities, namely:

i) $\pi_{ij} = 0$ for all $i, j \in I$;

ii) $\pi_{ij} > 0$ for some pair $i, j \in I$, but $\sum_{j \in I} \pi_{kj} < 1$ for some $k \in I$;

iii) $\sum_{j \in I} \pi_{ij} = 1$ for all $i \in I$.

A Markov chain is said to be *dissipative*, *semi-dissipative* or *non-dissipative* according as i), ii) or iii) holds.

The reader will be able to prove that the equality $\sum_{j \in I} \pi_{ij} = 1$ is equivalent to the relation

$$\lim_{n\to\infty} \sum_{j \in I} \left| \frac{1}{n} \sum_{m=1}^{n} p_{ij}^{(m)} - \pi_{ij} \right| = 0.$$

For various investigations concerning the π_{ij} see FOSTER (1951, 1952), HOLMES (1967), KENDALL (1951 a), KENDALL and REUTER (1957), VERE-JONES and KENDALL (1959).

1.1.2.4. Geometric ergodicity

A very important problem is that of determining the rate of convergence of the probabilities $p_{ij}^{(n)}$ to their limits. (For the sake of simplicity we shall consider only aperiodic states). Here we study the case of a geometric convergence. For an investigation of another rate of convergence see KARLIN (1955). The definition below was suggested by a property enjoyed by Markov chains with finite state space (Theorem 1.1.48).

Definition 1.1.16. An aperiodic state $i \in I$ is said to be *geometrically ergodic* iff there exists non-negative numbers $\rho_i < 1$ and A_i such that

$$|p_{ii}^{(n)} - \pi_{ii}| \leqslant A_i \, \rho_i^n$$

for all $n \in N$.

The concept of geometric ergodicity was introduced by KENDALL (1959) who proved

Theorem 1.1.17. *A state $i \in I$ is geometrically ergodic iff the series*

$$\sum_{n \in N^*} f_{ii}^{(n)} z^n$$

has a radius of convergence greater than unity.

Proof. In what follows $F_{ii}(z)$ and $P_{ii}(z)$ have the same meanings as in the proof of Proposition 1.1.13.

a) Necessity. If the state i is geometrically ergodic, then obviously the function $P_{ii}(z)$ can be continued analytically throughout some disk $|z| < a$ (where $a > 1$) save perhaps for a simple pole at $z = 1$, and on account of (1.1.17)

$$\operatorname{Re} P_{ii}(z) \geqslant \frac{1}{2} \text{ for } |z| < 1,$$

so that the function $P_{ii}(z)$ can have no zeros inside or on the perimeter of the unit disk, and a larger concentric disk inside which the function

$$1 - \frac{1}{P_{ii}(z)}$$

is analytic will actually exist. On account of (1.1.17) this last function coincides with $F_{ii}(z)$ near $z = 0$, and this completes the proof of necessity.

b) Sufficiency. Let $F_{ii}(z)$ have a radius of convergence greater than unity. We have $|F_{ii}(z)| < 1$ when $|z| < 1$ and $F_{ii}(z) = 1$ on the perimeter of the unit disk only at $z = 1$ (and not even if i is a nonrecurrent state), so that there will be a disk $|z| < b$ (where $b > 1$) throughout which the series $F_{ii}(z)$ converges and defines an analytic function which assumes the value 1 at the point $z = 1$ at most. It is clear that if $1 - F_{ii}(z)$ does have a zero at $z = 1$ then this zero must be simple, for $F'_{ii}(1) = \sum_{n \in N^*} n f_{ii}^{(n)} > 0$ in view of the aperiodicity of i (so it cannot be a nonreturn state). Relation (1.1.17), which holds for $|z| < 1$, now shows that the function $P_{ii}(z)$ can be continued analytically throughout a disk $|z| < c$ (where $c > 1$) save perhaps for a simple pole at $z = 1$. It follows that $p_{ii}^{(n)}$ can be expressed as the sum of a constant term (which will be absent if there is no pole) and a remainder which tends geometrically to zero when n tends to infinity. [8] ◇

Corollary. *A geometrically ergodic state $i \in I$ must be either nonrecurrent or positive; it cannot be null-recurrent.*

Proof. If i is geometrically ergodic, then by Theorem 1.1.17 $F_{ii}(z)$ can be continued analytically throughout a disk $|z| < a$ (where $a > 1$), and so the series $F'_{ii}(1) = \sum_{n \in N^*} n f_{ii}^{(n)}$ will be convergent and the state i must be positive if it is recurrent. ◇

---

[8] A careful scrutiny of the arguments in the above proof leads to the conclusion that a state $i \in I$ is geometrically ergodic iff the function

$$(1 - z) P_{ii}(z) = \frac{1 - z}{1 - F_{ii}(z)}$$

is holomorphic in an open disk containing the unit disk.

1.1.2.5. Essential states and classes of states

1.1.2.5.1. The types of states already introduced have been described in terms of returning. Now we shall consider two types of states whose definition is suggested by the motion from one state to another.

We say that state i *leads to* state j (or state j *can be reached from* state i) and write $i \to j$ iff there is an $n(i,j) = n \in N^*$ such that $p_{ij}^{(n)} > 0$. We say that states i and j *communicate* and write $i \leftrightarrow j$ iff $i \to j$ and $j \to i$.

Proposition 1.1.18. *$i \to j$ iff $f_{ij}^* > 0$; $i \leftrightarrow j$ iff $f_{ij}^* f_{ji}^* > 0$.*

Proof. The proof follows from the obvious inequalities

$$\sup_{n \in N^*} p_{ij}^{(n)} \leqslant f_{ij}^* \leqslant \sum_{n \in N^*} p_{ij}^{(n)}. \diamond$$

The relation $\to$ is transitive since

$$p_{ik}^{(m+n)} \geqslant p_{ij}^{(m)} p_{jk}^{(n)}$$

on account of the Chapman-Kolmogorov equation.

The relation $\leftrightarrow$ is thus symmetric and transitive. By virtue of Proposition 1.1.18 it is reflexive only on the set of return states. Therefore the relation $\leftrightarrow$ divides the set of return states into (equivalence) classes such that two states belong to the same *class* iff they communicate. We shall consider each nonreturn state as a class by itself.

Notice that the set of return states coincides with the set of states which communicate with some other state. Indeed, we have seen that if i is a return state then $i \leftrightarrow i$; if $i \leftrightarrow j$ then $i \to j$ and $j \to i$. Therefore, $i \leftrightarrow i$, that is, i is a return state.

1.1.2.5.2. Let us denote by $C(i)$ the class containing state i.

A property defined for all states from I is said to be a *class property* iff its possession for one state $i \in I$ implies its possession by all states from $C(i)$. Obviously, the negation of a class property is also a class property. We shall see that many interesting properties are class properties.

Definition 1.1.19. A state $i \in I$ is said to be *essential* iff $i \to j$ implies $j \to i$ for every state j. Otherwise it is called *inessential.*

Theorem 1.1.20. *An inessential state cannot be reached from an essential state.*

Proof. Let state i be essential and $i \to j$. For every state k such that $j \to k$ we have $i \to k$ by transitivity. Consequently $k \to i$ by the essentialness of i and then $k \to j$ by transitivity. Thus state j is essential by definition. $\diamond$

Corollary. *The property of being essential (or inessential) is a class property.*

Theorem 1.1.21. *The property of having a period equal to d is a class property.*

Proof. If a class contains more than one state, then the periods of its states will all be finite. Let $i \leftrightarrow j$. Then there are $m, n \in N^*$ such that $p_{ij}^{(m)} > 0$, $p_{ji}^{(n)} > 0$. If $p_{ii}^{(s)} > 0$, then

$$p_{jj}^{(n+s+m)} \geqslant p_{ji}^{(n)} p_{ii}^{(s)} p_{ij}^{(m)} > 0.$$

We have also

$$p_{jj}^{(n+2s+m)} > 0$$

because $p_{ii}^{(2s)} > 0$. It follows that d_j divides $(n + 2s + m) - (n + s + m) = s$, and therefore d_j divides d_i. By interchanging i and j, it follows that d_i divides d_j. Thus $d_i = d_j$. ◇

Theorem 1.1.22. *An inessential state is nonrecurrent (or, equivalently: a recurrent state is essential).*

Proof. By (1.1.3) we have

$$g_{ij} \leqslant \sum_{k \in I} p_{ik}^{(s)} f_{kj}^*$$

for all, $i, j \in I$, $s \in N^*$.

If i is an inessential state, then there exist $l \in I$, $m \in N^*$ such that $p_{il}^{(m)} > 0$ and $f_{li}^* = 0$. Then

$$g_{ii} \leqslant \sum_{k \in I} p_{ik}^{(m)} f_{ki}^* \leqslant \sum_{k \in I} p_{ik}^{(m)} - p_{il}^{(m)} < 1.$$

Therefore $g_{ii} = 0$, that is, state i is nonrecurrent. ◇

Theorem 1.1.23. *A nonrecurrent (null) state cannot be reached from a recurrent (positive) state.*

Proof. Let state i be recurrent and thus essential by Theorem 1.1.22. Therefore, if $i \to j$ then we must have $i \leftrightarrow j$. This means that there exist $s, r \in N^*$ such that $p_{ij}^{(s)} > 0$, $p_{ji}^{(r)} > 0$. The Chapman-Kolmogorov equation implies

$$p_{jj}^{(n+s+r)} \geqslant p_{ji}^{(r)} p_{ii}^{(n)} p_{ij}^{(s)}$$

for all $n \in N$. On account of Theorem 1.1.6 this inequality shows that state j is recurrent.

If state i is positive (thus recurrent), the same inequality yields

$$p_{jj}^{(nd+s+r)} \geqslant p_{ji}^{(r)} p_{ii}^{(nd)} p_{ij}^{(s)}$$

for all $n \in N$, where $d = d_i = d_j$ (by Theorem 1.1.21). Letting $n \to \infty$ and using Theorem 1.1.14 we get

$$\overline{\lim_{n \to \infty}} \; p_{jj}^{(n)} > 0 ,$$

and by Corollary to the same theorem state j is positive. ◇

Corollary. *The property of being recurrent (or nonrecurrent) is a class property. The property of being positive (or null) is a class property.*

We have also

Theorem 1.1.24. (D. G. KENDALL). *The property of being geometrically ergodic is a class property. If a class C is geometrically ergodic, then the non-diagonal transition probabilities $p_{ij}^{(n)}$, $i \neq j$ also converge geometrically to their ergodic limits π_{ij}, $i, j \in C$.*

Proof. See KENDALL (1959). ◇

We note that VERE-JONES (1962) has shown that for a geometrically ergodic class C there is a common rate of convergence $\rho < 1$, that is

$$| p_{ij}^{(n)} - \pi_{ij} | \leqslant A_{ij} \rho^n$$

whatever $i, j \in C, n \in N$. (This best common rate of convergence may not be the best possible for all transitions).

The question of what conditions the transition probabilities must satisfy to ensure geometric ergodicity of a class C is open. A sufficient condition for a positive class to be geometrically ergodic is that the elements in any one column of $(p_{ij})_{i,j \in C}$, apart from the diagonal term, be bounded away from zero (MILLER (1966, p. 372)). Two general methods for settling the question of geometric ergodicity were given by KENDALL (1960).

1.1.2.5.3. The above results concerning class properties allow us to speak of essential (inessential), recurrent (nonrecurrent), positive (null) or geometrically ergodic classes; also of the period of a class.

We have the following result concerning recurrent classes.

Theorem 1.1.25. *A class C is recurrent iff $f_{ij}^* = g_{ij} = 1$ for all $i, j \in C$. It is nonrecurrent iff $f_{ii}^* < 1$, $g_{ij} = 0$ for all $i, j \in C$.*

Proof. Let the class C be recurrent. If $i, j \in C$, then $g_{ii} = g_{jj} = 1$ and there is $m \in N^*$ such that $p_{ji}^{(m)} > 0$. If $f_{ij}^* < 1$, then by (1.1.3)

$$g_{jj} \leqslant \sum_{k \in I} p_{jk}^{(m)} f_{kj}^* < \sum_{k \in I} p_{jk}^{(m)} = 1,$$

and we are led to a contradiction. Thus $f_{ij}^*=1$, and (1.1.4) yields $g_{ij}=1$. Conversely, if $f_{ij}^* = g_{ij} = 1$ for all $i, j \in C$, then (1.1.4) yields $g_{ii} = 1$ for all $i \in C$ thus C is a recurrent class.

The second part of the theorem follows easily from the same relation (1.1.4). ◇

Using Theorem 1.1.25 and Doeblin formula (1.1.7) one can easily prove

Theorem 1.1.26. *A class C is recurrent or nonrecurrent according as the series* $\sum_{n \in N^*} p_{ij}^{(n)}$ *diverges or converges for all* $i, j \in C$.

We note that VERE-JONES (1962) has shown that if C is an aperiodic class, then the series $\sum_{n \in N^*} p_{ij}^{(n)} z^n$, for all $i, j \in C$, have the same radius of convergence, say ρ. When $z = \rho$, all the series again behave in the same way, for all $i, j \in C$, and there are now just three possibilities:

(i) the series converge, so that $p_{ij}^{(n)} \rho^n \to 0$ as $n \to \infty$;

(ii) the series diverge, but still $p_{ij}^{(n)} \rho^n \to 0$ as $n \to \infty$;

(iii) the series diverge, and the sequences $(p_{ij}^{(n)} \rho^n)_{n \in N^*}$ all tend to finite, positive limits.

It is only in case (i) that we can have $\rho = 0$. For a recurrent aperiodic class C (when $\rho = 1$) we obtain no additional information. These results suggest the following terminology: a class C is called *ρ-nonrecurrent* in case (i), *ρ-null-recurrent* in case (ii), and *ρ-positive-recurrent* in case (iii).

1.1.2.5.4. Now we shall show that any class of period $d > 1$ splits into d subclasses.

Proposition 1.1.27. *Let $C(i)$ be a class of period $d > 1$. To every $j \in C(i)$ there corresponds a unique residue class r_j modulo d such that $p_{ij}^{(n)} > 0$ implies that $n \equiv r_j \pmod d$. Conversely, if $n \in N^*$ is sufficiently large and $n \equiv r_j \pmod d$, then $p_{ij}^{(n)} > 0$ for $j \in C(i)$.*

Proof. Let $p_{ij}^{(n)} > 0$ and $p_{ij}^{(n')} > 0$. There is an $m \in N^*$ such that $p_{ji}^{(m)} > 0$. Consequently,

$$p_{ii}^{(m+n)} \geqslant p_{ij}^{(n)} p_{ji}^{(m)} > 0,$$

$$p_{ii}^{(m+n')} \geqslant p_{ij}^{(n')} p_{ji}^{(m)} > 0.$$

It follows that d divides $(m+n) - (m+n') = n - n'$, and the first assertion is proved.

To prove the second assertion, notice that by Proposition 1.1.12 we have $p_{ii}^{(md)} > 0$ for all $m \in N^*$ sufficiently large. To state $j \in C(i)$ there corresponds an $n' \equiv r_j \pmod d$ such that $p_{ij}^{(n')} > 0$. We then have

$$p_{ij}^{(n)} \geqslant p_{ii}^{(md)} p_{ij}^{(n')} > 0$$

for $m \in N^*$ sufficiently large with $n = md + n'$. ◇

According to the above proposition we may consider the non-void set of states $j \in C(i)$ corresponding to the same residue class r modulo d. This set is called the *subclass* $C_r(i)$ of the class $C(i)$. We shall extend this notation to all $r \in Z$ by setting

$$C_r(i) = C_s(i) \text{ if } r = s \pmod{d}.$$

Clearly, the subclasses $C_r(i)$, $0 \leqslant r \leqslant d - 1$, are disjoint and their union is $C(i)$. Moreover, if $j \in C_r(i)$ then $C_s(j) = C_{r+s}(i)$ for all $r, s \in N$. It follows that the subclasses $C_r(i)$, $0 \leqslant r \leqslant d - 1$, do not depend on the choice of state i in the class $C(i)$ except for a cyclic permutation. Consequently, we shall denote them simply by C_r.

Theorem 1.1.28. *If C is an essential class of period d with the subclasses C_r, $0 \leqslant r \leqslant d - 1$, then*

$$\sum_{j \in C_{r+n}} p_{ij}^{(n)} = 1, i \in C_r.$$

Proof. Since C is essential, we have by Theorem 1.1.20

$$\sum_{j \in C} p_{ij}^{(n)} = 1$$

for all $i \in C$, $n \in N^*$. It follows from previous remarks that if $i \in C_r$ and $p_{ij}^{(n)} > 0$, then $j \in C_{r+n}$. ◇

Note that Theorem 1.1.28 amounts to the fact that

$$\mathsf{P}(\xi_{m+n} \in C_{r+n} \mid \xi_m = i) = 1$$

for all $m, n \in N$, $i \in C_r$, $0 \leqslant r \leqslant d - 1$. It follows that given an essential class C and an initial probability distribution $\mathbf{p} = (p_i)_{i \in I}$ on I such that $p_i = 0$ for $i \notin C_r$, the sequence of random variables $(\xi_{nd})_{n \in N}$ is a Markov chain with state space C_r and transition matrix $(p_{ij}^{(d)})_{i,j \in C_r}$, $0 \leqslant r \leqslant d - 1$. The state space C_r for this reduced Markov chain is an essential class of period 1. Thus the case of a period $d > 1$ can be reduced to the case where the period equals one.

1.1.2.5.5. We shall now give a characterization of essential classes.

Definition 1.1.29. A non-void set A of states is said to be *(stochastically) closed* iff $\sum_{j \in A} p_{ij} = 1$ for all $i \in A$ [9] (equivalently, if $p_{ij} = 0$ what-

[9] It is easy to see that this implies

$$\sum_{j \in A} p_{ij}^{(n)} = 1, \; i \in A,$$

for all $n \in N^*$.

ever $i \in A$ and $j \notin A$). A closed set is said to be *minimal* iff it contains no proper closed subset. If a closed set contains only a single state i (therefore $p_{ii} = 1$), then i is said to be an *absorbing* state.

For a closed set A we have

$$\mathsf{P}(\xi_{n+1} \in A \mid \xi_n = i) = 1$$

whatever $n \in N$, $i \in A$. It follows that given a closed set A and an initial probability distribution $\mathbf{p} = (p_i)_{i \in I}$ on I such that $p_i = 0$ for $i \notin A$, the Markov chain $(\xi_n)_{n \in N}$ moves inside A, that is its state space is A and its transition matrix $(p_{ij})_{i,j \in A}$. In other words, the motion of the chain inside A can be studied independently of all other states.

Trivially, the set I of all states is closed but not necessarily minimal. From the proof of Theorem 1.1.28 we know that any essential class is a closed set. Now we prove more.

Proposition 1.1.30. *A set is minimal closed iff it is an essential class.*

Proof. Let C be an essential class and $A \subset C$ a closed set. If $i \in A$, then there is r, $0 \leqslant r \leqslant d-1$, such that $i \in C_r$. On account of Theorem 1.1.28 we have

$$\sum_{j \in C_{r+1}} p_{ij} = 1,$$

whence $C_{r+1} \subset A$. Hence $C_{r+2} \subset A$, etc., that is $A = C$. Therefore an essential class is minimal closed.

If the set A is minimal closed and contains an essential state, then by the preceding it consists of the corresponding essential class. Assume A contains only inessential states and let i be one of them. Let B be the set of all $j \in A$ such that $i \to j$ but not $j \to i$. Clearly, $i \notin B$. Then B is non-empty, closed and a proper subset of A. Hence A cannot be minimal closed. ◇

The above results suggest

Definition 1.1.31. A Markov chain is said to be *irreducible* iff its state space is minimal closed (it thus consists of one essential class). It is said to be *indecomposable* iff its state space does not contain two disjoint closed set.

Further, we have

Proposition 1.1.32. *The union of a finite number of inessential classes is not closed.*

Proof. Let $A_1, \ldots, A_m$ be distinct inessential classes. Consider the Cartesian products $\prod_{\nu=1}^{r} A_{s_\nu}$, $1 \leqslant r \leqslant m$, $1 \leqslant s_\nu \leqslant m$, $s_\nu \neq s_\mu$, $\nu \neq \mu$, $1 \leqslant \nu, \mu \leqslant r$, and retain only their elements $(i_{s_1}, \ldots, i_{s_r})$ for which $i_{s_1} \to$

$\ldots \to i_{s_r}$. Let $(i_1, \ldots, i_n)$, $n \leqslant m$, be one of the longest elements retained. Since the state i_n is inessential there is $j \in I$ such that $i_n \to j$ but not $j \to i_n$. By the choice of n we have $j \notin \bigcup_{n=1}^{m} A_m$, that is this union is not closed. ◇

1.1.2.6. Conditional motion

1.1.2.6.1. Proposition 1.1.32 implies that if the set of inessential states is finite, then it is not closed. We shall treat this case further as a corollary of Theorem 1.1.33 below. From now on, we denote by F either the set of inessential, nonrecurrent or null states.

Theorem 1.1.33. *Let $S \subset I$ be an arbitrary set of states and let $\mathfrak{r}_i$, $i \in S$, denote the probability that the Markov chain starting in state i does not leave the set S. Then*

$$\mathfrak{r}_i = \sum_{j \in S} p_{ij} \mathfrak{r}_j \tag{1.1.18}$$

for all $i \in S$. Further we have $\mathfrak{r}_i = 0$, $i \in S$, iff the system of equations

$$z_i = \sum_{j \in S} p_{ij} z_j, \quad i \in S, \tag{1.1.19}$$

admits of no nonzero bounded solution.

Proof. Denote by $\mathfrak{r}_i^{(n)}$, $i \in S$, the probability that the Markov chain starting in state i does not leave S during the first n steps, i.e.,

$$\mathfrak{r}_i^{(n)} = \mathsf{P}(\xi_{m+r} \in S,\ 1 \leqslant r \leqslant n \mid \xi_m = i)\ [10]$$

so that

$$\mathfrak{r}_i = \lim_{n \to \infty} \mathfrak{r}_i^{(n)}$$

the convergence being monotone. It is obvious that

$$\mathfrak{r}_i^{(n+1)} = \sum_{j \in S} p_{ij} \mathfrak{r}_j^{(n)}$$

for all $i \in S$, $n \in N^*$, thus letting $n \to \infty$ we obtain (1.1.18).

---

[10] Notice that in case $S = F$ on account of Theorems 1.1.20 and 1.1.23 we have

$$\mathfrak{r}_i^{(n)} = \mathsf{P}(\xi_{m+n} \in F \mid \xi_m = i) = \sum_{j \in F} p_{ij}^{(n)}.$$

More generally, if $S \subset F$, then

$$\mathfrak{r}_i^{(n)} \leqslant \sum_{j \in S} p_{ij}^{(n)}.$$

Now, suppose that there is a bounded nonzero solution z_i, $|z_i| \leqslant 1$, $i \in S$, to the system (1.1.19). We deduce easily that $|z_i| \leqslant \mathrm{r}_i^{(1)}$, and then, by induction, $|z_i| \leqslant \mathrm{r}_i^{(n)}$ for all $i \in S$, $n \in N^*$. It follows that $\mathrm{r}_i = 0$ for all $i \in S$ iff the system (1.1.19) admits of no nonzero bounded solution. ◇

Corollary. *If S is a finite subset of F, then $\mathrm{r}_i = 0$ for all $i \in S$, i.e., the probability that the Markov chain does not leave a finite subset of F equals zero.*

First proof. Suppose that the system (1.1.19) admits of a bounded nonzero solution with $z_{i_1} = \ldots = z_{i_r} = a > 0$ and $z_i < a$ for $i \neq i_j$, $1 \leqslant j \leqslant r$. Then we have

$$a = \sum_{j=1}^{r} p_{ii_j} a + \sum_{k \neq i_1, \ldots, i_r} p_{ik} z_k$$

for $i = i_1, \ldots, i_r$. These equalities are possible only if $p_{ik} = 0$ and $\sum_{j=1}^{r} p_{ii_j} = 1$ with $i = i_1, \ldots, i_r$, $k \neq i_1, \ldots, i_r$, so that the set $\{i_1, \ldots, i_r\}$ is closed. Thus on account of Proposition 1.1.32 we are led to a contradiction, which shows that $\mathrm{r}_i = 0$, $i \in S$. ◇

Second proof. Since the states from S are null, on account of Theorem 1.1.15 there is a $\mu \in N^*$ such that

$$\mathrm{r}_i^{(n)} \leqslant \sum_{j \in S} p_{ij}^{(n)} \leqslant \eta < 1$$

for all $n \geqslant \mu$, $i \in S$. Therefore

$$\sum_{j \in S} p_{ij}^{([s+1]\mu)} \leqslant \sum_{j,k \in S} p_{ik}^{(s\mu)} p_{kj}^{(\mu)} \leqslant \eta \sum_{k \in S} p_{ik}^{(s\mu)}, \quad i \in S, \quad s \in N^*,$$

whence

$$\sum_{j \in S} p_{ij}^{(s\mu)} \leqslant \eta^s, \quad i \in S, \quad s \in N^*,$$

and then

$$\sum_{j \in S} p_{ij}^{(n)} \leqslant \eta^{\frac{n}{\mu} - 1}, \qquad i \in S, \quad n \in N^*. \tag{1.1.20}$$

Thus $\mathrm{r}_i = \lim_{n \to \infty} \mathrm{r}_i^{(n)} = 0$, $i \in S$, the convergence being exponentially fast. ◇

Theorem 1.1.33 leads also to a criterion for nonrecurrence, namely

Theorem 1.1.34. *An essential class C is nonrecurrent iff for an arbitrary fixed state $k \in C$ the system of equations*

$$z_i = \sum_{j \in C - \{k\}} p_{ij} z_j, \quad i \in C - \{k\},$$

admits of a nonzero bounded solution.

Proof. Suppose that C is nonrecurrent. By Theorem 1.1.33 (with $S = C - \{k\}$) a nonzero bounded solution to the system considered is

$$z_i = \mathsf{P}(\xi_{m+n} \neq k,\ n \in N^* \mid \xi_m = i) = 1 - f_{ik}^*,$$

Indeed, we have $f_{ik}^* < 1$ for some $i \in C - \{k\}$ since supposing on the contrary that $f_{ik}^* = 1$ for all $i \in C - \{k\}$, (1.1.3) yields

$$g_{kk} = \lim_{m\to\infty} \sum_{j\in C} p_{kj}^{(m)} f_{jk}^*$$

$$= \lim_{m\to\infty} \sum_{j\in C} p_{kj}^{(m)} + (f_{kk}^* - 1) \lim_{m\to\infty} p_{kk}^{(m)} = 1^{11)}.$$

that is state k would be recurrent.

On the other hand, by the proof of Theorem 1.1.33, $z_i = 1 - f_{ik}^*$, $i \in C - \{k\}$ is the maximal bounded (by one) solution to our system. It follows on acount of Theorem 1.1.25 that if there is a nonzero bounded solution, then C is nonrecurrent. This ends the proof. ◇

1.1.2.6.2. Let S and A be two disjoint sets of states. Sometimes it is important to find the probability of entering the set A, the motion of the Markov chain being restricted to S. In case $S = F$ and A is a closed set, this probability may be called the *absorption probability* into A since transitions from $I - (F \cup A)$ to A are not possible and a closed set once entered cannot be left. Returning to the general case we note that if $\sum_{j\in A} p_{ij} > 0$ for at least one state $i \in S$, then there is a positive probability that the Markov chain starting in this state will not appear in S but in A at the next step. In this case we shall also say that the chain has been absorbed into A. More precisely, let $\mathfrak{a}_i^{(n)}(A)$ be the probability that, starting in $i \in S$, the Markov chain enters A for the first time at the nth step, in the intervening time remaining within S, i.e.,

$$\mathfrak{a}_i^{(n)}(A) = \mathsf{P}(\xi_{m+n} \in A,\ \xi_k \in S,\ m < k < m+n \mid \xi_m = i).$$

The absorption probability into A after moving inside S given that the Markov chain started in $i \in S$ will be

$$\mathfrak{a}_i(A) = \sum_{n\in N^*} \mathfrak{a}_i^{(n)}(A).$$

Note that if A is a recurrent class then we may write for $j \in A$

$$f_{ij} = \sum_{n\in N^*} \sum_{k\in A} \mathsf{P}(\xi_{m+n} = k,\ \xi_{m+\nu} \notin A,\ 0 < \nu < n \mid \xi_m = i) f_{kj}^*$$

---

11) We used the fact that state k is null and Theorem 1.1.14.

and since $f^*_{kj} = 1$, $k \in A$, by Theorem 1.1.25 we get

$$\mathfrak{a}_i(A) = f^*_j.$$

Returning to the general case it is easy to see that

$$\mathfrak{a}_i^{(n+1)}(A) = \sum_{j \in S} p_{ij}\, \mathfrak{a}_j^{(n)}(A). \qquad (1.1.21)$$

Thus the probabilities $\mathfrak{a}_i^{(n)}(A)$ can be determined recursively. Summing over $n \in N^*$ in (1.1.21) we obtain the system of equations

$$\mathfrak{a}_i(A) = \sum_{j \in S} p_{ij}\, \mathfrak{a}_j(A) + \mathfrak{a}_i^{(1)}(A) \qquad (1.1.22)$$

with

$$\mathfrak{a}_i^{(1)}(A) = \sum_{j \in A} p_{ij},\ i \in S.$$

We have

Proposition 1.1.35. *If* $\mathfrak{r}_i = 0$, $i \in S$, *then the solution to the system* (1.1.22) *is unique.*

Proof. If there were two distinct solutions, their difference, which is bounded and non-null would satisfy the system (1.1.19). ◇

1.1.2.6.3. In several applications it is important to discuss the behaviour of the Markov chain after a large number of steps, given that it started within S, and conditional on its remaining within S. When S is finite, this problem has been studied by BRENY (1962 b), DARROCH and SENETA (1965), MANDL (1959), and SENETA (1966). For infinite S, the corresponding questions were taken up in SENETA and VERE-JONES (1966). Here we give only an outline. For details the reader is referred to the quoted papers.

As in the above, denote by $\mathfrak{r}_i$ the probability of remaining within S given that the Markov chain started in $i \in S$. Thus $1 - \mathfrak{r}_i$ is the absorption probability into $I - S$ after moving within S given that the Markov chain started in $i \in S$.

Of particular interest are two conditional distributions and their limiting values. Consider the probability $\alpha_{ij}(n)$ that the chain is in state $j \in S$ at the nth step, given that it started in state $i \in S$, and conditional on absorption ultimately occurring, on or after the nth step:

$$\alpha_{ij}(n) = \frac{p_{ij}^{(n)}(1 - \mathfrak{r}_j)}{\sum_{k \in S} p_{ik}^{(n)}(1 - \mathfrak{r}_k)}. \quad {}^{12)}$$

---

12) To avoid complications in notation, in this subparagraph $p_{ij}^{(n)}$ will denote the entries of the nth power of the matrix $(p_{ij})_{i,j \in S}$

Consider also the probability $\beta_{ij}(m, n)$ that the chain will be in state $j \in S$ at the mth step given that it started in state $i \in S$, and conditional on absorption ultimately occurring, on or after the $(m+n)$th step:

$$\beta_{ij}(m, n) = \frac{p_{ij}^{(m)} \sum_{k \in S} p_{jk}^{(n)} (1 - \mathfrak{r}_k)}{\sum_{k \in S} p_{ik}^{(m+n)} (1 - \mathfrak{r}_k)}.$$

Obviously we have

$$\sum_{j \in S} \alpha_{ij}(n) = \sum_{j \in S} \beta_{ij}(m, n) = 1$$

for all $i \in S$, $m, n \in N^*$.

When S is finite and its states communicate [13], there exist the positive limits

$$\lim_{n \to \infty} \alpha_{ij}(n) = \alpha_j,$$

$$\lim_{m \to \infty} \lim_{n \to \infty} \beta_{ij}(m, n) = \beta_j \alpha_j$$

which are independent of the initial state $i \in S$. In this case the second limit coincides with the limiting value of the expected proportion of time spent in state $j \in S$ given that the chain started in state $i \in S$, and conditional on its remaining within S, namely

$$\lim_{n \to \infty} \frac{\frac{1}{n} \sum_{s=1}^{n} p_{ij}^{(s)} \left(\sum_{k \in S} p_{jk}^{(n-s)} \right)}{\sum_{k \in S} p_{ik}^{(n)}}.$$

The vectors $(\alpha_j)_{j \in S}$ and $(\beta_j)_{j \in S}$ are just the left and right eigenvectors associated with the maximum eigenvalue of $(p_{ij})_{i,j \in S}$.

1.1.3. Taboo and stationarity

This subsection is concerned with two closely related topics leading to deep results for Markov chains.

---

[13] The discussion of this case can be based on the spectral representation of the matrix $(p_{ij})_{i,j \in S}$, using the Perron-Frobenius theory (see Paragraph 1.1.4.2).

1.1.3.1. Taboo transition probabilities

1.1.3.1.1. For further developments it is useful to introduce the so-called *taboo transition probabilities*. Let

$$_kp_{ij}^{(n)} = \mathsf{P}(\xi_{m+n} = j,\ \xi_{m+\nu} \neq k,\ 0 < \nu < n \mid \xi_m = i),\ n \in N^*,$$

be the probability that the chain starting in state i reaches state j in n steps without entering the *taboo* state k in the intervening time [14]. Obviously $_jp_{ij}^{(n)} = f_{ij}^{(n)}$. For convenience we set

$$_kp_{ij}^{(0)} = \begin{cases} \delta_{ij} & \text{if } i \neq k \\ 0 & \text{if } i = k. \end{cases}$$

Define similarly

$$_kf_{ij}^{(n)} = \mathsf{P}(\xi_{m+n} = j,\ \xi_{m+\nu} \notin \{j, k\},\ 0 < \nu < n \mid \xi_m = i),\ n \in N^*,$$

the probability that the chain starting in state i reaches state j for the first time at the nth step without entering state k in the intervening time. Obviously $_kf_{ij}^{(n)} \leqslant f_{ij}^{(n)}$ whatever k. Finally we set

$$_kp_{ij}^* = \sum_{n \in N} {_kp_{ij}^{(n)}}, \quad _kf_{ij}^* = \sum_{n \in N^*} {_kf_{ij}^{(n)}}.$$

We have $_kf_{ij}^* \leqslant f_{ij}^*$ for all k and $_jp_{ij}^* = f_{ij}^*$. It may happen that $_kp_{ij}^* = \infty$ for $k \neq j$ but if $j \to k$ this possibility cannot occur. Namely we have

Proposition 1.1.36. *If $f_{jk}^* > 0$ (i.e. j leads to k), then $_kp_{ij}^* < \infty$ for all $i \in I$.*

Proof. Suppose $j \neq k$. Since $f_{jk}^* > 0$ we have $f_{jk}^{(m)} > 0$ for some $m \in N^*$. Clearly,

$$_kp_{ij}^{(n)}\ {_kp_{jk}^{(m)}} \leqslant {_kp_{ik}^{(n+m)}} = f_{ik}^{(n+m)}, \quad n \in N,$$

whence

$$_kp_{ij}^{(n)} \leqslant [f_{jk}^{(m)}]^{-1} f_{ik}^{(n+m)},$$

so that

$$_kp_{ij}^* = \sum_{n \in N} {_kp_{ij}^{(n)}} \leqslant [f_{jk}^{(m)}]^{-1} \sum_{n \in N} f_{ik}^{(n+m)} \leqslant [f_{jk}^{(m)}]^{-1}. \diamond$$

---

[14] More generally, we may consider transition avoiding a given set H of *taboo* states. In our case $H = \{k\}$; in Paragraph 1.1.2.6 we were in fact concerned with $H = I - S$. For details see CHUNG (1967).

By applying the method of first entrance into state j, we easily deduce the formulas

$$ {}_kp_{ij}^{(n)} = \sum_{m=1}^{n} {}_kf_{ij}^{(m)}\, {}_kp_{jj}^{(n-m)}, \quad k \neq j, \tag{1.1.23} $$

$$ p_{ii}^{(n)} = {}_jp_{ii}^{(n)} + \sum_{m=1}^{n-1} f_{ij}^{(m)}\, p_{ji}^{(n-m)}, \quad i \neq j. \tag{1.1.24} $$

We have also

$$ f_{ij}^{(n)} = \sum_{m=0}^{n-1} {}_jp_{ii}^{(m)}\, {}_if_{ij}^{(n-m)}, \quad i \neq j, \tag{1.1.25} $$

the derivation of which consists of considering the last time prior to n at which state i is entered (This may be called the *method of last exit* from a given state and is a natural dual of the method of first entrance).

Summing over $n \in N^*$ in (1.1.23) (with $k = i$) and (1.1.25) we get

$$ {}_ip_{ij}^* = {}_if_{ij}^*\, {}_ip_{jj}^*, \quad i \neq j, $$

$$ f_{ij}^* = {}_jp_{ii}^*\, {}_if_{ij}^*, \quad i \neq j. $$

It follows from these equations and Theorem 1.1.25 that within a recurrent class we have

$$ {}_ip_{ij}^* = \frac{{}_ip_{jj}^*}{{}_jp_{ii}^*}. \tag{1.1.26} $$

1.1.3.1.2. Let us consider the generating functions

$$ P_{ij}(z) = \sum_{n \in N} p_{ij}^{(n)} z^n, \quad F_{ij}(z) = \sum_{n \in N^*} f_{ij}^{(n)} z^n, $$

$$ {}_kP_{ij}(z) = \sum_{n \in N} {}_kp_{ij}^{(n)} z^n, \quad {}_kF_{ij}(z) = \sum_{n \in N^*} {}_kf_{ij}^{(n)} z^n. $$

with $|z| < 1$. Relation (1.1.6) implies that

$$ P_{ij}(z) - \delta_{ij} = F_{ij}(z)\, P_{jj}(z), \quad i, j \in I, $$

whence

$$ P_{ij}(z) = F_{ij}(z)\, P_{jj}(z), \quad i \neq j, \tag{1.1.27} $$

$$ P_{ii}(z) = \frac{1}{1 - F_{ii}(z)} \tag{1.1.28} $$

(The latter relation was already encountered — see (1.1.17)). For the sequel it is important that the alternative representation

$$P_{ii}(z) = \frac{{}_jP_{ii}(z)}{1 - F_{ij}(z)\,F_{ji}(z)} \tag{1.1.29}$$

holds for $j \neq i$. To prove (1.1.29) it suffices to remark that (1.1.24) implies

$$P_{ii}(z) = {}_jP_{ii}(z) + F_{ij}(z)\,P_{ji}(z), \quad i \neq j,$$

and to substitute there $P_{ji}(z)$ from (1.1.27).

We note also that (1.1.23) leads to

$${}_jP_{ii}(z) - 1 = {}_jF_{ii}(z)\,{}_jP_{ii}(z), \quad i \neq j,$$

whence

$${}_jP_{ii}(z) = \frac{1}{1 - {}_jF_{ii}(z)} \qquad i \neq j. \tag{1.1.30}$$

1.1.3.2. Ratio limit theorems

1.1.3.2.1. A typical example of a so-called *ratio limit theorem* is Doeblin's formula (1.1.7). In the following we need other similar results.

Proposition 1.1.37. *Within a recurrent class we have*

$$\lim_{m\to\infty} \frac{\sum_{n=0}^{m} p_{ii}^{(n)}}{\sum_{n=0}^{m} p_{jj}^{(n)}} = \frac{{}_jp_{ii}^*}{{}_ip_{jj}^*} = {}_jp_{ji}^*.$$

Proof. (FELLER (1966 a)). On account of (1.1.28) and (1.1.30) relation (1.1.29) may be rewritten in the form

$$\frac{1}{1 - F_{ij}(z)\,F_{ji}(z)} - \frac{P_{jj}(z)}{{}_ip_{jj}^*} = \Phi_{ij}(z), \tag{1.1.31}$$

where

$$\Phi_{ij}(z) = \frac{{}_if_{jj}^* - {}_iF_{jj}(z)}{1 - F_{jj}(z)}, \quad i \neq j.$$

Put

$$C_{ij}(z) = \frac{1}{1 - F_{ij}(z)\,F_{ji}(z)}, \quad i \neq j, \tag{1.1.32}$$

and denote the coefficients of $C_{ij}(z)$ by $c_{ij}^{(n)}$, $n \in N$.

Let us prove that

$$0 \leqslant \sum_{n=0}^{m} c_{ij}^{(n)} - \frac{1}{{}_ip_{jj}^{*}} \sum_{n=0}^{m} p_{jj}^{(n)} \leqslant 1. \tag{1.1.33}$$

By (1.1.31) the middle term in (1.1.33) is the mth coefficient of

$$\frac{\Phi_{ij}(z)}{1-z} = \frac{\sum\limits_{n \in N} \left(\sum\limits_{r \geqslant n} {}_if_{jj}^{(r+1)}\right) z^n}{1 - F_{jj}(z)} =$$

$$= \left(1 + \sum_{n \in N^*} [F_{jj}(z)]^n\right) \left[\sum_{n \in N} \left(\sum_{r \geqslant n} {}_if_{jj}^{(r+1)}\right) z^n\right],$$

therefore the left-inequality in (1.1.33) is obvious. On the other hand the coefficients in the power series $\Phi_{ij}(z)\,(1-z)^{-1}$ are dominated by the coefficients in the power series

$$\frac{\sum\limits_{n \in N} \left(\sum\limits_{r \geqslant n} f_{jj}^{(r+1)}\right) z^n}{1 - F_{jj}(z)} = \frac{1}{1-z} = \sum_{n \in N} z^n.$$

Thus we have also proved the right inequality in (1.1.33).

The proof now follows trivially on comparing (1.1.33) with the relation obtained by interchanging i and j, remembering that $c_{ij}^{(n)} = c_{ji}^{(n)}$, and making use of Theorem 1.1.6 and relation (1.1.27). ◇

Corollary. *Within a recurrent class we have*

$$\lim_{m \to \infty} \frac{\sum\limits_{n=0}^{m} p_{ji}^{(n)}}{\sum\limits_{n=0}^{m} p_{jj}^{(n)}} = {}_jp_{ji}^{*}.$$

Proof. We may write

$$\frac{\sum\limits_{n=0}^{m} p_{ji}^{(n)}}{\sum\limits_{n=0}^{m} p_{jj}^{(n)}} = \frac{\sum\limits_{n=0}^{m} p_{ji}^{(n)}}{\sum\limits_{n=0}^{m} p_{ii}^{(n)}} \cdot \frac{\sum\limits_{n=0}^{m} p_{ii}^{(n)}}{\sum\limits_{n=0}^{m} p_{jj}^{(n)}}.$$

Since both i and j are in the same recurrent class, by Theorem 1.1.25 we have $f^*_{ij} = f^*_{ji} = 1$. According to Doeblin formula (1.1.7), the first ratio on the right-hand side tends to $f^*_{ji} = 1$ and on account of Proposition 1.1.37 the second ratio approaches ${}_jp^*_{ji}$. ◇

In fact Proposition 1.1.37 implies that within a recurrent class we have

$$\lim_{m\to\infty} \frac{\sum_{n=0}^{m} p_{ij}^{(n)}}{\sum_{n=0}^{m} p_{kl}^{(n)}} = {}_lp^*_{ij},$$

In the following, however, we shall use only the above corollary.

If in a recurrent class we have also

$$\lim_{n\to\infty} \frac{p_{ij}^{(n+m)}}{p_{kl}^{(n)}} = {}_lp^*_{ij}$$

for all $m \in Z$, we say that the *strong ratio limit property* holds. For details see Folkman and Port (1966), Kingman and Orey (1964), Molčanov (1967), Orey (1961), and Papangelou (1967).

1.1.3.2.2. According to Corollary to Theorem 1.1.15, the Cesaro limit

$$\pi_{ij} = \lim_{n\to\infty} \frac{1}{n} \sum_{m=0}^{n} p_{ij}^{(m)}$$

exists and equals $f^*_{ij}\nu_{jj}$ for all $i, j \in I$.

Let us set

$$\pi_{ii} = \pi_i, \quad i \in I.$$

We have $\pi_i = 0$ or > 0 according as state i is null or positive and thus $\pi_i = \nu_{ii}$ for all $i \in I$.

On account of Proposition 1.1.37 we may assert that

$${}_jp^*_{ji} = \frac{\pi_i}{\pi_j}$$

for every recurrent state $j \in I$. Indeed, we have

$$\sum_{i\in C(j)} {}_jp_{ji}^{(n)} = \mathsf{P}(\xi_\nu \neq j,\ 1 \leqslant \nu \leqslant n \mid \xi_0 = j) = \sum_{\nu\geqslant n} f_{jj}^{(\nu)},$$

thus

$$\sum_{i\in I} {}_jp^*_{ji} = \sum_{i\in I} \sum_{n\in N^*} {}_jp^{(n)}_{ji} = \sum_{n\in N^*} \sum_{i\in I} {}_jp^{(n)}_{ji} =$$

$$= \sum_{n\in N^*} \sum_{\nu\geq n} f^{(\nu)}_{jj} = \sum_{\nu\in N^*} \nu f^{(\nu)}_{jj} = \frac{1}{\nu_{ji}}.$$

Therefore we may state

Proposition 1.1.38. *The series* $\sum_{i\in C(j)} {}_jp^*_{ji}$ *converges in a positive class and diverges in a null-recurrent class.*

1.1.3.3. Stationary distributions

1.1.3.3.1. Among the initial probability distributions for a Markov chain the most important are those which make it a strictly stationary chain.

Definition 1.1.39. A probability distribution $\mathbf{p} = (p_i)_{i\in I}$ is said to be *stationary* for the Markov chain $(\xi_n)_{n\in N}$ iff $p_i = \sum_{j\in I} p_j p_{ji}$ for all $i \in I$.

It follows from Definition 1.1.39 that if $\mathbf{p} = (p_i)_{i\in I}$ is a stationary distribution, then

$$p_i = \sum_{j\in I} p_j p^{(n)}_{ji}$$

for all $i \in I$, $n \in N^*$. Therefore we have

$$\mathsf{P}_{\mathbf{p}}(\xi_n = i) = p_i$$

for all $i \in I$, $n \in N^*$. Moreover, it is easy to prove that the joint distribution of $(\xi_{n+h}, \ldots, \xi_{n+m+h})$ for all $n \in N^*$, $m \in N$ is independent of $h \in N$ so that the chain is a strictly stationary one in the established sense.

For applications of the spectral representation of stationary processes to Markov chains see Loynes (1966).

1.1.3.3.2. As we shall see, the quantities π_i, $i \in I$, introduced in Subparagraph 1.1.3.2.2. satisfy a certain system of equations and are related to the stationary distributions of the Markov chain.

Theorem 1.1.40. *Let C be an essential class of period d. The only solutions $\{u_i,\ i \in C\}$ of the system of equations*

$$u_i = \sum_{j\in C} u_j p_{ji}, \quad i \in C, \tag{1.1.34}$$

such that $\sum_{i \in C} |u_i| < \infty$ *are given by*

$$u_i = c\pi_i$$

where c is an arbitrary constant (W. FELLER).

If C is a positive class with subclasses C_r, $0 \leqslant r \leqslant d-1$, *then we have for every r*

$$\sum_{i \in C_r} \pi_i = \frac{1}{d}, \tag{1.1.35}$$

thus $\sum_{i \in I} \pi_i = 1$ (A. N. KOLMOGOROV).

Proof. (CHUNG, (1967, p. 35)). We prove successively that

I. $\{\pi_i,\ i \in C\}$ is a solution of (1.1.34) such that $\sum_{i \in C} \pi_i < \infty$.

II. This solution is unique except for a constant factor.

III. Relation (1.1.35) holds.

Proof of I. It suffices to consider only the case where C is a positive class. According to Theorem 1.1.28

$$p_{ii}^{(nd)} = \sum_{j \in C} p_{ij}^{(nd-1)} p_{ji} = \sum_{j \in C_{-1}(i)} p_{ij}^{(nd-1)} p_{ji}.$$

Letting $n \to \infty$ we have by Theorem 1.1.15 and Fatou's lemma

$$d\pi_i = \lim_{n\to\infty} p_{ii}^{(nd)} \geqslant \sum_{j \in C_{-1}(i)} \lim_{n\to\infty} p_{ij}^{(nd-1)} p_{ji} =$$

$$= \sum_{j \in C_{-1}(i)} d f_{ij}^*(d-1)\, \nu_{jj} p_{ji} = d \sum_{j \in C} \pi_j p_{ji},$$

because $f_{ij}^*(d-1) = f_{ij}^* = 1$ for $j \in C_{d-1}(i) = C_{-1}(i)$. Thus,

$$\pi_i \geqslant \sum_{j \in C} \pi_j p_{ji}. \tag{1.1.36}$$

On the other hand, we have for every r, $0 \leqslant r \leqslant d-1$,

$$1 = \lim_{n\to\infty} \sum_{j \in C_r(i)} p_{ij}^{(nd+r)} \geqslant \sum_{j \in C_r(i)} \lim_{n\to\infty} p_{ij}^{(nd+r)} = \sum_{j \in C_r(i)} d\pi_j.$$

Therefore

$$\sum_{j \in C_r(i)} \pi_j \leqslant \frac{1}{d}$$

and

$$\sum_{j\in C} \pi_j \leqslant 1. \tag{1.1.37}$$

Summing (1.1.36) over all $i \in C$ we obtain

$$\sum_{i\in C} \pi_i \geqslant \sum_{i\in C}\sum_{j\in C} \pi_j p_{ji} = \sum_{j\in C} \pi_j \sum_{i\in C} p_{ji} = \sum_{j\in C} \pi_j \tag{1.1.38}$$

(the interchange of summations is justified by (1.1.37)). Relations (1.1.36) and (1.1.38) imply

$$\pi_i = \sum_{j\in C} \pi_j p_{ji}, \quad i \in C. \tag{1.1.39}$$

Proof of II. Suppose that $\{u_i, i \in C\}$ is a solution of (1.1.34) such that $\sum_{i\in C} |u_i| < \infty$. Iterating (1.1.34) and interchanging summations we get

$$u_i = \sum_{j\in C_{-r}(i)} u_j p_{ji}^{(nd+r)}$$

for all $i \in C$, $n \in N^*$, $0 \leqslant r \leqslant d-1$. Letting $n \to \infty$ we have by Theorem 1.1.15

$$u_i = \Big(\sum_{j\in C_{-r}(i)} u_j\Big)\, d\pi_i. \tag{1.1.40}$$

Since equation (1.1.40) is true for every r, the bracketed sum does not depend on r. Hence it does not depend on i either, since $C_{-r}(i)$, $0 \leqslant r \leqslant d-1$, is simply an enumeration of the subclasses of C. It follows that $u_i = c\pi_i$, $i \in C$.

Proof of III. If C is a positive class, then $\pi_i > 0$ for $i \in C$. The results already proved allow us to take $u_i = \pi_i$ in (1.1.40). We thus obtain (1.1.35). ◇

Corollary. *If C is a positive class, then we have for every $i \in I$,*

$$\sum_{j\in C} \pi_{ij} = \alpha_i(C).$$

In particular, the sum equals one if $i \in C$.

Proof. By Subparagraph 1.1.2.6.2. we have $\alpha_i(C) = f_{ij}$ whatever $j \in C$ and $i \in I$. Therefore,

$$\sum_{j\in C} \pi_{ij} = \sum_{j\in C} f_{ij}\nu_{jj} = \alpha_i(C)\sum_{j\in C} \nu_{jj} = \alpha_i(C)\sum_{j\in C} \pi_j = \alpha_i(C). \;\diamond$$

This corollary shows that a Markov chain is non-dissipative iff whatever the starting state, the probability of entering the set of positive states equals one.

The problem of characterizing stationary distributions is solved by

Theorem 1.1.41. (K. L. CHUNG). *Let $(D_\alpha)_{\alpha \in A}$, $D_{\alpha'} \neq D_{\alpha''}$ if $\alpha' \neq \alpha''$, be the positive classes of the Markov chain $(\xi_n)_{n \in N}$. In order that $(p_i)_{i \in I}$ be a stationary distribution it is necessary and sufficient that there exist a family of constants $\{\lambda_\alpha, \alpha \in A\}$ with $\lambda_\alpha \geqslant 0$, $\sum_{\alpha \in A} \lambda_\alpha = 1$ such that*

$$p_i = \begin{cases} 0 & \text{if } i \notin \bigcup_{\alpha \in A} D_\alpha \\ \lambda_\alpha \pi_i & \text{if } i \in D_\alpha, \ \alpha \in A. \end{cases}$$

Proof. i) Necessity. We have for all $r \in N^*$, $i \in I$,

$$p_i = \sum_{j \in I} p_j p_{ji}^{(r)}. \tag{1.1.41}$$

If $i \notin D = \bigcup_{\alpha \in A} D_\alpha$, we have by Theorem 1.1.15

$$\lim_{n \to \infty} p_{ij}^{(n)} = 0$$

for all $j \in I$. Hence, letting $r \to \infty$ in (1.1.41) we get $p_i = 0$ for all $i \notin D$. If $i \in D_\alpha$, then since $p_{ji}^{(r)} = 0$ for all $j \in D - D_\alpha$, $r \in N^*$, we have

$$p_i = \sum_{j \in D_\alpha} p_j p_{ji}^{(r)},$$

whence

$$p_i = \lim_{n \to \infty} \sum_{j \in D_\alpha} p_j \left(\frac{1}{n} \sum_{r=1}^{n} p_{ji}^{(r)} \right) = \left(\sum_{j \in D_\alpha} p_j \right) \pi_i.$$

(The passage to the limit under the summation sign is justified by the fact that the series is dominated by $\sum_{j \in D_\alpha} p_j \leqslant 1$). Thus we can take $\lambda_\alpha = \sum_{j \in D_\alpha} p_j$.

ii) Sufficiency. Let $i \notin D$; then if $j \in D$, $p_{ji} = 0$. Thus

$$\sum_{j \in I} p_j p_{ji} = 0 = p_i$$

for all $i \notin D$. Further, let $i \in D_\alpha$; then if $j \in D - D_\alpha$, $p_{ji} = 0$. By (1.1.40) (with $C = D_\alpha$) we have

$$\sum_{j \in I} p_j p_{ji} = \sum_{j \in D_\alpha} \lambda_\alpha \pi_j p_{ji} = \lambda_\alpha \pi_i = p_i$$

for all $i \in D_\alpha$. This completes the proof. ◇

Corollary. *Suppose that the state space I is a positive class. The probability distribution* $(p_i)_{i \in I}$ *is stationary iff* $p_i = \pi_i$, $i \in I$.

1.1.3.3.3. In recent years many investigations have been devoted to the so-called *quasi-stationary* distributions for Markov chains. The starting point was BARTLETT's (1957) paper which suggested an investigation of distributions describing the long-term behaviour within an inessential class of a Markov chain. Actually, the papers by DARROCH and SENETA (1965), EWENS (1964), SENETA and VERE-JONES (1966), VERE-JONES (1962, 1967, 1968) permit to conclude (see KENDALL (1966c)) that the quasi-stationary distributions for a Markov chain appear to be stationary distributions for another suitable Markov chain. Thus the distribution $(\beta_j \alpha_j)_{j \in S}$ encountered in Subparagraph 1.1.2.6.3. is a stationary one for a Markov chain with transition probabilities

$$p^*_{ij} = \frac{\alpha_j p_{ji}}{\lambda_1 \alpha_i}, \quad i, j \in S,$$

where λ_1 is the maximum eigenvalue of $(p_{ij})_{i,j \in S}$.

1.1.3.4. Stationary measures

1.1.3.4.1. Theorem 1.1.41 shows that if there are no positive states, then there is no stationary distribution. In such a case we shall consider a more general concept than that of a stationary distribution, namely that of a stationary measure.

Definition 1.1.42. A family $(u_i)_{i \in I}$ of non-negative numbers is said to be a *stationary measure* for the Markov chain $(\xi_n)_{n \in N}$ iff $u_i = \sum_{j \in I} u_j p_{ji}$ for all $i \in I$ [15].

An interesting probabilistic interpretation of stationary measures has been given by DERMAN (1955).

1.1.3.4.2. The following result is an extension of Theorem 1.1.40.

---

[15] No conditions are imposed on the convergence of the series $\sum_{i \in I} u_i$.

Theorem 1.1.43 (C. DERMAN). *Let C be a recurrent class. The only non-negative solutions* $\{u_i, i \in C\}$ *of the system of equations*

$$u_i = \sum_{j \in C} u_j p_{ji}, \; i \in C, \tag{1.1.42}$$

are given by $u_i = c \; _kp^*_{ki}$ *where k is any state in C and c an arbitrary non-negative constant* [16].

Proof. We have, by the very definition of $_kp^*_{ki}$,

$$\sum_{j \neq k} {_kp^{(n)}_{kj}} p_{ji} = {_kp^{(n+1)}_{ki}}.$$

Hence (here and below the interchange of the order of summation is justified since all terms are non-negative)

$$\sum_{j \in C} {_kp^*_{kj}} p_{ji} = \sum_{j \in C} \sum_{n \in N^*} {_kp^{(n)}_{kj}} p_{ji} = \sum_{n \in N^*} \sum_{j \in C} {_kp^{(n)}_{kj}} p_{ji} =$$

$$= \sum_{n \in N^*} \{ {_kp^{(n+1)}_{ki}} + {_kp^{(n)}_{kk}} p_{ki} \} = {_kp^*_{ki}} - {_kp^{(1)}_{ki}} + {_kp^*_{kk}} p_{ki} = {_kp^*_{ki}}.$$

We took into account here that $_kp^{(1)}_{ki} = p_{ki}$ and $_kp^*_{kk} = f^*_{kk} = 1$. Therefore $u_i = c \; _kp^*_{ki}$, $i \in C$, is a solution.

Passing to the uniqueness, let $\{u_i, i \in C\}$ be a non-negative solution. It is easily seen that

$$u_i = \sum_{j \in C} u_j p^{(n)}_{ji}, \; i \in C,$$

for every $n \in N^*$. It follows that if one $u_i = 0$, then all $u_i = 0$. Thus suppose that all $u_i > 0$.

Let us introduce the quantities

$$q_{ij} = \frac{u_j}{u_i} p_{ji}, \; i, j \in C.$$

Since

$$q_{ij} \geqslant 0, \sum_{j \in C} q_{ij} = 1,$$

---

[16] Compare with Proposition 1.1.38.

we may regard q_{ij}, $i, j \in C$, as transition probabilities of a Markov chain [17]. It is easily seen that

$$q_{ij}^{(n)} = \frac{u_j}{u_i} p_{ji}^{(n)},\ i, j \in C.$$

By Theorem 1.1.3 there exists a Markov chain with state space C and transition matrix $(q_{ij})_{i,\, j \in C}$[18]. It is obvious that this Markov chain is an irreducible one and since $\sum_{n \in N^*} q_{ii}^{(n)} = \sum_{n \in N^*} p_{ii}^{(n)} = \infty$ it is recurrent by Theorem 1.1.6. We have

$$\frac{\sum_{n=0}^{m} q_{ik}^{(n)}}{\sum_{n=0}^{m} q_{kk}^{(n)}} = \frac{u_k}{u_i} \cdot \frac{\sum_{n=0}^{m} p_{ki}^{(n)}}{\sum_{n=0}^{m} p_{kk}^{(n)}}$$

Using Doeblin formula (1.1.7) for the new Markov chain and the Corollary to Proposition 1.1.37 for the old Markov chain we deduce

$$1 = \frac{u_k}{u_i} {}_k p_{ki}^{*}.$$

For an arbitrary fixed $k \in C$ this proves the stated uniqueness. We notice that by (1.1.26), for two different choices of k the corresponding solutions are proportional. ◇

1.1.3.4.3. The modern point of view on the study of the system of equations (1.1.42) defining stationary measures as well as of the system of equations appearing in Theorem 1.1.34 is to present it as a corollary of potential theory of Markov chains alluded to at the end of Subparagraph 1.1.2.1.2.

1.1.3.4.4. The problem of characterizing stationary measures for a nonrecurrent class is still open. Here we give a necessary and sufficient condition for the existence of a stationary measure for such a class.

Theorem 1.1.44. (T. E. HARRIS and W. VEECH). *Let C be a nonrecurrent class (essential or not) whose states are labelled by the non-*

---

[17] These probabilities are called the *inverse probabilities* for the Markov chain considered. If C is a positive class, then

$$q_{ij} = \mathsf{P}_{\pi}(\xi_n = j \mid \xi_{n+1} = i),\ n \in N,$$

where $\pi = (\pi_i)_{i \in C}$.

[18] Sometimes it is called the *reverse chain*.

negative integers. In order that the system (1.1.42) *admit a not identical null solution it is necessary and sufficient that there exist an infinite set* $K \subset C$ *such that*

$$\lim_{\substack{j\to\infty,\ k\to\infty \\ (k\in K)}} \frac{\sum\limits_{r\geqslant j}\sum\limits_{n\in N} {}_i p_{kr}^{(n)}\, p_{ri}}{\sum\limits_{r\in N}\sum\limits_{n\in N} {}_i p_{kr}^{(n)}\, p_{ri}} = 0 \tag{1.1.43}$$

for all $i \in N$.

Proof. See HARRIS (1957) and VEECH (1963). ◇

We note that the numerator in (1.1.43) is the probability that the Markov chain starting in state k reaches state i with the first visit preceded by a visit to a state with index $\geqslant j$, while the denominator is the probability of ever reaching state i starting in state k.

For examples concerning stationary measures for nonrecurrent classes we refer the reader to DERMAN (1955). An interesting application of Theorem 1.1.44 may be found in PRUITT (1964).

1.1.3.5. Integral representations

1.1.3.5.1. Previous considerations show that the computation of the n-step transition probabilities is of fundamental importance. It was proved by KENDALL (1959) that a Fourier representation can be found for the sequence $(p_{ij}^{(n)})_{n\in N}$ whenever i and j are identical states or distinct communicating states. Kendall's approach is based on Hilbert space techniques which, despite their formal elegance, are not natural for the problem; his arguments were subsequently greatly simplified by FELLER (1966 a) who also improved the results.

In what follows we present Feller's treatment. The use of methods from functional analysis will be illustrated in special cases (see Subparagraphs 1.2.1.5.2 and 1.2.2.5.1).

1.1.3.5.2. Let us first consider the case of a nonrecurrent class. We have

Theorem 1.1.45. *Let* C *be a nonrecurrent class and let* $v_i > 0$, $i \in C$ *be arbitrary. There exist bounded complex-valued continuous functions* m_{ij} *defined on* $[-\pi, \pi]$ *such that* $m_{ii} \geqslant 0$, $i, j \in C$, *and*

$$\int_{-\pi}^{\pi} e^{in\theta}\, m_{ij}(\theta)\, d\theta = \begin{cases} \left(\dfrac{v_i}{v_j}\right)^{1/2} p_{ij}^{(n)} & \text{for } n \in N \\[2ex] \left(\dfrac{v_j}{v_i}\right)^{1/2} p_{ji}^{(-n)} & \text{for } n \in -N \end{cases} \tag{1.1.44}$$

(implying $m_{ij}(-\theta) = m_{ji}(\theta) = \overline{m_{ij}(\theta)}$).

Proof. By Theorem 1.1.26, the power series $P_{ij}(z)$, $i, j \in C$, converges uniformly for $|z| = 1$, and hence, when $i \neq j$,

$$\frac{1}{2\pi}\int_{-\pi}^{\pi} e^{-in\theta} P_{ij}(e^{i\theta})\, d\theta = \begin{cases} p_{ij}^{(n)} & \text{for } n \in N \\ 0 & \text{for } n \in -N. \end{cases}$$

An analogous relation holds with i and j interchanged, and so (1.1.44) holds with

$$m_{ij}(\theta) = \frac{1}{2\pi}\left[\left(\frac{v_i}{v_j}\right)^{1/2} P_{ij}(e^{-i\theta}) + \left(\frac{v_j}{v_i}\right)^{1/2} P_{ji}(e^{i\theta})\right].$$

The case $i = j$ is covered by the result of Subparagraph 1.1.2.2.3. ◇

1.1.3.5.3. For a recurrent class we shall prove

Theorem 1.1.46. *Let C be an aperiodic recurrent class. Let $(u_i)_{i\in C}$ stand for the (essentially unique) stationary measure. There exists a unique system of integrable complex-valued functions m_{ij} defined on $[-\pi, \pi]$ such that $m_{ij} = \bar{m}_{ji}$ (implying $m_{ii} \geqslant 0$), $i, j \in C$ and*

$$\left(\frac{u_i}{u_j}\right)^{1/2} (p_{ij}^{(n)} - \pi_{ij}) = \int_{-\pi}^{\pi} e^{in\theta} m_{ij}(\theta)\, d\theta. \tag{1.1.45}$$

The functions m_{ij}, $i, j \in C$, are continuous outside every neighborhood of $\theta = 0$ and

$$\frac{1}{u_i} m_{ii} - \frac{1}{u_j} m_{jj} \tag{1.1.46}$$

is square integrable.

Proof. Consider first the diagonal terms. It was proved in Subparagraph 1.1.2.2.3. that (1.1.45) holds with

$$m_{ii}(\theta) = \frac{1}{2\pi}\,[2 \operatorname{Re} P_{ii}(e^{i\theta}) - 1] \geqslant 0, \quad i \in C.$$

On account of (1.1.31), Theorem 1.1.43, and (1.1.26), it is clear that to prove the square integrability of (1.1.46) it suffices to prove that the boundary values $\Phi_{ij}(e^{i\theta})$ are square integrable over $[-\pi, \pi]$. In order to do this, put $z = \rho\, e^{i\theta}$ and note that by Schwarz's inequality

$$|{}_if_{jj}^* - {}_iF_{jj}(z)|^2 \leqslant {}_if_{jj}^* \sum_{n\in N^*} {}_if_{jj}^{(n)} |1 - \rho^n e^{in\theta}|^2 \leqslant$$

$$\leqslant 2 \sum_{n\in N^*} {}_if_{jj}^{(n)}(1 - \rho^n \cos\theta) \leqslant 2 \operatorname{Re}[1 - F_{jj}(z)]. \tag{1.1.47}$$

Thus

$$|\Phi_{ij}(z)|^2 \leqslant 2\,\frac{\operatorname{Re}[1-F_{jj}(z)]}{|1-F_{jj}(z)|^2} = 2\operatorname{Re}\frac{1}{1-F_{jj}(z)}, \qquad (1.1.48)$$

and, as in Subparagraph 1.1.2.2.3, the right-hand side is a non-negative function whose integral over $[-\pi, \pi]$ does not exceed 2.

Turning to the non-diagonal terms, put

$$u_{ij}(z) = \frac{1}{2\pi}\left[\left(\frac{u_i}{u_j}\right)^{1/2}\overline{P_{ij}(z)} + \left(\frac{u_j}{u_i}\right)^{1/2} P_{ji}(z)\right]$$

for $i \neq j$. For fixed $\rho < 1$, it is clear that

$$\int_{-\pi}^{\pi} e^{in\theta} u_{ij}(\rho e^{i\theta})\, d\theta = \begin{cases} \left(\dfrac{u_i}{u_j}\right)^{1/2} \rho^n p_{ij}^{(n)} & \text{for } n \in N \\[2ex] \left(\dfrac{u_j}{u_i}\right)^{1/2} \rho^{-n} p_{ji}^{(-n)} & \text{for } n \in -N. \end{cases}$$

Let

$$m_{ij}(\theta) = u_{ij}(e^{i\theta}).$$

Then $m_{ij} = \overline{m_{ji}}$ and, to prove the representation (1.1.45) it suffices to prove that u_{ij} is integrable over $|z| = 1$. To do this we refer again to (1.1.31) and recall that Φ_{ij} has integrable boundary values. Using (1.1.27), Theorem 1.1.43 and (1.1.26) it is clear that it suffices to prove that the function

$$F_{ij}C_{ij} + \overline{F}_{ji}\overline{C}_{ij} = (F_{ij} - \overline{F}_{ji})\,\overline{C}_{ij} + 2\overline{F}_{ji}\operatorname{Re} C_{ij}$$

(C_{ij} was defined by (1.1.32)) is integrable over $|z| = 1$. As in (1.1. 7) and (1.1.48) we get, for $|z| < 1$,

$$|(F_{ij} - \overline{F}_{ji})\, C_{ij}|^2 \leqslant 2(1 - \operatorname{Re} F_{ij}F_{ji})\, |C_{ij}|^2 = 2\operatorname{Re} C_{ij}\,.$$

To end the proof we need only remark that $F_{ij}F_{ji}$ may be considered as the generating function of the interarrival time distribution of a regenerative phenomenon. Thus the argument used to prove the integrability of the right-hand side of (1.1.48) applies also for $\operatorname{Re} C_{ij}$. ◇

1.1.4. Finite state Markov chains

This subsection is concerned with the special case in which the state space I is a finite set.

1.1.4.1. Specific properties

1.1.4.1.1. The results which have been obtained so far hold whether the number of states is finite or not. However, as we shall see in the case of a finite state space, the situation is simpler in various respects. For a detailed study of finite state Markov chains the reader is referred to KEMENY and SNELL (1960).

Proposition 1.1.47. *In a finite state Markov chain there are no essential null states, and not all states can be inessential.*

Proof. Suppose to the contrary that there is an essential null state and let C be its essential class. Since C is closed,

$$\sum_{j\in C} p_{ij}^{(n)} = 1$$

for all $i \in C$, $n \in N^*$. According to Theorem 1.1.14 this leads to

$$1 = \lim_{n\to\infty} \sum_{j\in C} p_{ij}^{(n)} = 0,$$

and we reach a contradiction; thus the first assertion is proved [19]. The second assertion follows in an analogous manner from

$$\sum_{j\in I} p_{ij}^{(n)} = 1, \quad i \in I, \quad n \in N^*. \quad \diamond$$

It is easy to see that Proposition 1.1.47 implies that in a finite state Markov chain the three dichotomies positive-null, recurrent-nonrecurrent, essential-inessential coincide. In the following we shall use only the last dichotomy, and F will mean the set of inessential states.

We notice that the transition matrix of an arbitrary finite state Markov chain may be written in the form

$$\mathbf{P} = \begin{pmatrix} \mathbf{P}_1 & \mathbf{O} & \dots & \mathbf{O} & \mathbf{O} \\ \mathbf{O} & \mathbf{P}_2 & \dots & \mathbf{O} & \mathbf{O} \\ \cdot & \cdot & \dots & \cdot & \cdot \\ \cdot & \cdot & \dots & \cdot & \cdot \\ \cdot & \cdot & \dots & \cdot & \cdot \\ \mathbf{O} & \mathbf{O} & \dots & \mathbf{P}_r & \mathbf{O} \\ \mathbf{F}_1 & \mathbf{F}_2 & \dots & \mathbf{F}_r & \mathbf{F} \end{pmatrix}$$

[19] This proof shows that in a denumerable state Markov chain an essential null class is either empty or infinite.

where the square matrices $\mathbf{P}_1, \ldots, \mathbf{P}_r$ correspond to the essential classes and the matrices $\mathbf{F}_1, \ldots, \mathbf{F}_r$, $\mathbf{F}$ to the inessential states.

1.1.4.1.2. Let us characterize the case of an indecomposable finite state Markov chain whose essential states are aperiodic. By Theorem 1.1.15 we have

$$\lim_{n\to\infty} p_{ij}^{(n)} = q_j, \quad i, j \in I, \tag{1.1.49}$$

where $q_j = \pi_j$ or zero according as state j is essential or inessential. Therefore this limit does not depend on the starting state i. Conversely, if (1.1.49) holds, then there is a single aperiodic essential class $C = \{j : q_j > 0\}$.

We have now the following result essentially due to A. A. MARKOV himself.

Theorem 1.1.48. *In order that* (1.1.49) *holds it is necessary and sufficient that there are an* $n_0 \in N^*$ *and a* $j_0 \in I$ *such that*

$$p_{ij_0}^{(n_0)} > 0$$

for all $i \in I$. *Moreover, if there is a set* $I_1 \subset I$ *consisting of* s *elements such that*

$$\min_{i \in I,\, j \in I_1} p_{ij}^{(n_0)} \geqslant \delta > 0,$$

then

$$|p_{ij}^{(n)} - q_j| \leqslant (1 - s\delta)^{\frac{n}{n_0} - 1}$$

for all $i, j \in I$, $n \in N^*$.

Proof. See e.g. DOOB (1953, p. 173). ◇

Analogously the case of an irreducible aperiodic finite state Markov chain is characterized by the existence of an $n_0 \in N^*$ such that

$$p_{ij}^{(n_0)} > 0, \qquad i, j \in I.$$

1.1.4.1.3. Concerning the motion of a finite state Markov chain within the set F of inessential states, we have

Proposition 1.1.49. *The probability that a finite state Markov chain moves into* F *infinitely many times is zero.*

Proof. Apply Corollary to Theorem 1.1.33. ◇

Note that the second proof of Corollary to Theorem 1.1.33 implies the existence of numbers $a > 0$, $0 < b < 1$ such that $p_{ij}^{(n)} \leqslant ab^n$, $n \in N$, for any $i, j \in F$. In combination with Theorem 1.1.48 this remark yields

Theorem 1.1.50. *Any indecomposable finite state Markov chain whose essential states are aperiodic is geometrically ergodic.*

1.1.4.1.4. Concerning the matrix $\mathbf{\Pi} = (\pi_{ij})_{i,j \in I}$, introduced in Subparagraph 1.1.2.3.3 and which is, in general, substochastic, we have

Proposition 1.1.51. *For finite Markov chains the matrix* $\mathbf{\Pi}$ *is stochastic.*
Proof. We can write

$$\mathbf{\Pi} = \lim_{n \to \infty} \frac{1}{n} \sum_{m=1}^{n} \mathbf{P}^n,$$

that is, $\mathbf{\Pi}$ is the limit of a sequence of finite order stochastic matrices. ◇

In Subparagraph 1.1.4.2.5 we shall give a method for computing $\mathbf{\Pi}$

1.1.4.1.5. We shall now consider an interesting special case, namely that of a *doubly stochastic* matrix $\mathbf{P}$ of finite order characterized by

$$\sum_{j \in I} p_{ij} = \sum_{i \in I} p_{ij} = 1, \qquad i, j \in I.$$

If $\mathbf{P}$ is doubly stochastic, then its powers are also. Therefore $\mathbf{\Pi}$ will be also a doubly stochastic matrix. It follows that there are no inessential states (otherwise the matrix $\mathbf{\Pi}$ would contain a null column). Since in an essential class π_{ij} does not depend on i, we deduce easily that

$$\pi_{ij} = \frac{1}{\text{number of elements in } C}, \qquad i, j \in C,$$

for every essential class C.

In particular, the matrix $\mathbf{P}$ is doubly stochastic if it is symmetric ($p_{ij} = p_{ji}$, $i, j \in I$). In this case no inessential states are possible and the period of an essential class can be at most two. Thus

$$\lim_{n \to \infty} p_{ij}^{(n)} = \frac{1}{\text{number of elements in } C}, \qquad i, j \in C$$

for every aperiodic essential class C [20]. Conversely, if the above relation holds for an aperiodic essential class C, then $p_{ij} = p_{ji}$, $i, j \in C$.

For more details about doubly stochastic matrices, the reader is referred to MIRSKY (1962/63).

---

[20] It is easily seen that an infinite essential class with doubly stochastic matrix is either nonrecurrent or null-recurrent.

1.1.4.1.6. For finite state Markov chains the problem of finding the probabilities of absorption $\mathfrak{a}_i(C)$ into an essential class C given that the Markov chain started in $i \in F$ is completely solved by the system (1.1.22). Indeed, by the remark made in Subparagraph 1.1.4.1.3, the matrix $\mathbf{I} - \mathbf{F}$, where $\mathbf{F} = (p_{ij})_{i,j \in F}$ and $\mathbf{I} = (\delta_{ij})_{i,j \in F}$, has an inverse (which is easily seen to be $\sum_{k \in N} \mathbf{F}^k$). Thus Propositions 1.1.35 and 1.1.49 and relation (1.1.22) imply that

$$(\mathfrak{a}_i(C))_{i \in F} = (\mathbf{I} - \mathbf{F})^{-1}(\sum_{j \in C} p_{ij})_{i \in F}.$$

The matrix $(\mathbf{I} - \mathbf{F})^{-1}$ is known as the *fundamental* matrix associated with $\mathbf{P}$. It was shown by KEMENY and SNELL (1960, Chapter III) that moments of many quantities describing the motion of the chain within suitable sets of states may be expressed in terms of the corresponding fundamental matrices. We have

Theorem 1.1.52. *Let S be a set of s states such that no essential class is a subset of it. Let $\mathbf{S} = (p_{ij})_{i,j \in S}$, $\mathbf{I} = (\delta_{ij})_{i,j \in S}$, and denote by $\boldsymbol{\rho}_k$, $k \in I - S$, and $\mathbf{1}$ the s-component column vectors with entries p_{ik}, $i \in S$, respectively with all entries 1. Then*

i) *The ij-th entry of $\widetilde{\mathbf{N}} = (\mathbf{I} - \mathbf{S})^{-1}$ is the mean number of visits to the state $j \in S$ before leaving S, given that the Markov chain started in $i \in S$.*

ii) *The ij-th entry of $\widetilde{\mathbf{N}}\,(2\widetilde{\mathbf{N}}_{dg} - \mathbf{I}) - \widetilde{\mathbf{N}}_{sq}$*[21] *is the variance of the same function under the same initial condition.*

iii) *The i-th entry of $\widetilde{\mathbf{N}}\mathbf{1}$ is the mean number of steps taken before leaving S, given that the Markov chain started in $i \in S$.*

iv) *The i-th entry of $(2\widetilde{\mathbf{N}} - \mathbf{I})\,\widetilde{\mathbf{N}}\mathbf{1} - (\widetilde{\mathbf{N}}\mathbf{1})_{sq}$ is the variance of the same function under the same initial condition.*

v) *The i-th entry of $\widetilde{\mathbf{N}}\boldsymbol{\rho}_k$ is the probability that starting in $i \in S$ the Markov chain goes to $k \in I - S$ when it leaves S.*

Proof. Consider a new Markov chain $(\tilde{\xi}_n)_{n \in N}$ with transition probabilities

$$\tilde{p}_{ij} = \begin{cases} p_{ij} & \text{if} \quad (i,j) \in S \times S \\ \delta_{ij} & \text{if} \quad (i,j) \notin S \times S \end{cases}$$

(thus all the states in $I - S$ are made absorbing). It is easy to see that the set of inessential states for $(\tilde{\xi}_n)_{n \in N}$ is just S. Moreover, if $\xi_0 = \tilde{\xi}_0 \in S$, then before leaving S the motions of the two Markov chains coincide. Consequently, i) and v) follow from a remark made in Subparagraph 1.1.2.1.2 and the beginning of this subparagraph. Straightforward

[21] $\mathbf{A}_{dg}$ and $\mathbf{A}_{sq}$ are the matrices resulting from setting off-diagonal entries of $\mathbf{A}$ equal to 0 and by squaring each entry of $\mathbf{A}$ respectively.

computations lead to ii), iii), and iv). Details may be found in KEMENY and SNELL (1960, p. 49—54). ◇

1.1.4.1.7. By virtue of Proposition 1.1.47 and Theorem 1.1.41, every finite state Markov chain has at least one stationary distribution.

Theorem 1.1.53. *Every stationary distribution of a finite state Markov chain is a linear convex combination of probability distributions* $(\pi_{ij})_{j \in I}$, $i \in I - F$.

Proof. Apply Theorem 1.1.41. ◇

Let us consider an essential class C. We know that there is a unique stationary distribution for C. DECELL JR. and ODELL (1967 a, b) have shown that a simple form for it may be obtained using the theory of generalized matrix inversion (see e.g., PENROSE (1955)). In this way one avoids the calculation of powers of the transition matrix.

1.1.4.2. The matrix method

1.1.4.2.1. In Subsections 1.1.1—1.1.3 we studied Markov chains using the so-called direct method initiated by A. N. KOLMOGOROV in 1936. It is clear that the (spectral) theory of matrices may be an important tool of investigation in this area. Finite state Markov chains can be exhaustively studied by means of purely algebraic methods based on the Perron-Frobenius theory of finite order non-negative matrices. This approach to finite state Markov chains was initiated by V. I. ROMANOVSKI in 1935. Unfortunately, a completely analogous treatment of Markov chains with infinitely denumerable state space is not possible (for details see SARYMSAKOV (1954); an interesting attempt to develop a theory analogous to the Perron-Frobenius theory for infinite order non-negative matrices was made by VERE-JONES (1967, 1968)). It must be noted that although the algebraic methods offer some useful algorithms, they can obscure the probabilistic aspects. In what follows we shall give without proofs [22] the most important algebraic results concerning non-negative finite order matrices.

1.1.4.2.2. Let $\mathbf{A} = (a_{ij})_{1 \leqslant i, j \leqslant n}$ be a complex matrix of order n. A complex number λ is called an *eigenvalue* of the matrix $\mathbf{A}$ if there exists an n-dimensional (column) vector $\mathbf{u} \neq \mathbf{0}$, such that $\mathbf{u}'\mathbf{A} = \lambda\mathbf{u}'$. If λ is an eigenvalue of $\mathbf{A}$, then the set $\mathfrak{U}_\lambda$ consisting of all vectors $\mathbf{u}$ which satisfy the equation $\mathbf{u}'\mathbf{A} = \lambda\mathbf{u}'$ is called the *left eigenspace* of $\mathbf{A}$ corresponding to the eigenvalue λ, and the elements of $\mathfrak{U}_\lambda$ are called *left eigenvectors* for λ. The set $\mathfrak{U}_\lambda$ is a linear manifold and its dimension is called the *geometric multiplicity* of λ. The eigenvalues of $\mathbf{A}$ are precisely the roots of the nth degree algebraic equation (the *characteristic equation*)

$$\det(\lambda\mathbf{I} - \mathbf{A}) = 0,$$

[22] The proofs can be found, e.g., in GANTMACHER (1959) or KARLIN (1966, Ch. IV and Appendix). See also WIELANDT (1950) for a very elegant treatment.

where **I** is the identity matrix of order n. Thus, **A** can have only a finite number (at most n) of eigenvalues and left eigenspaces. The geometric multiplicity of an eigenvalue is at most its order of multiplicity as root of the characteristic equation.

The values of λ such that $\mathbf{A}\mathbf{v} = \lambda\mathbf{v}$ for suitable n-dimensional (column) vectors $\mathbf{v} \neq \mathbf{0}$ are precisely the eigenvalues of **A** as defined above. Moreover, the dimension of the manifold $\mathfrak{V}_\lambda$ of vectors $\mathbf{v}$ satisfying $\mathbf{A}\mathbf{v} = \lambda\mathbf{v}$ (the so-called *right eigenvectors* for λ) is just the geometric multiplicity of λ.

If $\mathfrak{U}_{\lambda_i}$ ($\mathfrak{V}_{\lambda_i}$), $1 \leqslant i \leqslant r$ are distinct left (right) eigenspaces of the matrix **A** and $\mathbf{u}_1^{(i)}, \ldots, \mathbf{u}_{m_i}^{(i)}$ ($\mathbf{v}_1^{(i)}, \ldots, \mathbf{v}_{m_i}^{(i)}$) is a basis of $\mathfrak{U}_{\lambda_i}$ ($\mathfrak{V}_{\lambda_i}$), then

$$\mathbf{u}_1^{(1)}, \ldots, \mathbf{u}_{m_1}^{(1)}, \ldots, \mathbf{u}_1^{(r)}, \ldots, \mathbf{u}_{m_r}^{(r)}$$

$$(\mathbf{v}_1^{(1)}, \ldots, \mathbf{v}_{m_1}^{(1)}, \ldots, \mathbf{v}_1^{(r)}, \ldots, \mathbf{v}_{m_r}^{)r)})$$

is a linearly independent set.

1.1.4.2.3. From now on we assume that **A** is real. Suppose that the sum of geometric multiplicities of distinct eigenvalues of **A** equals n. From the above remark it follows that we can construct a basis for R^n using left (right) eigenvectors of **A**. These bases $\mathbf{u}_1, \ldots, \mathbf{u}_n$ and $\mathbf{v}_1, \ldots, \mathbf{v}_n$ may be chosen to be biorthogonal, i.e., $\mathbf{u}_i'\mathbf{v}_j = \delta_{ij}$, $1 \leqslant i, j \leqslant n$. [23]

Using this result we can obtain the so-called *spectral representation* of **A**. Let λ_i be the eigenvalue corresponding to $\mathbf{u}_i$ (the λ_i need not be distinct), $1 \leqslant i \leqslant n$. We have

$$\mathbf{A} = \sum_{i=1}^{n} \lambda_i \mathbf{v}_i \mathbf{u}_i' \tag{1.1.50}$$

and since $\mathbf{u}_i'\mathbf{v}_j = \delta_{ij}$, $1 \leqslant i, j \leqslant n$,

$$\mathbf{A}^m = \sum_{i=1}^{n} \lambda_i^m \mathbf{v}_i \mathbf{u}_i' .$$

Therefore, once the spectral representation (1.1.50) is known, it is not difficult to compute the powers of **A**.

1.1.4.2.4. We close these considerations with the Frobenius theorems. We shall write $\mathbf{M} > \mathbf{0}$ or $\mathbf{M} \geqslant \mathbf{0}$ according as the elements of the matrix **M** are all positive or nonnegative.

---

[23] In particular, if the roots of the characteristic equation are distinct, then there are n distinct unidimensional left (right) eigenspaces and the two bases are biorthogonal after a suitable normalization.

Let $\mathbf{A} \geqslant \mathbf{0}$, and let Λ be the set of real numbers λ to each of which corresponds a vector $\mathbf{x}' = (x_1, \ldots, x_n)$ such that

$$\sum_{i=1}^{n} x_i = 1, \quad \mathbf{x} \geqslant \mathbf{0} \quad \text{and} \quad \mathbf{A}\mathbf{x} \geqslant \lambda\mathbf{x}.$$

Put $\lambda_0(\mathbf{A}) = \sup\{\lambda : \lambda \in \Lambda\}$. It is easy to see that $\lambda_0(\mathbf{A})$ is finite and if $\mathbf{A} > \mathbf{0}$, then $\lambda_0(\mathbf{A})$ is positive.

The first Frobenius theorem is

Theorem 1.1.54. *If* $\mathbf{A}^r > \mathbf{0}$ *for some* $r \in N^*$, *then* i) *there is* $\mathbf{u}_0 > \mathbf{0}$ *(resp.* $\mathbf{v}_0 > \mathbf{0}$*) such that* $\mathbf{u}_0' \mathbf{A} = \lambda_0(\mathbf{A})\, \mathbf{u}_0'$ *(resp.* $\mathbf{A}\mathbf{v}_0 = \lambda_0(\mathbf{A})\, \mathbf{v}_0$*)*; ii) *if* $\lambda \neq \lambda_0(\mathbf{A})$ *is any other eigenvalue of* $\mathbf{A}$, *then* $|\lambda| < |\lambda_0(\mathbf{A})|$; iii) $\lambda_0(\mathbf{A})$ *is a simple root of the characteristic equation.*

Using this theorem one may deduce

Proposition 1.1.55. *If* $\mathbf{A}^r > \mathbf{0}$ *for some* $r \in N^*$, *then*

$$\lim_{m \to \infty} \frac{\mathbf{A}^m}{[\lambda_0(\mathbf{A})]^m} = \mathbf{v}_0 \mathbf{u}_0',$$

the convergence being geometrically fast.

The main Frobenius theorem is

Theorem 1.1.56. *If* $\mathbf{A} \geqslant \mathbf{0}$ *then* i) $\lambda_0(\mathbf{A})$ *is an eigenvalue of* $\mathbf{A}$ *with a left (right) eigenvector* $\mathbf{u}_0 \geqslant \mathbf{0}$ $(\mathbf{v}_0 \geqslant \mathbf{0})$; ii) *if* λ *is any other eigenvalue of* $\mathbf{A}$, *then* $|\lambda| \leqslant |\lambda_0(\mathbf{A})|$; iii)

$$\frac{1}{m} \sum_{i=1}^{m} \frac{\mathbf{A}^i}{[\lambda_0(\mathbf{A})]^i}$$

converges as $m \to \infty$ *if* $\mathbf{v}_0 > \mathbf{0}$; iv) *if* λ *is an eigenvalue of* $\mathbf{A}$ *and* $|\lambda| = |\lambda_0(\mathbf{A})|$, *then* $\eta = \lambda/\lambda_0(\mathbf{A})$ *is a root of unity and* $\eta^m \lambda_0(\mathbf{A})$ *is an eigenvalue of* $\mathbf{A}$ *for* $m \in N$.

1.1.4.2.5. If $\mathbf{P}$ is a finite n-order stochastic matrix, then it is easily seen that $\lambda_0(\mathbf{P}) = 1$. The (either geometric or algebraic) multiplicity of the eigenvalue 1 equals the number of essential classes associated with $\mathbf{P}$. Let $C_1, \ldots, C_r$ be these classes. Then the vectors

$$\mathbf{u}_k = (\pi_j \delta_j(C_k))_{1 \leqslant j \leqslant n}, \qquad 1 \leqslant k \leqslant r,$$

where $\delta_x(A)$ is 1 or 0 according as $x \in A$ or $x \notin A$, form a basis for the left eigenspace corresponding to the eigenvalue 1. A basis for the corresponding right eigenspace is formed by the vectors

$$\mathbf{v}_k = (a_j(C_k))_{1 \leqslant j \leqslant n}, \qquad 1 \leqslant k \leqslant r,$$

(the quantities $\mathfrak{a}_j(C_k)$ were defined in Subparagraph 1.1.2.6.2). These two bases are biorthogonal.

The general theorem characterizing finite order stochastic matrices is

Theorem 1.1.57. *Let* **P** *be a finite order stochastic matrix. Any eigenvalue of* **P** *of modulus* 1 *is a root of unity. The dth roots of unity are eigenvalues of* **P** *iff* **P** *has an essential class with period d. The multiplicity of each dth root of unity is just the number of essential classes of period d.*

The characteristic polynomial $f(\lambda) = \det(\lambda\mathbf{I} - \mathbf{P})$ of a stochastic finite order matrix **P** can be used to obtain the matrix $\mathbf{\Pi}$ as already stated in Subparagraph 1.1.4.1.4. Let g be the minimal monic polynomial such that $g(\mathbf{P}) = 0$; g will be a factor of f, and may be a proper factor. Let $f(\lambda) = a(\lambda - 1)^d \varphi(\lambda)$, $g(\lambda) = b(\lambda - 1)\psi(\lambda)$ where the constants a and b are such that $\varphi(1) = \psi(1) = 1$.

Proposition 1.1.58. (VERE-JONES and KENDALL (1959)). *We have* $\mathbf{\Pi} = \varphi(\mathbf{P}) = \psi(\mathbf{P})$.

The proof follows from the properties of the Jordan canonical form for **P**.

1.1.4.3. The Ehrenfest model

1.1.4.3.1. Consider the Markov chain with states $-a, -a+1, \ldots, 0, \ldots, a-1, a$, where $a \in N^*$, and transition probabilities

$$p_{i,i+1} = p_{-i,-i-1} = \frac{a-i}{2a}$$
$$p_{i,i-1} = p_{-i,-i+1} = \frac{a+i}{2a} \tag{1.1.51}$$

for $0 \leqslant i \leqslant a-1$, $p_{a,a-1} = p_{-a,-a+1} = 1$ and $p_{ij} = 0$, otherwise. The transition matrix is therefore (for $a > 3$)

$$\mathbf{P} = \begin{array}{c|cccccccccc} & -a & -a+1 & -a+2 & -a+3 & \ldots & a-3 & a-2 & a-1 & a \\ -a & 0 & 1 & 0 & 0 & \ldots & 0 & 0 & 0 & 0 \\ -a+1 & 1/2a & 0 & 1-1/2a & 0 & \ldots & 0 & 0 & 0 & 0 \\ -a+2 & 0 & 2/2a & 0 & 1-2/2a & \ldots & 0 & 0 & 0 & 0 \\ \vdots & \vdots & \vdots & \vdots & \vdots & & \vdots & \vdots & \vdots & \vdots \\ a-2 & 0 & 0 & 0 & 0 & \ldots & 1-2/2a & 0 & 2/2a & 0 \\ a-1 & 0 & 0 & 0 & 0 & \ldots & 0 & 1-1/2a & 0 & 1/2a \\ a & 0 & 0 & 0 & 0 & \ldots & 0 & 0 & 1 & 0 \end{array}$$

This Markov chain is irreducible and of period two. It may be interpreted in terms of the *Ehrenfest model of diffusion*, describing the stochastic motion of molecules between two connected containers (I and II). Imagine $2a$ molecules (balls) numbered consecutively from 1 to $2a$, distributed in the two containers. At time n we choose at random an integer between 1 and $2a$ (all these integers are assumed to be equiprobable) and move the ball whose number has been drawn from its container into the other. We identify the state i of this system with the number of balls $a + i$ in I $(-a \leqslant i \leqslant a)$. It is easily seen that the transition probabilities describing this model are given by (1.1.51).

Following KAC (1947) and COX and MILLER (1965, p. 129) we shall obtain the eigenvalues of $\mathbf{P}$ and the corresponding left and right eigenvectors. The left eigenvectors $\mathbf{u}' = (u_0, \ldots, u_{2a})$ satisfy $\mathbf{u}'\mathbf{P} = \lambda\mathbf{u}'$ or

$$\begin{aligned}
&\frac{u_1}{2a} = \lambda u_0 \\
&\frac{2u_2}{2a} + \frac{2au_0}{2a} = \lambda u_1 \\
&\frac{3u_3}{2a} + \frac{(2a-1)u_1}{2a} = \lambda u_2 \\
&\quad\vdots \qquad\qquad \vdots \\
&\frac{2au_{2a}}{2a} + \frac{2u_{2a-1}}{2a} = \lambda u_{2a-1} \\
&\frac{u_{2a-1}}{2a} = \lambda u_{2a}\,.
\end{aligned} \tag{1.1.52}$$

If we introduce the polynomial

$$u(z) = \sum_{i=0}^{2a} u_i z^i,$$

then the system (1.1.52) can be written as

$$\sum_{i=0}^{2a} i u_i z^{i-1} + \sum_{i=0}^{2a} (2a - i)\, u_i z^{i+1} = 2a\lambda\, u(z)$$

or

$$(1 - z)\, u'(z) + 2a(z - \lambda)\, u(z) = 0. \tag{1.1.53}$$

The general solution to this differential equation is

$$u(z) = C(1+z)^{(1+\lambda)a}(1-z)^{(1-\lambda)a},$$

where C is an arbitrary constant. We are interested only in solutions $u(z)$ which are polynomials of degree $2a$ in z. Notice that the values

$$\lambda = \frac{i}{a}, \qquad -a \leqslant i \leqslant a,$$

furnish $2a+1$ such polynomials. These values are thus seen to be eigenvalues of $\mathbf{P}$ and, since there are $2a+1$ of them, we conclude that we have found all the eigenvalues. It also follows that the components of the left eigenvector corresponding to $\lambda = i/a$ are proportional to the coefficients of $1, z, \ldots, z^{2a}$ in

$$(1+z)^{a+i}(1-z)^{a-i}. \tag{1.1.54}$$

The unique stationary distribution is obtained by setting $i = a$ (i.e. $\lambda = 1$) in (1.1.54) and normalizing so that the coefficients sum to 1. We get $2^{-2a}(1+z)^{2a} = (1/2+z/2)^{2a}$, that is, a symmetric binomial distribution. We can therefore say that, whatever the initial number of molecules in the first container, after a long time the probability of finding a specified number of balls in it is nearly the same as if the $2a$ balls had been distributed at random, each ball having probability 1/2 of being in I.

Let

$$p_i(z) = 2^{-2a}z^{-a}(1+z)^{a+i}(1-z)^{a-i} = \sum_{j=-a}^{a} c_{ij}z^j, \qquad -a \leqslant i \leqslant a.$$

The left eigenvector corresponding to $\lambda = i/a$ is now given by the coefficients of $z^{-a}, \ldots, z^a$ in $p_i(z)$. Consider the matrix $\mathbf{C} = (c_{ij})_{-a\leqslant i,\, j\leqslant a}$. It is obvious that the right eigenvectors of $\mathbf{P}$ are the columns of $\mathbf{C}^{-1} = (c^{ij})_{-a\leqslant i,\, j\leqslant a}$. Straightforward calculations yield

$$c^{ij} = (-1)^{a+i}\, 2^{2a} c_{-i,\, j}$$

On account of Subparagraph 1.1.4.2.3 we have

$$p_{ij}^{(n)} = 2^{2a}(-1)^{a+i} \sum_{k=-a}^{a} \left(\frac{k}{a}\right)^n c_{-i,\, k}\, c_{kj}\,. \tag{1.1.55}$$

1.1.4.3.2. Following again KAC (1947) we note first that the Ehrenfest model is a simple and convenient model of the heat exchange between two isolated bodies of unequal temperatures. The temperatures are represented by the numbers of balls in the containers and the heat exchange is not an orderly process, as in classical thermodynamics, but a random one as in the kinetic theory of matter. The realistic value of the model is greatly enhanced by the fact that the average excess over a of the number of balls in I, namely, the quantity

$$\mu_i^{(n)} = \sum_{k=-a}^{a} k p_{ik}^{(n)}$$

can be shown to be equal to

$$i\left(1 - \frac{1}{a}\right)^n \tag{1.1.56}$$

which in the limit as $a \to \infty$, $1/a\Delta \to \gamma$, $n\Delta = t$, yields

$$i e^{-\gamma t},$$

or the Newton law of cooling. One could prove (1.1.56) using (1.1.55) but we shall proceed by a simple calculation analogous to that leading to (1.1.53). It is easily seen that the moment generating function

$$M_i^{(n)}(\theta) = \sum_{k=-a}^{a} p_{ik}^{(n)} e^{-k\theta}$$

satisfies

$$a M_i^{(n)}(\theta) \cosh\theta - \left(\frac{d}{d\theta} M_i^{(n)}(\theta)\right) \sinh\theta = a M_i^{(n+1)}(\theta).$$

Differentiating this equation with respect to θ and remembering that

$$\left.\frac{d}{d\theta} M_i^{(n)}(\theta)\right|_{\theta=0} = \mu_i^{(n)},$$

we obtain the recurrence relation

$$\mu_i^{(n+1)} = \left(1 - \frac{1}{a}\right) \mu_i^{(n)}.$$

This leads to (1.1.56) since $\mu_i^{(0)} = i$.

The Ehrenfest model is also particularly suited for the discussion of a famous paradox which at the turn of this century nearly wrecked L. BOLTZMANN's inspired efforts to explain thermodynamics on the basis of the kinetic theory. In classical thermodynamics the process of heat exchange of two isolated bodies of unequal temperatures is irreversible. On the other hand, if the bodies are treated as a dynamical system, the famous "Wiederkehrsatz" of H. POINCARÉ asserts that "almost every" state of the system (except for a set of states which, when interpreted as points in phase space, form a set of Lebesgue measure 0) will be achieved arbitrarily closely. Thus, argued E. ZERMELO, the irreversibility postulated in thermodynamics and the "recurrence" properties of dynamical systems are irreconciliable. Boltzmann then replied that the "Poincaré cycles" (time until states "nearly recur" for the first time, — the word "nearly" requiring further specification) are so long compared to time intervals involved in ordinary experiences that predictions based on classical thermodynamics can be fully trusted. This explanation, though correct in principle, was set forth in a manner which was not quite convicing, and the controversy raged on. It was mainly through the efforts of EHRENFEST and M. SMOLUCHOWSKI that the situation became completely clarified, and the irreversibility interpreted in a proper statistical manner.

It is easy to discuss this explanation by appealing to the Ehrenfest model. We know that every state is recurrent or, in other words: *each state is bound to recur with probability one.* This is the statistical analogue of the "Wiederkehrsatz". Furthermore, by Theorem 1.1.14 the mean recurrence time $\mathsf{E}(\tau_{ii})$ of state i equals

$$\frac{2}{\lim\limits_{n\to\infty} p_{ii}^{(2n)}} = \frac{2}{2^{2a}(-1)^{a+i}[c_{-i,a}\, c_{ai} + c_{-i,-a}\, c_{-a,i}]} = 2^{2a}\,\frac{(a+i)!\,(a-i)!}{(2a)!}.$$

This is the statistical analogue of a "Poincaré cycle", and it tells us, roughly speaking, how long, on the average, one will have to wait for state i to recur.

If $a+i$ and $a-i$ differ considerably, then $\mathsf{E}(\tau_{ii})$ is enormous. For example, if $a = 10{,}000$ and $i = 10{,}000$, we get $\mathsf{E}(\tau_{ii}) = 2^{20{,}000}$ steps (of the order of $10^{6{,}000}$ years if one step is performed in a second). If, on the other hand, $a+i$ and $a-i$ are nearly equal, $\mathsf{E}(\tau_{ii})$ is rather small. If in the above example we set $i = 0$ we get (using Stirling's formula) $\mathsf{E}(\tau_{ii}) \approx 100\sqrt{\pi} \approx 175$ steps.

It was SMOLUCHOWSKI who suggested that if one starts in a state with a long recurrence time the process will appear to be irreversible. In our example if one starts with 20,000 balls in one container and none in the other, one would observe, for a long time, an essentially irreversible flow of balls. On the other hand, if the mean recurrence time is short, it makes no sense to speak of irreversibility.

We close by noticing that the Ehrenfest model appears to be a special case of a general urn model due to B. FRIEDMAN (1949) [24]. A modified Ehrenfest model with a continuous time parameter has been discussed by BELLMAN and HARRIS (1951).

1.2. Noteworthy classes of denumerable Markov chains

Even in simple cases it is very difficult to decide, for example, whether a Markov chain is recurrent or nonrecurrent; and the computation of basic probabilistic quantities is almost always cumbersome. The purpose of this section is to present some classes of Markov chains occurring in various contexts, to classify their states, and to compute (certain) fundamental quantities for them [25].

1.2.1. Random walk

1.2.1.1. Homogeneous random walk

1.2.1.1.1. Let Z be the set of all integers. Consider for a given natural number s the set Z^s of s-tuples whose elements belong to Z. If $\mathbf{i} = (i_1, \ldots, i_s)$ and $\mathbf{j} = (j_1, \ldots, j_s) \in Z^s$, we shall denote by $\mathbf{i} - \mathbf{j}$ the s-tuple $(i_1 - j_1, \ldots, i_s - j_s) \in Z^s$. A Markov chain $(\xi_n)_{n \in N}$ with state space $I = Z^s$ is said to be a *homogeneous s-dimensional random walk* if the transition probabilities $p_{\mathbf{ij}}$ depend only on $\mathbf{j} - \mathbf{i}$, that is, $p_{\mathbf{ij}} = p(\mathbf{j} - \mathbf{i})$, $\mathbf{i}, \mathbf{j} \in Z^s$, the function p satisfying the relations

$$p(\mathbf{i}) \geqslant 0, \qquad \sum_{\mathbf{i} \in Z^s} p(\mathbf{i}) = 1.$$

A simple induction shows that the same is true for the n-step transition probabilities $p_{\mathbf{ij}}^{(n)}$. It is also easily seen that the quantities $f_{\mathbf{ij}}^{(n)}$, $n \in N^*$, and $f_{\mathbf{ij}}^*$ depend only on $\mathbf{j} - \mathbf{i}$.

We shall prove that a chain $(\xi_n)_{n \in N}$ is a homogeneous s-dimensional random walk iff it is a Z^s — valued chain with independent and identically distributed increments. Indeed, if $p_{\mathbf{ij}} = p(\mathbf{j} - \mathbf{i})$, let us put

$$y_0 = \xi_0, \quad y_m = \xi_m - \xi_{m-1}, \qquad m \in N^*.$$

[24] We note that the so-called stimulus-sampling learning models (see e.g. ATKINSON and ESTES (1963)) may also be regarded as special cases of Friedman's urn. For the application of the theory of finite state Markov chains to these models see KEMENY and SNELL (1960, pp. 182–191).

[25] For another class of Markov chains see KEMENY (1966 a).

We have for $\mathbf{i}_\nu \in Z^s$, $0 \leqslant \nu \leqslant n+1$, $n \in N$

$$\mathsf{P}(y_{n+1} = \mathbf{i}_{n+1} | y_\nu = \mathbf{i}_\nu, \ 0 \leqslant \nu \leqslant n) =$$

$$= \mathsf{P}\left(\xi_{n+1} = \sum_{\nu=0}^{n+1} \mathbf{i}_\nu \middle| \xi_\nu = \sum_{r=0}^{\nu} \mathbf{i}_r, \ 0 \leqslant \nu \leqslant n\right) =$$

$$= \mathsf{P}\left(\xi_{n+1} = \sum_{\nu=0}^{n+1} \mathbf{i}_\nu \middle| \xi_n = \sum_{r=0}^{n} \mathbf{i}_r\right) = p(\mathbf{i}_{n+1}).$$

Therefore $(y_n)_{n \in N}$ is a sequence of independent random variables, the distributions of the variables y_m, $m \in N^*$, being identical :

$$\mathsf{P}(y_m = \mathbf{i}) = p(\mathbf{i}), \qquad \mathbf{i} \in Z^s,$$

and

$$\xi_{n+1} = \xi_0 + \sum_{r=1}^{n+1} y_r, \qquad n \in N.$$

It follows in particular that

$$p_{\mathbf{0i}}^{(m)} = Pr(S_m = \mathbf{i}), \qquad \mathbf{i} \in Z^s, \qquad m \in N^*, \tag{1.2.1}$$

where S_m, $m \in N^*$, are the partial sums of an arbitrary sequence of independent and identically distributed random variables with the same distribution as y_1.

Conversely, if

$$\xi_{n+1} = \xi_0 + \sum_{r=1}^{n+1} y_r, \qquad n \in N,$$

the variable ξ_0 being independent of y_m, $m \in N^*$, and the variables y_m, $m \in N^*$, being independent and identically distributed, then

$$\mathsf{P}(\xi_{n+1} = \mathbf{j} | \xi_n = \mathbf{i}) =$$
$$= \mathsf{P}(y_{n+1} = \mathbf{j} - \mathbf{i} | \xi_n = \mathbf{i}) = \mathsf{P}(y_{n+1} = \mathbf{j} - \mathbf{i}),$$

for $\mathbf{i}, \mathbf{j} \in Z^s$, $n \in N$, that is $(\xi_n)_{n \in N}$ is a homogeneous s-dimensional random walk.

1.2.1.1.2. Since $f^*_{\mathbf{ii}} = f^*_{\mathbf{00}}$ for all $\mathbf{i} \in Z^s$, we can assert that *all* states $\mathbf{i} \in Z^s$ are either recurrent (iff $f^*_{\mathbf{00}} = 1$) or nonrecurrent (iff $f^*_{\mathbf{00}} < 1$). In this connection one speaks about a *recurrent* or a *nonrecurrent random* *walk*. It is easily seen that if an irreducible homogeneous random walk is recurrent, it can only be null recurrent.

A general criterion for determining the recurrence of homogeneous random walk is available. To give it we need the so-called *characteristic function* of the random walk, namely the function

$$\varphi(\mathbf{t}) = \sum_{\mathbf{j} \in Z^s} p(\mathbf{j}) \exp \mathrm{i}\, (j_1 t_1 + \ldots + j_s t_s),$$

where $\mathbf{t} = (t_1, \ldots, t_s) \in R^s$. It is clear that φ is the characteristic function of $y_1 = \xi_1 - \xi_0$. Relation (1.2.1) implies then that

$$\varphi^n(\mathbf{t}) = \sum_{\mathbf{j} \in Z^s} p_{\mathbf{0j}}^{(n)} \exp \mathrm{i}\, (j_1 t_1 + \ldots + j_s t_s)$$

for all $n \in N^*$, whence by multiplying each side by $\exp\{-\mathrm{i}(k_1 t_1 + \ldots + k_s t_s)\}$ and integrating over the cube C in R^s with center at the origin and sides of length 2π, we get

$$p_{\mathbf{0k}}^{(n)} = (2\pi)^{-s} \int_C \exp\{-\mathrm{i}\,(k_1 t_1 + \ldots + k_s t_s)\}\, \varphi^n(\mathbf{t})\, \mathrm{d}(\mathbf{t}). \qquad (1.2.2)$$

We can now prove

Theorem 1.2.1 (K. L. Chung, W. H. J. Fuchs). *A homogeneous s-dimensional random walk is nonrecurrent iff*

$$\lim_{x \to 1-} \int_C \mathrm{Re} \frac{1}{1 - x\varphi(\mathbf{t})}\, \mathrm{d}\mathbf{t} < \infty.$$

Proof. (Spitzer (1964, p. 83)). The random walk is nonrecurrent, according to Theorem 1.1.6, iff

$$G(\mathbf{0}, \mathbf{0}) = \sum_{n \in N} p_{\mathbf{00}}^{(n)} < \infty.$$

But, by Abel's lemma

$$G(\mathbf{0}, \mathbf{0}) = \lim_{x \to 1-} \sum_{n \in N} p_{\mathbf{00}}^{(n)} x^n,$$

whether this limit is finite or not. Using (1.2.2) we have

$$(2\pi)^s G(\mathbf{0}, \mathbf{0}) = \lim_{x \to 1-} \sum_{n \in N} x^n \int_C \varphi^n(\mathbf{t})\, \mathrm{d}\mathbf{t} = \lim_{x \to 1-} \int_C \frac{\mathrm{d}\mathbf{t}}{1 - x\varphi(\mathbf{t})}.$$

Noticing that for $0 \leqslant x < 1$,

$$\int_C \frac{d\mathbf{t}}{1 - x\varphi(\mathbf{t})} = \operatorname{Re} \int_C \frac{d\mathbf{t}}{1 - x\varphi(\mathbf{t})} = \int_C \operatorname{Re} \frac{1}{1 - x\varphi(\mathbf{t})}\, d\mathbf{t},$$

we find

$$(2\pi)^s G(\mathbf{0}, \mathbf{0}) = \lim_{x \to 1-} \int_C \operatorname{Re} \frac{1}{1 - x\varphi(\mathbf{t})}\, d\mathbf{t},$$

which proves the theorem. ◇

We remark that SPITZER (1964, p. 84—85) has shown that it is possible to justify the interchange of limits in Theorem 1.2.1 and obtain the more elegant criterion: *homogeneous random walk is nonrecurrent or recurrent according as the real part of* $[1 - \varphi(\mathbf{t})]^{-1}$ *is Lebesgue integrable on the cube* C *or not.*

It is possible to deduce from Theorem 1.2.1 that a genuinely [26] s-dimensional random walk is nonrecurrent when $s \geqslant 3$.

It is difficult to handle the above criterion without a thorough investigation of the characteristic function φ, which may be found in SPITZER (1964, Ch. 2). In what follows we shall obtain by a direct way some simple sufficient conditions for recurrence or nonrecurrence.

Let us set

$$\mu = \sum_{\mathbf{i} \in Z^s} |\mathbf{i}|\, p(\mathbf{i}),$$

and, if $\mu < \infty$,

$$\mathbf{m} = \sum_{\mathbf{i} \in Z^s} \mathbf{i}\, p(\mathbf{i})$$

Here we denote by $|\mathbf{i}|$ the Euclidean distance from $\mathbf{i} = (i_1, \ldots, i_s)$ to the origin, that is $\left(\sum_{k=1}^{s} i_k^2\right)^{1/2}$.

Proposition 1.2.2. *In the one-dimensional case* ($s = 1$), *if* $\mu < \infty$ *and* $m = 0$, *then the random walk is recurrent.*

---

[26] The term "genuinely" is used to avoid an artificial increase of the state space of the random walk by imbedding it in a space of higher dimension and defining p to be zero wherever it had not been previously defined.

Proof. Relation (1.2.1) allows us to write (independently of the value of m)

$$\lim_{n\to\infty} \sum_{\left[i:\left|\frac{i}{n}-m\right|>\varepsilon\right]} p_{0i}^{(n)} = 0 \tag{1.2.3}$$

for all $\varepsilon > 0$. Indeed, (1.2.3) is a transcription of Hinčin's weak law of large numbers for independent and identically distributed random variables.

By virtue of (1.1.6′) we can write for all $n \in N^*$ and $C > 0$

$$\sum_{k=0}^{n} p_{00}^{(k)} \geqslant \frac{1}{2C+1} \sum_{|i|\leqslant C} \sum_{k=0}^{n} p_{0i}^{(k)}. \tag{1.2.4}$$

By using (1.2.3) we obtain

$$\sum_{|i|\leqslant C} \sum_{k=0}^{n} p_{0i}^{(k)} = \sum_{k=0}^{n} \sum_{|i|\leqslant C} p_{0i}^{(k)} \geqslant \sum_{k=0}^{n} \sum_{\left[i:\left|\frac{i}{k}\right|\leqslant\left|\frac{C}{n}\right|\right]} p_{0i}^{(k)}. \tag{1.2.5}$$

By taking $C = an$, $a > 0$, from (1.2.4) and (1.2.5) we obtain

$$\sum_{k=0}^{n} p_{00}^{(k)} \geqslant \frac{1}{2an+1} \sum_{k=0}^{n} \sum_{[i:\,|i|\leqslant ka]} p_{0i}^{(k)}. \tag{1.2.6}$$

Taking into account the fact that $m=0$ and $\sum_{i\in Z} p_{0i}^{(k)} = 1$, we deduce from (1.2.3) that

$$\lim_{k\to\infty} \sum_{[i:\,|i|\leqslant ka]} p_{0i}^{(k)} = 1$$

for all $a > 0$. Relation (1.2.6) then gives

$$\varliminf_{n\to\infty} \sum_{k=0}^{n} p_{00}^{(k)} \geqslant \frac{1}{2a},$$

whence, a being arbitrary, we obtain

$$\sum_{k=0}^{\infty} p_{00}^{(k)} = \infty,$$

and it remains only to apply Theorem 1.1.6. ◇

Proposition 1.2.3. *Whatever the dimension* s, *if* $\mu < \infty$ *and* $\mathbf{m} \neq \mathbf{0}$, *then the random walk is nonrecurrent.*

Proof. According to the strong law of large numbers we have

$$Pr\left(\lim_{n\to\infty} \frac{S_n}{n} = \mathbf{m}\right) = 1,$$

which, if we define

$$A_n = \left(\left|\frac{S_n}{n} - \mathbf{m}\right| > \frac{|\mathbf{m}|}{2}\right), \qquad n \in N^*,$$

implies

$$Pr\,(\overline{\lim_{n\to\infty}}\, A_n) = 0.$$

Note that

$$(S_n = \mathbf{0}) \subset A_n, \qquad n \in N^*.$$

Thus

$$Pr\,[\overline{\lim_{n\to\infty}}\,(S_n = \mathbf{0})] = 0,$$

that is

$$\mathsf{P}_{\mathbf{0}}[\overline{\lim_{n\to\infty}}\,(\xi_n = \mathbf{0})] = 0.$$

It follows that state $\mathbf{0}$ is nonrecurrent, and so also the random walk. ◇

Corollary. *The one-dimensional random walk with* $\mu < \infty$ *is recurrent iff* $m = 0$.

1.2.1.2. Some important special cases

1.2.1.2.1. A particularly important random walk is the so-called *simple* (or *symmetric*) random walk defined by

$$p(\mathbf{i}) = \begin{cases} \dfrac{1}{2s} & \text{if} \quad |\mathbf{i}| = 1 \\ 0 & \text{if} \quad |\mathbf{i}| \neq 1. \end{cases}$$

Thus the probabilities $p(\mathbf{i})$ are nonnull (and equal) only for the $2s$ points $\mathbf{i} = \pm\,(\delta_{kl})_{1\leqslant l\leqslant s}$, $1 \leqslant k \leqslant s$. The simple random walk is obviously an irreducible Markov chain of period two.

1.2.1.2.2. A somewhat more general random walk is the *Bernoulli* random walk for which the probabilities $p(\mathbf{i})$, $|\mathbf{i}| = 1$, are no longer equal, the only condition imposed on them being that they sum to one.

The one-dimensional Bernoulli random walk is thus defined by the probabilities

$$p(1) = p, \qquad p(-1) = 1 - p = q,$$

where $0 < p < 1$. Historically this was the first random walk studied in probability theory. In fact, the corresponding Markov chain is the sequence of partial sums of a sequence $(\beta_n)_{n \in N^*}$ of independent random variables taking on the values 1 and -1 with probabilities p and $1 - p$.

The Bernoulli random walk has been used by physicists as a coarse approximation to the one-dimensional unrestricted diffusion of particles[27].

The corollary to Proposition 1.2.3. allows us to assert that the one-dimensional Bernoulli random walk is recurrent iff $p = 1/2$, i.e. the simple random walk is the only recurrent Bernoulli random walk. A direct proof is as follows: we have

$$p_{ij}^{(n)} = Pr(\beta_1 + \dots + \beta_n = j - i) =$$

$$= \begin{cases} \dbinom{n}{(n+j-i)/2} p^{(n+j-i)/2} q^{(n-j+i)/2} & \text{if } |j-i| \leqslant n \text{ and } |j-i|+n \text{ is even} \\ 0 & \text{otherwise.} \end{cases} \tag{1.2.7}$$

Consequently,

$$\begin{aligned} P_{00}(z) &= \sum_{n \in N} p_{00}^{(n)} z^n = \\ &= 1 + \sum_{n \in N^*} \binom{2n}{n} (pq)^n z^{2n} = (1 - 4pqz^2)^{-1/2}, \end{aligned} \tag{1.2.8}$$

$$F_{00}(z) = 1 - \frac{1}{P_{00}(z)} = 1 - (1 - 4pqz^2)^{1/2},$$

$$f_{00}^* = F_{00}(1) = 1 - |1 - 2p|.$$

Therefore $f_{00}^* = 1$ iff $p = \frac{1}{2}$[28],

---

[27] We note that every random walk may be thought of as the motion of a "particle". The language and ideas associated with such a motion are intuitive and convenient.

[28] The two-dimensional simple random walk is also recurrent, but as we have already noted, the three (or more)-dimensional simple random walk is not. This was first proved by G. Polya. For a simple proof see e.g. Feller (1968, p. 360).

We record another interesting property of the simple one-dimensional random walk. It is obvious that we have for $i, j > 0$

$$p_{ij}^{(n)} = {}_0p_{ij}^{(n)} + \sum_{m=0}^{n-1} p_{i0}^{(m)}\, {}_0p_{0j}^{(n-m)}$$

$$p_{-i,j}^{(n)} = \sum_{m=0}^{n-1} p_{-i,0}^{(m)}\, {}_0p_{0j}^{(n-m)} = \sum_{m=1}^{n-1} p_{i0}^{(m)}\, {}_0p_{0j}^{(n-m)},$$

whence

$${}_0p_{ij}^{(n)} = p_{ij}^{(n)} - p_{-i,j}^{(n)}. \tag{1.2.9}$$

This relation is called the *reflection principle* of D. André. It expresses the fact that there is a one-to-one correspondence between all "paths" from i to j containing 0 and all "paths" from $-i$ to j. For details see Feller (1968, Ch. 3).

1.2.1.3. Barriers

1.2.1.3.1. In various applications (to physics, chemistry, genetics, etc.) we are led to study Markov chains obtained by restricting the motion of a "particle" which performs a random walk. Usually this is done by introducing so-called *barriers*. Consider, for example, the one-dimensional Bernoullian case, and define a Markov chain with state space Z as follows: B being a given subset of Z (the elements of B will be called barriers), the transition probabilities are given by

$$p_{i,i+1} = p, \qquad p_{i,i-1} = 1 - p \qquad \text{if} \quad i \notin B,$$

$$p_{i,i+1} = p_i, \qquad p_{i,i-1} = q_i, \qquad p_{ii} = r_i \quad \text{if} \quad i \in B,$$

where $0 \leqslant p_i,\ q_i,\ r_i \leqslant 1$, $p_i q_i r_i = 0$, and $p_i + q_i + r_i = 1$. A state $i \in B$ is said to be an *absorbing, right-reflecting, left-reflecting, right-elastic* or *left-elastic* barrier according as $r_i = 1$, $p_i = 1$, $q_i = 1$, $q_i = 0$ and $p_i r_i > 0$, or $p_i = 0$ and $q_i r_i > 0$. The Markov chain defined in this manner is referred to as a *random walk with barriers*[29]. It no longer has independent increments. Moreover, the state space of a random walk with barriers is a proper subset of Z. Thus, in the case of a unique absorbing barrier at 0, the state space is N or $-N$ according as the "particle" starts at the right or at the left of the origin; in the case of two barriers 0 and $l > 0$ with $q_0 = p_l = 0$, the state space is the finite set $\{0, 1, \ldots, l\}$ assuming that the particle starts in i, $0 \leqslant i \leqslant l$.

---

[29] One can easily imagine numerous other types of barriers. For example, assume that the "particle" returns to its initial position after reaching a barrier: this is another type of reflection.

In what follows we give some examples. For a more detailed treatment of random walks with barriers the reader is referred to Cox and MILLER (1965, Ch. 2). For some special problems see ONICESCU (1964).

1.2.1.3.2. Let us first consider the case of an absorbing barrier at 0. If the "particle" starts in $i > 0$, then the state space will be $I = N$. The transition matrix is

$$\mathbf{P} = \begin{array}{c} \\ 0 \\ 1 \\ 2 \\ \cdot \\ \cdot \\ \cdot \end{array} \begin{array}{c} \begin{array}{cccccccc} 0 & 1 & 2 & 3 & 4 & \cdot & \cdot & \cdot \end{array} \\ \left(\begin{array}{cccccccc} 1 & 0 & 0 & 0 & 0 & \cdot & \cdot & \cdot \\ q & 0 & p & 0 & 0 & \cdot & \cdot & \cdot \\ 0 & q & 0 & p & 0 & \cdot & \cdot & \cdot \\ \cdot & \cdot & \cdot & \cdot & \cdot & \cdot & \cdot & \cdot \\ \cdot & \cdot & \cdot & \cdot & \cdot & \cdot & \cdot & \cdot \\ \cdot & \cdot & \cdot & \cdot & \cdot & \cdot & \cdot & \cdot \end{array}\right) \end{array}.$$

All the states except 0 form an inessential class of period 2; state 0 being absorbing forms a recurrent class by itself. If $i > j > 0$, then f_{ij}^* is the same as for the (unrestricted) Bernoulli random walk since starting in i the "particle" must enter j before it can enter 0, and the transition probabilities for $i, j \in N^*$ are the same in the two cases. Furthermore, it is easily seen that the first passage time τ_{ij}, $i > j > 0$, is the sum of $i - j$ independent random variables each with the same distribution as τ_{10}. This implies that

$$F_{ij}(z) = F_{10}^{i-j}(z), \quad i > j > 0.$$

Thus it suffices to find $F_{10}(z)$ for the Bernoulli random walk. By (1.2.7) and (1.2.8) we have

$$P_{10}(z) = \sum_{n \in N^*} p_{10}^{(n)} z^n = \sum_{n \in N} \binom{2n+1}{n} p^n q^{n+1} z^{2n+1} =$$

$$= \frac{1}{2pz} \sum_{n \in N} \binom{2n+2}{n+1} (pq)^{n+1} z^{2n+2} = \frac{1}{2pz} (P_{00}(z) - 1),$$

thus on account of (1.1.27),

$$F_{10}(z) = \frac{1 - (1 - 4pqz^2)^{1/2}}{2pz},$$

and then

$$F_{ij}(z) = \left[\frac{1 - (1 - 4pqz^2)^{1/2}}{2pz}\right]^{i-j}, \qquad i > j > 0.$$

It follows that $f_{10}^{*} = F_{10}(1)$ equals 1 or $1 - 1/p$ according as $p \leqslant \frac{1}{2}$ or $p > \frac{1}{2}$. The absorption probability $\alpha_i(\{0\}) = f_{i0}^{*}$ will be 1 or $\left(1 - \frac{1}{p}\right)^i$ according as $p \leqslant \frac{1}{2}$ or $p > \frac{1}{2}$. If absorption is certain, then the mean time to absorption is

$$\mathsf{E}(\tau_{i0}) = F_{i0}'(1) = iF_{10}'(1) = \begin{cases} \dfrac{i}{1-2p} & \text{if } p < \dfrac{1}{2} \\[2ex] \infty & \text{if } p = \dfrac{1}{2}. \end{cases}$$

We close by noticing that for $p = 1/2$, the n-step transition probabilities $p_{ij}^{(n)}$, $i, j > 0$ may be obtained from the reflection principle (1.2.9). Indeed, in this case $p_{ij}^{(n)}$, $i, j > 0$ coincides with the taboo probability ${}_0p_{ij}^{(n)}$ for the simple random walk.

1.2.1.3.3. Now consider the case of two absorbing barriers at 0 and l. If the "particle" starts in i, $0 \leqslant i \leqslant l$, then the state space is finite. The transition matrix is

$$\mathbf{P} = \begin{array}{c} \\ 0 \\ 1 \\ 2 \\ \cdot \\ \cdot \\ \cdot \\ l-1 \\ l \end{array} \begin{array}{c} \begin{array}{cccccccccc} 0 & 1 & 2 & 3 & . & . & . & l-2 & l-1 & l \end{array} \\ \left(\begin{array}{cccccccccc} 1 & 0 & 0 & 0 & . & . & . & 0 & 0 & 0 \\ q & 0 & p & 0 & . & . & . & 0 & 0 & 0 \\ 0 & q & 0 & p & . & . & . & 0 & 0 & 0 \\ . & . & . & . & . & . & . & . & . & . \\ . & . & . & . & . & . & . & . & . & . \\ . & . & . & . & . & . & . & . & . & . \\ 0 & 0 & 0 & 0 & . & . & . & q & 0 & p \\ 0 & 0 & 0 & 0 & . & . & . & 0 & 0 & 1 \end{array}\right) \end{array}.$$

All the states except 0 and 1 form an inessential class of period 2; states 0 and 1 being absorbing form two essential classes by themselves. A classical illustration of this random walk is the '*gambler's ruin*' *problem*: consider two persons playing a certain game repeatedly and starting off with i and $l - i$ units of capital. Assume that one unit is bet each time and the probabilities of winning for the two players are p and $1 - p$. What is the probability of ruin of each of the players? It is clear that the winnings of the first player are described by the random walk we have just considered and that the desired "ruin" probabilities are the absorption probabilities $\mathfrak{a}_i(\{0\})$ and $\mathfrak{a}_i(\{l\})$. To obtain them we can use Theorem 1.1.52. In this case it is easily verified that the entries of the fundamental matrix $\mathbf{N} = (n_{ij})_{1 \leqslant i, j \leqslant l-1}$ are given by

$$n_{ij} = \frac{1}{(2p-1)(a^l - 1)} \cdot \begin{cases} (a^j - 1)(a^{l-i} - 1) & \text{if } j \leqslant i \\ (a^i - 1)(a^{l-i} - a^{j-i}) & \text{if } j > i \end{cases}$$

for $p \neq \frac{1}{2}$, where $a = p/q$, and

$$n_{ij} = \frac{2}{l} \cdot \begin{cases} j(l-i) & \text{if } j \leqslant i \\ i(l-j) & \text{if } j > i \end{cases}$$

for $p = \frac{1}{2}$. According to v) Theorem 1.1.52. we have

$$\mathfrak{a}_i(\{0\}) = \sum_{j=1}^{l-1} n_{ij} p_{j0} = \begin{cases} \dfrac{a^{l-i} - 1}{a^l - 1} & \text{if } p \neq \dfrac{1}{2} \\ 1 - i/l & \text{if } p = \dfrac{1}{2}, \end{cases}$$

$$\mathfrak{a}_i(\{l\}) = 1 - \mathfrak{a}_i(\{0\}).$$

By iii) of the same theorem the mean time to absorption at either 0 or l (i.e. the mean duration of game) is

$$\sum_{j=1}^{l-1} n_{ij} = \begin{cases} \dfrac{1}{2p-1}\left(l\dfrac{a^l - a^{l-i}}{a^l - 1} - i\right) & \text{if } p \neq \dfrac{1}{2} \\ i(l-i) & \text{if } p = \dfrac{1}{2}. \end{cases}$$

For a detailed study of the gambler's ruin problem the reader is referred to FELLER (1968, Ch. XIV).

1.2.1.3.4. Consider the case of a right-elastic barrier at 0. If the "particle" starts in $i > 0$, then the state space will be $I = N$.

The transition matrix is

$$\mathbf{P} = \begin{array}{c} \\ 0 \\ 1 \\ 2 \\ \cdot \\ \cdot \\ \cdot \end{array} \begin{array}{c} \begin{array}{cccccccc} 0 & 1 & 2 & 3 & 4 & \cdot & \cdot & \cdot \end{array} \\ \left(\begin{array}{cccccccc} r_0 & p_0 & 0 & 0 & 0 & \cdot & \cdot & \cdot \\ q & 0 & p & 0 & 0 & \cdot & \cdot & \cdot \\ 0 & q & 0 & p & 0 & \cdot & \cdot & \cdot \\ \cdot & \cdot & \cdot & \cdot & \cdot & \cdot & \cdot & \cdot \\ \cdot & \cdot & \cdot & \cdot & \cdot & \cdot & \cdot & \cdot \\ \cdot & \cdot & \cdot & \cdot & \cdot & \cdot & \cdot & \cdot \end{array}\right) \end{array}.$$

All the states form an aperiodic essential class. To classify its states we use Theorem 1.1.34. Consider the system of equations

$$z_i = \sum_{j \in N^*} p_{ij} z_j, \qquad i \in N^*.$$

These reduce to

$$z_1 = pz_2, \quad z_i = pz_{i+1} + qz_{i-1}, \quad i \geqslant 2,$$

which can be written as

$$p(z_{i+1} - z_i) = q(z_i - z_{i-1}), \quad i \geqslant 2.$$

We obtain

$$z_{i+1} - z_i = \left(\frac{q}{p}\right)^i z_1, \quad i \in N^*,$$

so that

$$z_i = z_1 \sum_{k=0}^{i-1} \left(\frac{q}{p}\right)^k, \quad i \in N^*.$$

Therefore, a bounded solution exists iff $p > q$, that is, iff $p > \frac{1}{2}$. Thus by Theorem 1.1.34, the states are nonrecurrent or recurrent according as $p > \frac{1}{2}$ or $p \leqslant \frac{1}{2}$.

To decide when the states are positive, we consider the system of equations determining the stationary measure in the case $p \leqslant \frac{1}{2}$, namely

$$u_0 = r_0 u_0 + qu_1$$

$$u_1 = p_0 u_0 + qu_2$$

$$u_i = pu_{i-1} + qu_{i+1}, \quad i \geqslant 2.$$

Solving these successively, we obtain

$$u_i = \frac{p_0}{q}\left(\frac{p}{q}\right)^{i-1} u_0, \quad i \in N^*.$$

The series $\sum_{i \in N^*} u_i$ converges iff $p < \frac{1}{2}$. Therefore by Theorem 1.1.43 and Proposition 1.1.38, the states are positive for $p < \frac{1}{2}$. For $p = \frac{1}{2}$, they are null-recurrent. It follows also that the stationary distribution is given by

$$\pi_0 = \frac{q-p}{q+p_0-p}, \quad \pi_i = \frac{p_0}{p}\left(\frac{p}{q}\right)^{i-1} \pi_0, \quad i \in N^*.$$

1.2.1.3.5. The case of a right-reflecting barrier may be obtained from the preceding one by taking $r_0 = 0$, $p_0 = 1$. All considerations concerning classification of states remain valid for the new case except for the period which is 2. KAC (1947) was able to derive an integral representation for the n-step transition probabilities. It is obvious that we have the following equations

$$\begin{aligned} p_{ij}^{(n+1)} &= p p_{i,j-1}^{(n)} + q p_{i,j+1}^{(n)} \quad n, i \in N,\ j \geqslant 2, \\ p_{i1}^{(n+1)} &= p_{i0}^{(n)} + q p_{i2}^{(n)},\ p_{i0}^{(n+1)} = q p_{i1}^{(n)}, \quad n, i \in N. \end{aligned} \tag{1.2.10}$$

Let $\mathbf{p}_{n,i}$ be the infinite column vector with entries $p_{ij}^{(n)}$, $j \in N$, and $\mathbf{A}$ the infinite matrix

$$\mathbf{A} = \begin{array}{c} \\ 0 \\ 1 \\ 2 \\ 3 \\ \cdot \\ \cdot \\ \cdot \end{array}\!\!\begin{array}{c} \begin{array}{cccccccc} 0 & 1 & 2 & 3 & 4 & \cdot & \cdot & \cdot \end{array} \\ \left(\begin{array}{cccccccc} 0 & q & 0 & 0 & 0 & \cdot & \cdot & \cdot \\ 1 & 0 & q & 0 & 0 & \cdot & \cdot & \cdot \\ 0 & p & 0 & q & 0 & \cdot & \cdot & \cdot \\ 0 & 0 & p & 0 & q & \cdot & \cdot & \cdot \\ \cdot & \cdot & \cdot & \cdot & \cdot & \cdot & \cdot & \cdot \\ \cdot & \cdot & \cdot & \cdot & \cdot & \cdot & \cdot & \cdot \\ \cdot & \cdot & \cdot & \cdot & \cdot & \cdot & \cdot & \cdot \end{array}\right) \end{array}$$

Then equations (1.2.10) can be written in matrix form as

$$\mathbf{p}_{n+1,i} = \mathbf{A}\mathbf{p}_{n,i}, \quad n, i \in N,$$

whence

$$\mathbf{p}_{n,i} = \mathbf{A}^n \mathbf{p}_{0,i}, \quad n, i \in N.$$

It follows that $p_{ij}^{(n)}$ is the ij th entry of $\mathbf{A}^n$. KAC showed that when $p < 1/2$

$$p_{ij}^{(n)} = \frac{1-2p}{2pq}\left(\frac{p}{q}\right)_*^j [1+(-1)^{i+j+n}] +$$

$$+\frac{2}{\pi}\left(\frac{q}{p}\right)^{\frac{i}{2}}\left(\frac{p}{q}\right)_*^{\frac{j}{2}}(2\sqrt{pq})^n \int_0^\pi \cos^n\theta \frac{\tan^2\theta}{(1-2p)^2+\tan^2\theta} f_i(\theta) f_j(\theta)\, d\theta,$$

where

$$\left(\frac{p}{q}\right)_*^\alpha = \begin{cases} \left(\frac{p}{q}\right)^\alpha & \text{if } \alpha > 0 \\ p & \text{if } \alpha = 0, \end{cases}$$

and

$$f_m(\theta) = \cos m\theta - (1-2p)\frac{\sin m\theta \cos\theta}{\sin\theta}, \quad m \in N.$$

In case $p > \frac{1}{2}$ it was shown by SHENTON (1955) that $p_{ij}^{(n)}$ is given by the second term in the above expression. For the case $p = \frac{1}{2}$ see LEHNER (1963).

1.2.1.4. Generalizations

1.2.1.4.1. A first generalization of the homogeneous random walk is obtained by abandoning the restriction that the "particle" move in steps which are integral multiples of unity. This is done by retaining the basic property, namely that the increments be independent and identically distributed. To be precise, let $(\eta_n)_{n \in N^*}$ be a sequence of independent and identically distributed random variables with values in the Euclidean space R^s. Then the sequence $(S_n)_{n \in N^*}$ of their partial sums is said to be a *generalized* homogeneous s-dimensional random walk. From a point of view differing from that of Markov chains these generalized random walks were studied in the *fluctuation theory* initiated by E. SPARRE ANDERSEN at the beginning of the fifties. Basically, the problems investigated in this theory consists of finding the distribution of various functions which are definable in terms of the sums S_n and which give a measure (in some sense) of the amount of oscillation which the sums undergo. Typical quantities investigated in fluctuation theory are:

a) the number N_n of non-negative sums among the first n sums ($N_0 = 0$).

b) the value of the maximum $\overline{M}_n$ and of the minimum $\underline{M}_n$ of the first n sums ($\overline{M}_0 = \underline{M}_0 = 0$).

c) L_n — the smallest subscript k $(0 \leqslant k \leqslant n)$ for which $S_k = \overline{M}_n$.

d) the value R_{nk} of the sum which falls kth from the bottom when the sums $S_0 = 0, S_1, \ldots, S_n$ are arranged in increasing order.

For various aspects of fluctuation theory we refer the reader to CARTIER (1962/1963), FELLER (1968, Ch. 3), PORT (1963), PRABHU (1965, Ch. 6) and SPITZER (1964, § 20).

The study of the generalized homogeneous random walk from the point of view of Markov chains was carried out by CHUNG and FUCHS (1951). (See also FELLER (1966 b, Ch. 6) and ORNSTEIN (1969)). In this context the gambler's ruin problem leads to consideration of the random variables $\nu = \nu(a, b)$ defined by

$$\nu = \min \{n: S_n \leqslant -b \text{ or } S_n \geqslant a\}, \quad 0 < a, b < \infty.$$

Obviously, ν is the first exit time from the interval $(-b, a)$. It is of fundamental importance in Wald's sequential analysis. Details may be found in COX and MILLER (1965, Ch. 2) and KEMPERMAN (1961).

Of the numerous applications of generalized homogeneous random walks (to insurance risk theory, storage theory, etc.) we shall briefly describe that to the escape of comets from the solar system due to KENDALL (1961 a, b): During one revolution around the earth, the energy of a comet undergoes a change brought about by the disposition of the planets. In successive revolutions the changes in the energy of the comet are assumed to be independent and identically distributed random variables $\eta_1, \eta_2, \ldots$. If initially the comet has positive energy e_0, which is taken as origin on the scale of the η_n, then after n revolutions the energy will be $S_n = \eta_1 + \ldots + \eta_n$. If at any stage the energy S_n becomes 0 or negative, the comet escapes from the solar system. Thus the energy level of the comet performs a generalized homogeneous random walk with an absorbing barrier at 0. Problems of interest are determining the probability of escape and the distribution of the time to escape.

1.2.1.4.2. The point of view in the previous subparagraph was naturally extended by considering "random walks" $(S_n)_{n \in N^*}$ associated with a sequence $(\eta_n)_{n \in N^*}$ of independent random variables (i.e. $S_n = \eta_1 + \ldots + \eta_n$) which are no longer identically distributed. Consequently, the sequence $(S_n)_{n \in N^*}$ still has independent increments but these are no longer identically distributed. The theory of these random walks, which will be called *nonhomogeneous*, in general coincides with that for chains with independent increments. In special cases the nonhomogeneous random walk can be investigated from the Markov chain point of view. Such a case will be examined below.

In closing we mention that in recent years certain Markov chains which do not have independent increments have been abusively designated as random walks. It thus appears that the term "random walk" has tended to lose its initial meaning and has become a synonym for "Markovian motion".

1.2.1.5. Integral representations for certain nonhomogeneous random walks

1.2.1.5.1. Let us consider with HARRIS (1952) a nonhomogeneous random walk [30] with state space $I = N$ and right-elastic barrier at 0 whose transition probabilities are given by

$$p_{ij} = 0 \quad \text{if} \quad |i-j| > 1,$$

$$p_{i,i-1} = q_i, \; p_{ii} = r_i, \; p_{i,i+1} = p_i,$$

$$p_i + q_i + r_i = 1, \qquad i \in N^*,$$

$$p_{00} = r_0, \, p_{01} = p_0 = 1 - r_0.$$

We assume that $p_i > 0$, $q_{i+1} > 0$, $r_i \geqslant 0$, $i \in N$.

It is obvious that all states form an essential class. To classify them we use Theorem 1.1.34. The system of equations

$$z_i = \sum_{j \in N^*} p_{ij} z_j, \quad i \in N^*,$$

reduces to

$$z_1 = r_1 z_1 + p_1 z_2, \; z_i = p_i z_{i+1} + r_i z_i + q_i z_{i-1}, \; i \geqslant 2,$$

which can be written as

$$p_i(z_{i+1} - z_i) = q_i(z_i - z_{i-1}), \quad i \geqslant 2,$$

whence

$$z_{i+1} - z_i = \frac{q_1 \cdots q_i}{p_1 \cdots p_i} z_1, \quad i \in N.$$

We see that a bounded solution exists iff $\sum_{i \in N^*} \rho_i < \infty$, where

$$\rho_i = \frac{q_1 \cdots q_i}{p_1 \cdots p_i}.$$

Therefore by Theorem 1.1.34, the states are nonrecurrent or recurrent according as the series $\sum_{i \in N} \rho_i$ converges or diverges.

---

[30] This can be considered as a generalization of the random walk studied in Subparagraph 1.2.1.3.4.

To decide when the states are positive, we write the system of equations

$$u_j = \sum_{i \in N} u_i p_{ij} \quad , \quad j \in N,$$

determining the stationary measure. This system reduces to

$$u_0 = u_0 r_0 + u_1 q_1$$

$$u_j = u_{j-1} p_{j-1} + u_j r_j + u_{j+1} q_{j+1}, \quad j \in N^*.$$

Solving these equations successively yields

$$u_j = \frac{p_0 \cdots p_{j-1}}{q_1 \cdots q_j} u_0 \quad , \quad j \in N^*,$$

that is,

$$u_j = \frac{p_0 u_0}{p_j \rho_j} \quad , \quad j \in N^*. \tag{1.2.11}$$

By Theorem 1.1.43 and Proposition 1.1.38, the states are positive iff $\sum_{i \in N^*} \rho_i = \infty$ and $\sum_{i \in N^*} 1/p_i \rho_i < \infty$ and null-recurrent iff $\sum_{i \in N^*} \rho_i = = \sum_{i \in N} 1/p_i \rho_i = \infty$.

1.2.1.5.2. Now, following Karlin and McGregor (1959), we shall obtain an integral representation for the n-step transition probabilities associated with the random walk considered. Their approach is based on classical results from functional analysis. The matrix $\mathbf{P} = (p_{ij})_{,i\,j \in N}$ determines a linear transformation in the space of all sequences $\mathbf{x} = (x_i)_{i \in N}$ of complex numbers, by means of the formula

$$(\mathbf{P}\,\mathbf{x})_i = \sum_{j \in N} p_{ij} x_j \quad , \quad i \in N.$$

The series on the right has at most three nonzero terms. The solution of $\mathbf{P}\mathbf{x} = \lambda \mathbf{x}$, where λ is a complex constant, is unique up to a constant factor. If we normalize the sequence $\mathbf{x} = (x_i)_{i \in N}$ by setting $x_0 = 1$, then the nth term x_n, $n \in N^*$, is a polynomial $Q_n(\lambda)$ in λ of degree exactly n and the recurrence relations

$$Q_0(\lambda) \equiv 1,$$

$$\lambda\, Q_0(\lambda) = r_0\, Q_0(\lambda) + p_0\, Q_1(\lambda), \tag{1.2.12}$$

$$\lambda Q_n(\lambda) = q_n\, Q_{n-1}(\lambda) + r_n\, Q_n(\lambda) + p_n\, Q_{n+1}(\lambda) \quad , \quad n \in N^*.$$

are satisfied.

Let us consider the Hilbert space $L_2(\mathbf{u})$ of sequences $(x_i)_{i \in N}$ such that

$$\| \mathbf{x} \|^2 = \sum_{i \in N} | x_i |^2 u_i < \infty,$$

the inner product in $L_2(\mathbf{u})$ being defined by

$$(\mathbf{x}, \mathbf{y}) = \sum_{i \in N} x_i \bar{y}_i u_i$$

Here $\mathbf{u} = (u_i)_{i \in N}$ is the stationary measure (1.2.11) normalized by $u_0 = 1$. Notice that we have

$$p_{ij} u_i = p_{ji} u_j \quad , \quad i, j \in N. \tag{1.2.13}$$

Proposition 1.2.4. *The linear transformation* $\mathbf{x} \to \mathbf{P}\mathbf{x}$ *induces in* $L_2(\mathbf{u})$ *a bounded self-adjoint linear operator* U *with norm one.*

Proof. The self-adjointness follows from the symmetry equations (1.2.13) [31]), and the fact that $\| U \| = 1$ from the stationarity of the measure $(u_i)_{i \in N}$. ◇

Notice that for $\mathbf{x} \in L_2(u)$ we have

$$(U^n \mathbf{x})_i = \sum_{j \in N} p_{ij}^{(n)} x_j \quad , \quad n \in N^*.$$

In order to express the entries of $\mathbf{P}^n$ in terms of the operator U, we consider the sequences $\mathbf{e}_i = (e_{ij})_{j \in N}$ defined by

$$e_{ij} = \frac{1}{u_i} \delta_{ij} \quad , \quad i, j \in N.$$

Since $p_i u_i = q_{i+1} u_{i+1}$, we find by direct computation that

$$U\mathbf{e}_i = q_i \mathbf{e}_{i-1} + r_i \mathbf{e}_i + p_i \mathbf{e}_{i+1}$$

with an obvious modification for $i = 0$, and hence by an inductive argument based on (1.2.12).

$$Q_n(U) \mathbf{e}_0 = \mathbf{e}_n \quad , \quad n \in N.$$

---

31) In fact (1.2.13) (which amounts to the reversibility with respect to $\mathbf{u}$ of the chain) and the self-adjointness of U are equivalent. Kendall (1959) characterized irreducible Markov chains satisfying (1.2.13) by the fact that

$$p_{ii_1} p_{i_1 i_2} \cdots p_{i_r i} = p_{ii_r} p_{i_r i_{r-1}} \cdots p_{i_1 i}$$

for each $r \geqslant 2$ and for all sets of distinct states $i, i_1, \ldots, i_r$.

Now, the inner product $(U^n \mathbf{e}_j, \mathbf{e}_i)$ is equal to $p_{ij}^{(n)}/u_j$, and therefore

$$p_{ij}^{(n)} = u_j (U^n Q_j(U) \mathbf{e}_0, Q_i(U) \mathbf{e}_0).$$

The polynomials Q_n have real coefficients so the operator polynomials $Q_n(U)$ are self-adjoint, and since they commute with U,

$$p_{ij}^{(n)} = u_j (U^n Q_i(U) Q_j(U) \mathbf{e}_0, \mathbf{e}_0).$$

Consequently, if (E_λ) is the resolution of the identity corresponding to the self-adjoint operator U, and if $\psi(\lambda) = (E_\lambda \mathbf{e}_0, \mathbf{e}_0)$, then

$$p_{ij}^{(n)} = u_j \int_{-1}^{1} \lambda^n Q_i(\lambda) Q_j(\lambda) \, d\psi(\lambda), \tag{1.2.14}$$

where the integral includes any jumps which may be present at $\lambda = 1$ or at $\lambda = -1$.

Theorem 1.2.5. *There is a unique positive regular distribution function* ψ *on* $[-1, 1]$ *such that* (1.2.14) *is valid for all* $i, j, n \in N$.

Proof. The existence of one such ψ has been proved. But (1.2.14) with $i = j = 0$ determines all the moments of ψ and hence determines ψ uniquely. ◇

Setting $n = 0$ in (1.2.14) we see that the polynomials Q_n are the orthogonal polynomials belonging to ψ, in fact,

$$u_j \int_{-1}^{1} Q_i(\lambda) Q_j(\lambda) \, d\psi(\lambda) = \delta_{ij}.$$

In the special case $r_i = 0$, $i \in N$, the distribution function ψ is symmetric about $\lambda = 0$. To see this, notice that in this case $p_{00}^{(n)}$ is zero or positive according as n is odd or even, and hence from (1.2.14), all the odd order moments of ψ are zero so that ψ is symmetric. Conversely, one can prove that every distribution function on $[-1, 1]$ symmetric about $\lambda = 0$, not supported by a finite set of points, and with total mass one, is the spectral function ψ of a nonhomogeneous random walk with $r_i = 0$, $i \in N$.

The representation formula (1.2.14) provides a tool for relating recurrence properties of the random walk to the properties of ψ. Thus it follows from (1.1.8) that

$$F_{00}(z) = 1 - \frac{1}{\displaystyle\int_{-1}^{1} \frac{d\psi(\lambda)}{1 - z\lambda}}.$$

The recurrence of the random walk is equivalent to the divergence of $\int_{-1}^{1} (1-\lambda)^{-1}\,\mathrm{d}\psi(\lambda)$. Using the analytical tools of the theory of orthogonal polynomials one may prove that

$$\int_{-1}^{1} \frac{\mathrm{d}\psi(\lambda)}{1-\lambda} = \sum_{n\in N} \frac{1}{p_n u_n},$$

that is, the random walk is recurrent iff the series $\sum_{n\in N} 1/p_n u_n$ diverges. This result is in accordance with the results of the previous subparagraph.

We note that the random walk is positive iff $\overline{\lim_{n\to\infty}}\ p_{00}^{(n)} > 0$. Since $\lambda^{2n} \to 0$ monotonically on $(-1, 1)$ as $n \to \infty$, we see from (1.2.14) that the random walk is positive iff ψ has a jump at either $\lambda = 1$ or $\lambda = -1$. If ψ has no jump at $\lambda = 1$, then the magnitude of the jump at $\lambda = -1$ is

$$-\lim_{n\to\infty} \int_{-1}^{1} \lambda^{2n+1}\,\mathrm{d}\psi(\lambda) = -\lim_{n\to\infty} p_{00}^{(2n+1)} \leqslant 0;$$

so there is no jump at $\lambda = -1$. Consequently, the random walk is positive iff $\lambda = 1$ is an eigenvalue of the self-adjoint operator U, and this is the case iff

$$\sigma^{-1} = \sum_{n\in\infty} u_n < \infty,$$

where σ is the magnitude of the jump.

For the important problem of computing the spectral distribution function ψ we refer the reader to KARLIN and MCGREGOR (1959). These authors also apply the above method to the nonhomogeneous random walk with two elastic barriers and nonhomogeneous unrestricted random walk.

1.2.2. Galton-Watson chains

1.2.2.1. Basic properties

1.2.2.1.1. We shall now study a class of Markov chains which arose in a demography problem considered in 1873 by F. GALTON and partially solved in 1874 by H. W. WATSON [32]. These are the so-called *bran-*

[32] For the history of Galton-Watson chains see the delightful (1966 b) paper of KENDALL.

ching chains of Galton-Watson type which may be characterized as follows: let $(\eta_n)_{n\in N^*}$ be a sequence of independent and identically distributed random variables such that $Pr(\eta_1 = k) = \mathfrak{p}_k$, $k \in N$, with $\mathfrak{p}_k \geqslant 0$, $\sum_{k\in N} \mathfrak{p}_k = 1$. A homogeneous Markov chain $(\xi_n)_{n\in N}$ with state space $I = N$ is a *Galton-Watson chain* if

$$p_{ij} = \mathsf{P}(\xi_{n+1} = j \mid \xi_n = i) = Pr(\eta_1 + \ldots + \eta_i = j)$$

for $i \in N^*$, $j \in N$, and $p_{00} = 1$.

This definition may be interpreted as follows: a particle (organism, bacterium, neutron, etc.) at the end of its lifetime produces a random number η_1 of offspring constituting the first generation. All of these offspring act independently of each other and at the end of their lifetime (the lifespans of all particles are assumed to be the same) individually produce progeny in accordance with the probability distribution of η_1. The particles of the first generation give rise in this way to the second generation, and so on. If we denote by ξ_n, $n \in N$, the number of particles in the nth generation, then $\xi_0 = 1$ and for $n \geqslant 1$ the conditional distribution of ξ_n, given $\xi_{n-1} = i \neq 0$, coincides with the distribution of the sum $\eta_1 + \ldots + \eta_i$.

Unless otherwise stated, we shall always assume that for a Galton-Watson chain $(\xi_n)_{n\in N}$ we have $\xi_0 = 1$. Since the families of the initial particles develop independently of each other, the appropriate adjustments when $\xi_0 \neq 1$ are made easily.

1.2.2.1.2. We remark initially that the transition probabilities p_{ij} of the Galton-Watson chain have no simple expressions (See, e.g., SELIVANOV (1969)). For this reason, the main tool in investigating these chains will be the probability generating function of η_1:

$$f(z) = \sum_{k\in N} \mathfrak{p}_k z^k, \; |z| \leqslant 1.$$

We define the functions $f_n(z)$, $n \in N$ by

$$f_0(z) = z, f_1(z) = f(z),$$

$$f_{n+1}(z) = f(f_n(z)) \quad , \quad n \in N^*.$$

It is easy to verify that

$$f_{m+n}(z) = f_m(f_n(z)) \quad , \quad m, n \in N,$$

and, thus, in particular

$$f_{n+1}(z) = f_n(f(z)). \tag{1.2.15}$$

The interpretation of the functions $f_n(z)$, $n \geqslant 1$ follows from

Theorem 1.2.6 (H. WATSON). *The generating function of* ξ_n is $f_n(z)$, $n \in N$.

Proof. The theorem is trivially verified for $n = 0$. Let $f_{(n)}(z)$ be the generating function of ξ_n, $n \in N$. The conditional distribution of ξ_{u+1} given $\xi_n = k$ is, by definition, $[f(z)]^k$, $k \in N$. Consequently,

$$f_{(n+1)}(z) = \sum_{k \in N} \mathsf{P}(\xi_n = k) [f(z)]^k = f_{(n)}(f(z)), \ n \in N. \quad (1.2.16)$$

Since $f_{(0)}(z) = f_0(z)$, it follows by induction from (1.2.15) and (1.2.16) that $f_{(n)}(z) = f_n(z)$[33], $n \in N$. ◇

Corollary. *The n-step transition probabilities are given by the relations*

$$p_{ij}^{(n)} = \text{coefficient of } z^j \text{ in } [f_n(z)]^i,$$

$i, j \in N$, $n \in N^*$.

Define the function K_n by

$$K_n(u) = \log f_n(z) \quad , \quad z = e^u, \ n \in N.$$

By using (1.2.16) one can also prove

Theorem 1.2.7 (P. WOODWARD). *The cumulant generating functions* K_n *satisfy the relation*

$$K_{n+1}(u) = K_1(K_n(u)) \quad , \quad n \in N.$$

1.2.2.1.3. Theorem 1.2.6 will allow us to compute the moments of the random variables ξ_n, $n \in N^*$. Put

$$m = \mathsf{E}\,\xi_1 = \sum_{k \in N} k\mathfrak{p}_k = f'(1),$$

$$\sigma^2 = \mathsf{D}\,\xi_1 = \sum_{k \in N} k^2\mathfrak{p}_k - m^2 = f''(1) + m - m^2, \text{[34]}$$

and suppose that m and σ^2 are finite. To obtain the mean value of ξ_n, we differentiate (1.2.15) and let $z \to 1-$. We have

$$f'_{n+1}(1) = f'_n[f(1)]f'(1) = mf'_n(1) \quad , \quad n \in N^*,$$

[33] It is clear that if $\xi_0 = i$ then the generating function of ξ_n is $[f_n(z)]^i$.

[34] The symbols $f'(1)$, $f''(1)$, etc., will usually refer to left-hand derivatives at $z = 1$.

from which we deduce

$$f'_n(1) = m^n \quad , \quad n \in N^*.$$

Consequently,

$$\mathsf{E}\,\xi_n = m^n \quad , \quad n \in N^*.$$

Differentiating (1.2.15) again yields

$$f''_{n+1}(1) = f''_n(1)\,[f'(1)]^2 + f''(1)\,f'_n(1). \tag{1.2.17}$$

Similarly, differentiating $f_{n+1}(z) = f(f_n(z))$ twice produces

$$f''_{n+1}(1) = f''(1)[f'_n(1)]^2 + f'(1)\,f''_n(1). \tag{1.2.18}$$

If $m \neq 1$, then from (1.2.17) and (1.2.18) we deduce

$$f''_n(1) = \sigma^2 m^n \frac{m^n - 1}{m^2 - m} + m^n(m^n - 1)$$

and

$$\mathsf{D}\xi_n = \frac{\sigma^2 m^n (m^n - 1)}{m^2 - m} \quad , \quad n \in N^*.$$

If $m = 1$ then (1.2.17) and (1.2.18) yield directly $f''_n(1) = n\sigma^2$, and therefore $\mathsf{D}\,\xi_n = n\sigma^2$, $n \in N^*$.

If higher order moments of ξ_1 exist, then the corresponding moments of ξ_n may be obtained as above.

1.2.2.14. Concerning the state classification for Galton-Watson chains, we note that this is rendered difficult by the hindrance already mentioned regarding the transition probabilities; it is however obvious that the state $i = 0$ is absorbing (and thus positive).

Proposition 1.2.8. *If* $\mathfrak{p}_1 \neq 1$, *then all states* $k \neq 0$ *are nonrecurrent.*

Proof. It suffices to prove that $f^*_{kk} < 1$. We have

$$f^*_{kk} = \mathsf{P}\Big(\bigcup_{n \in N^*} (\xi_{m+n} = k) \mid \xi_m = k\Big) = \mathfrak{p}_1^k < 1$$

if $\mathfrak{p}_0 = 0$, and

$$f^*_{kk} \leqslant 1 - p_{k0} = 1 - \mathfrak{p}_0^k < 1$$

if $\mathfrak{p}_0 > 0$. ◇

Corollary. *If* $\mathfrak{p}_1 \neq 1$, *then*

$$\lim_{n\to\infty} \mathsf{P}(\xi_n = k) = 0$$

for all states $k \neq 0$.

Proof. Note that

$$\mathsf{P}(\xi_n = k) = p_{1k}^{(n)}$$

and apply Theorem 1.1.15. ◇

The reader can prove without difficulty that, assuming $0 < \mathfrak{p}_0 < 1$, the states $k \neq 0$ form a single aperiodic (nonrecurrent) class if $\mathfrak{p}_1 > 0$ and $\mathfrak{p}_i > 0$ for some $i > 1$.

1.2.2.1.5. We shall now briefly describe an extension of Galton-Watson chains which arises by permitting an immigration contribution. More precisely, besides the probability generating function $f(z)$ corresponding to the progeny distribution per particle per generation consider another probability generating function $g(z) = \sum_{k \in N} g_k z^k$ whose coefficients g_k, $k \in N$, are the probabilities that k new particles immigrate into the population of particles. The contributions arising from birth and death of the present population and new immigration are assumed to be independent. Moreover, newly arriving particles undergo growth following the laws of the Galton-Watson chain induced by $f(z)$.

Then, if there are i particles present in one generation, the probability q_{ij} that there will be j particles in the next generation is

$$q_{ij} = \text{coefficient of } z^j \text{ in } [f(z)]^i\, g(z).$$

A Markov chain with transition matrix $(q_{ij})_{i,j \in N}$ is said to be a *Galton-Watson chain with immigration.* One can show that

$$q_{ij}^{(n)} = \text{coefficient of } z^j \text{ in } [f_n(z)]^i \prod_{m=0}^{n-1} g(f_m(z)).$$

In a Galton-Watson chain with immigration, the state 0 is no longer absorbing. Moreover, it is possible that in such a chain all the states be recurrent. The interested reader is referred to Heathcote (1966) and Seneta (1968 a).

1.2.2.2. Extinction probability

1.2.2.2.1. We shall now consider the problem raised by Galton: what is the probability that the population will eventually die out, i.e.,

$$q = \mathsf{P}(\xi_n = 0 \text{ for some } n \in N^*) = \mathsf{P}\Big(\bigcup_{n \in N^*} (\xi_n = 0)\Big).$$

Since $\mathsf{P}(\xi_{n+1} = 0 \mid \xi_n = 0) = p_{00} = 1$, we have

$$q = \mathsf{P}\{(\xi_1 = 0) \cup (\xi_2 = 0) \cup \ldots\} = \lim_{n \to \infty} \mathsf{P}\{(\xi_1 = 0) \cup \ldots \cup (\xi_n = 0)\} =$$

$$= \lim_{n \to \infty} \mathsf{P}(\xi_n = 0) = \lim_{n \to \infty} f_n(0).$$

In what follows we suppose that the generating function does not coincide with the identity function, i.e., $p_1 \neq 1$. This means that either f is linear with $f(0) > 0$, or it is strictly convex on (0,1). In case $f(z) \equiv z$ we obviously have $q = 0$ since $\xi_n \equiv 1$, $n \in N$.

Our problem is solved by

Theorem 1.2.9 (H. Watson, J. Steffensen). *If* $m = \mathsf{E}\, \xi_1 \leqslant 1$ *then* $q = 1$; *if* $m > 1$, *then the extinction probability* q *is the unique solution in* [0,1) *of the equation* $x = f(x)$.

Proof. It is obvious that the sequence $(f_n(0))_{n \in N^*}$ is nondecreasing. Therefore, $0 \leqslant f_1(0) \leqslant f_2(0) \leqslant \ldots \leqslant q = \lim_{n\to\infty} f_n(0) \leqslant 1$. Since $f_{n+1}(0) = f(f_n(0))$, we have $q = f(q)$ and $0 \leqslant q \leqslant 1$. If $m \leqslant 1$, then $f(0) > 0$ and $f'(x) \leqslant f'(1) = m \leqslant 1$ for $x \in [0,1]$. The mean-value theorem then implies that $f(x) > x$ for $x \in [0,1)$, so that we must have $q = 1$.

Suppose now that $m > 1$. The mean-value theorem then implies that $f(x) < x$ for x sufficiently near 1. The strict convexity of f (in this case f cannot be linear) assures the uniqueness of the solution to the equation $x = f(x)$ in [0,1). To complete the proof it remains to show that $q \neq 1$. If q were equal to 1, then $f_n(0)$ would approach 1 as $n \to \infty$ and by the above considerations we would have $f_{n+1}(0) = f(f_n(0)) < f_n(0)$. Thus we reach a contradiction. ◇

It is easy to see that, whatever the value of m, we have

$$q = \lim_{n\to\infty} f_n(x)$$

for all $x \in [0,1)$.

Corollary. *Whatever the value of* m, *the sequence* $(\xi_n)_{n\in N}$ *converges as* $n \to \infty$ *to* ∞ *or* 0 *respectively with probabilities* $1 - q$ *and* q.

Proof. Since the states $k \neq 0$ are nonrecurrent, we can write $\mathsf{P}(\xi_n = k \text{ i. o.}) = 0$. Thus in the sequence $(\xi_n)_{n\in N}$ none of states $k \neq 0$ appear i. o. This means that this sequence approaches either 0 or ∞ as $n \to \infty$. By Theorem 1.2.9, the probability of approaching 0 is q. ◇

1.2.2.2.2. Concerning the rate of convergence of $f_n(0)$ to q we quote the following results

Theorem 1.2.10 (G. Koenigs, D. G. Kendall). *Assume that* $m > 1$ *and* $\mathfrak{p}_0 > 0$. *Then* $0 < f'(q) < 1$ *and*

$$f_n(0) = q - \left(\sum_{j\in N} \gamma_j q^j\right)[f'(q)]^n + O\{[f'(q)]^{2n}\}$$

when $n \to \infty$, *where the nonnegative* γ_j *are uniquely determined by the linear equations*

$$\sum_{i\in N^*} \gamma_i p_{ij} = f'(q)\gamma_j \quad , \quad j \in N^*,$$

with

$$\sum_{i\in N^*} \gamma_i\, i\, q^{i-1} = 1.$$

Proof. See HARRIS (1963, p. 16) and KENDALL (1966 a). ◇

One can verify without difficulty that for a fractional linear generating function (see Subparagraph 1.2.2.2.3) we have

$$\gamma_i = (1-q)^2 \quad , \quad i \in N^*.$$

Theorem 1.2.11 (A. N. KOLMOGOROV)[35]. *Assume that $m < 1$. Then*

$$\frac{m^n}{1-f_n(0)}$$

converges as $n \to \infty$ to a constant $K \neq 0$ which is finite or not according as the series $\sum_{i \in N^*} i \log i \mathfrak{p}_i$ (equivalently the integral $\int_0^1 \frac{1-mx-f(1-x)}{x^2} \mathrm{d}x$) is convergent or divergent.

Proof. See HEATHCOTE, SENETA and VERE-JONES (1967) and A. V. NAGAEV and BADALBAEV (1967). ◇

Bounds for K were given in HEATHCOTE and SENETA (1966—67).

Theorem 1.2.12 (A. N. KOLMOGOROV)[36]. *Assume that $m = 1$, $\mathfrak{p}_1 \neq 1$, $\sigma^2 < \infty$. Then*

$$f_n(0) = 1 - \frac{2}{n\sigma^2} + O(n^{-1})$$

as $n \to \infty$

Proof. See KESTEN, NEY and SPITZER (1966). ◇

1.2.2.2.3. LOTKA (1931) applied Theorem 1.2.9 to determine the probability of extinction of a surname descended from a single white American male.

Note first that in the special case

$$\mathfrak{p}_k = bc^{k-1} \quad , \quad k \in N^* \quad , \quad 0 < b, c \quad ; \quad b \leqslant 1 - c,$$

$$\mathfrak{p}_0 = 1 - \sum_{k \in N^*} \mathfrak{p}_k,$$

the generating function f is fractional linear, given by

$$f(z) = \frac{1-b-c}{1-c} + \frac{bz}{1-cz} \cdot$$

so that

$$m = f'(1) = \frac{b}{(1-c)^2}.$$

[35] KOLMOGOROV proved this theorem under the unnecessarily restrictive condition $\sigma^2 < \infty$. See also Theorem 1.2.25.

[36] KOLMOGOROV proved this theorem under the unnecessarily restrictive condition $f'''(1) < \infty$.

The equation $x = f(x)$ has the non-negative root

$$x_0 = \frac{1-b-c}{c(1-c)}.$$

For $m = 1$ we have $x_0 = 1$; for $m \neq 1$ x_0 is the unique solution in $[0,1)$. Thus $x_0 = q$ if $m \geqslant 1$.

From data for white males in the 1920 United States census LOTKA found that the probabilities $\mathfrak{p}_k$, $k \in N^*$ were well approximated by the geometric series $\mathfrak{p}_k = 0.2126 \times (0.5893)^{k-1}$ and $\mathfrak{p}_0 = 0.4825$. Hence, the probability of extinction of a surname descended from a single male is $q = x_0 = 0.819$, perhaps an unexpectedly high value. But, if initially there are i males with the same surname, and if their descendants reproduce independently of each other, then the probability that the surname becomes extinct is $(0.819)^i$ which is small for i large.

1.2.2.3. Time to extinction

1.2.2.3.1. We define the *time to extinction* ν as the smallest subscript n such that $\xi_n = 0$ (i.e. the first passage time to state 0). Obviously ν may be thought of as the number of generations to extinction. The distribution of ν is

$$\mathsf{P}(\nu = 0) = 0,$$

$$\mathsf{P}(\nu = n) = \mathsf{P}(\xi_{n-1} \neq 0\ ,\ \xi_n = 0) = \mathsf{P}(\xi_n = 0) - \mathsf{P}(\xi_{n-1} = 0,\ \xi_n = 0)$$

$$= \mathsf{P}(\xi_n = 0) - \mathsf{P}(\xi_{n-1} = 0) = f_n(0) - f_{n-1}(0)$$

for $n \in N^*$.

In case $m > 1$, we have from the corollary to Theorem 1.2.9

$$\mathsf{P}(\nu < \infty) = q < 1,$$

therefore $\mathsf{E}\,\nu = \infty$.

In case $m < 1$, by Theorem 1.2.11 the tail $\mathsf{P}(\nu > n) = 1 - f_n(0)$ approaches zero at least exponentially fast, therefore all moments of ν are finite.

It is sometimes erroneously stated that when $m = 1$, $\mathsf{E}\,\nu = \infty$. A sufficient condition for $\mathsf{E}\,\nu = \infty$ is $\sigma^2 < \infty$, since then $\mathsf{P}(\nu > n) \sim \sim 2/n\sigma^2$ as $n \to \infty$ by Theorem 1.2.12. The general case is treated by

Theorem 1.2.13 (H. BRENY, G. E. H. REUTER, E. SENETA). *If* $m = 1$, *then*

$$\mathsf{E}\,\nu \textit{ and } \mathfrak{J} = \int_0^1 \frac{1-x}{f(x)-x}\,\mathrm{d}x$$

are finite or infinite together.

Proof (SENETA (1967 a)). Put $h(x) = (1 - x) / (f(x) - x)$, $x \in [0,1)$ so that

$$h(x) = \frac{1}{1 - \dfrac{1 - f(x)}{1 - x}}.$$

Here since $f'(1) = m = 1$, $(1 - f(x)) / (1 - x)$ is a proper probability generating function (of the tail of the offspring distribution). Hence, h is positive, continuous and monotone increasing in $[0, 1)$ with $h(x) \to \infty$ as $x \to 1-$. Thus $\mathfrak{J}$ exists, and $\mathfrak{J} \leqslant \infty$.

Further, putting $a_n = f_n(0)$ — so that $a_0 = 0$ and $a_{n+1} = f(a_n)$ — we have

$$\Sigma_1 \equiv \sum_{n \in N} (a_{n+1} - a_n) h(a_n) \leqslant \mathfrak{J} \leqslant$$

$$\leqslant \sum_{n \in N} (a_{n+1} - a_n) h(a_{n+1}) = \sum_{n \in N^*} (a_n - a_{n-1}) h(a_n) \equiv \Sigma_2$$

on account of the positivity and monotonicity of h, and because $a_n \uparrow 1$. Now, $(a_{n+1} - a_n) = (a_n - a_{n-1}) f'(\theta_n)$ where $a_{n-1} < \theta_n < a_n$ by the mean-value theorem, so that for n sufficiently large, say $n \geqslant n_0$, $(a_{n+1} - a_n) \geqslant \geqslant \frac{1}{2}(a_n - a_{n-1})$ since $\theta_n \to 1$ as $n \to \infty$.

Hence it follows immediately that if Σ_1 is finite, then Σ_2 is also; and obviously the reverse implication is true. Thus Σ_1, Σ_2, and $\mathfrak{J}$ converge or diverge together. It remains to note that $\Sigma_1 = \sum_{n \in N} (1 - a_n) = = \mathsf{E}\nu$. ◇

1.2.2.3.2. It may sometimes be of interest to consider the total number γ of individuals which have existed up to the time to extinction. It is clear that

$$\gamma = \xi_0 + \ldots + \xi_\nu.$$

Let

$$c_j = \mathsf{P}(\gamma = j), \quad j \in N^*,$$

$$c(z) = \sum_{j \in N^*} c_j z^j.$$

We shall prove

Theorem 1.2.14 (D. HAWKINS, S. ULAM). *If* $\mathfrak{p}_0 > 0$, *then the generating function* c *is the unique root of the functional equation*

$$c(z) = zf(c(z))$$

such that $0 < c(z) \leqslant 1$ *for* $0 < z \leqslant 1$. *Moreover,*

$$\mathsf{P}(\gamma < \infty) = c(1) = \begin{cases} 1 & \textit{if} \quad m \leqslant 1 \\ q & \textit{if} \quad m > 1. \end{cases}$$

Proof. We have $c_1 = \mathfrak{p}_0$ and

$$c_{j+1} = \sum_{k \in N^*} \mathfrak{p}_k \sum_{(j_1+\cdots+j_k=j)} c_{j_1} \ldots c_{j_k}, \; j \in N^*,$$

from which we obtain

$$c(z) = z\,\mathfrak{p}_0 + z \sum_{k \in N^*} \mathfrak{p}_k [c(z)]^k = zf(c(z)).$$

The uniqueness of the solution of this functional equation can be established as follows: the function $g(u) = \dfrac{1}{u} f(u)$ is strictly convex on (0,1], and $g(0+) = \infty$, $g(1) = 1$. Therefore the equation $\dfrac{1}{z} = g(u)$ has a unique root $u = c(z)$ in (0,1] for all $z \in (0,1]$, and $c(1)$ will be the smallest positive root of the equation $f(u) = u$, that is 1 or q, according as $m \leqslant 1$ or $m > 1$. ◇

1.2.2.4. Stationary distributions and stationary measures

1.2.2.4.1. If $\mathfrak{p}_1 \neq 1$ we have $\pi_{ij} = 0$ for $i \in N$, $j \in N^*$ since by Proposition 1.2.8, the states $j \in N^*$ are nonrecurrent (therefore null). Obviously $\pi_{00} = 1$. For the sake of completeness we note that if $\mathfrak{p}_0 > 0$, then $\pi_{i0} > 0$ for $i \in N^*$. Indeed, we have $\pi_{i0} = f_{i0}^{'} \nu_{00} = f_{i0}^{*} > p_{i0} = \mathfrak{p}_0^i > 0$.

It follows from Theorem 1.1.41 that the unique stationary distribution of a Galton-Watson chain with $\mathfrak{p}_1 \neq 1$ is that concentrated at 0 i.e. (1, 0, 0, 0, ...).

We shall prove that under the same restriction, (1, 0, 0, 0, ...) is also the unique stationary measure. For, if $\mathfrak{p}_0 > 0$, then $p_{i0} = \mathfrak{p}_0^i > 0$, $i \in N^*$, and from the equation $u_0 = \sum_{j \in N} u_j p_{j0}$ and the fact that $p_{00} = 1$, we see that $u_i = 0$, $i \in N^*$. If $\mathfrak{p}_0 = 0$, then $p_{ij} = 0$ for $j < i$ and hence $u_i = u_1 p_{1i} + \ldots + u_i p_{ii}$, $i \in N^*$ Let u_n be the first of the numbers $u_1, u_2, \ldots$ differing from 0. It follows that $u_n = u_n p_{nn}$. Now p_{nn} is the coefficient of z^n in $[f(z)]^n$ and the assumption $\mathfrak{p}_1 \neq 1$ implies that $p_{nn} < 1$. Thus $u_n = 0$ and we arrive at a contradiction. Hence all the u_i must be zero [37].

---

[37] In fact, this proof may be considered as an alternate direct proof for the uniqueness of the stationary distribution.

1.2.2.4.2. A more interesting investigation is that of the existence and uniqueness of a stationary measure corresponding to the nonzero states, i.e. a sequence $(u_n)_{n \in N^*}$ of real nonnegative numbers satisfying

$$u_j = \sum_{i \in N^*} u_i p_{ij}, \quad j \in N^*.$$

The importance of this investigation for applications in genetics will become evident in II; 4.1.2.3.

Theorem 1.2.15 (T. E. HARRIS). *Let* $0 < \mathfrak{p}_0 < 1$. *Then there exists a stationary measure* $(u_n)_{n \in N^*}$ *such that* $\sum_{n \in N^*} u_n = \infty$. *The generating function*

$$u(z) = \sum_{n \in N^*} u_n z^n$$

is analytic for $|z| < q$ *(the extinction probability) and satisfies Abel's functional equation*

$$u(f(z)) = 1 + u(z), \quad |z| < q, \tag{1.2.19}$$

provided the numbers u_n *are normalized so that* $u(\mathfrak{p}_0) = 1$.

Proof. See HARRIS (1963, p. 24). ◇

Concerning the uniqueness of the solution of (1.2.19) it was pointed out by FATOU (1919, 1920) that if the coefficients u_n are not required to be nonnegative, then (1.2.19) has infinitely many linearly independent solutions. More recently, KINGMAN (1965) has shown that when $m \neq 1$, the solution of (1.2.19) can be nonunique, even when the coefficients are required to be nonnegative. His counterexample is the simple case $f(z) = (1 - z) / (1 - pz)$, $0 < p < 1$, $p \neq \frac{1}{2}$. Uniqueness of the stationary measure (up to a multiplicative constant) in the case $m = 1$, was proved by KARLIN and MCGREGOR (1967 a) under excessive assumptions on the moments. Here we state more general results.

Theorem 1.2.16 (H. KESTEN, P. NEY, F. SPITZER). *Let* $\mathfrak{p}_1 \neq 1$, $m = 1$, *and* $\sigma^2 < \infty$. *Then there is a unique stationary measure* $(u_i)_{i \in N^*}$ *satisfying* $\sum_{i \in N^*} u_i \mathfrak{p}_0^i = 1$ *given by*

$$u_j = \lim_{n \to \infty} \frac{\sigma^2}{2k} n^2 p_{kj}^{(n)} = \lim_{i \to \infty} \sum_{n \in N^*} p_{ij}^{(n)}, \quad k, i \in N^*.$$

We have $u_j > 0$ *iff* j *belongs to the additive semigroup generated by those* $k \in N^*$ *for which* $\mathfrak{p}_k > 0$. *Finally,*

$$\lim_{n \to \infty} \frac{1}{n} (u_1 + \ldots + u_n) = \frac{2}{\sigma^2}.$$

Proof. See KESTEN, NEY and SPITZER (1966). ◇

Theorem 1.2.17 (E. SENETA). *Let* $\mathfrak{p}_1 \neq 1$, $m = 1$. *Then as* $n \to \infty$

$$P(\xi_n = j \mid \nu = n + k) \to u_j(k)$$

for every fixed $j, k \in N^*$ *(here* ν *is the time to extinction), and* $\sum_{j \in N^*} u_j(k) = 1$, $k \in N^*$. *The numbers*

$$u_j = \frac{u_j(k)}{[f_k(0)]^j - [f_{k-1}(0)]^j}, \quad j \in N^*,$$

are independent of $k \in N^*$ *and represent the components of the unique stationary measure satisfying* $\sum_{j \in N^*} u_j \mathfrak{r}_0^j = 1$.

Proof. See SENETA (1967 a). ◇

For further considerations concerning the case $m = 1$ and $\sigma^2 = \infty$ see SLACK (1968).

A partial asymptotic expansion for the stationary measure $(u_i)_{i \in N^*}$ is provided by

Theorem 1.2.18 (S. KARLIN, J. MCGREGOR). *Let* $m = 1$ *and* $f^{(\mathrm{IV})}(1) < \infty$. *Assume that the greatest common divisor of the* $k \in N^*$ *for which* $\mathfrak{p}_k > 0$ *equals* 1[38]. *Then*

$$u_i = \frac{2}{\sigma^2}\left[1 - \frac{a}{i} + \frac{e_i}{i}\right], \quad i \in N^*,$$

where $\sum_{i \in N^*} e_i^2 < \infty$ *and*

$$a = \frac{2}{\sigma^2}\left[\frac{\sigma^4}{4} - \frac{f'''(1)}{6}\right].$$

Proof. See KARLIN and MCGREGOR (1967 b). ◇

1.2.2.5. Spectral theory of Galton-Watson chains

1.2.2.5.1. Using methods from functional analysis, KARLIN and MCGREGOR (1966 a, b) have obtained very interesting results concerning the spectral representation of the n-step transition probabilities of Galton-Watson chains. In what follows we reproduce without proof some of these results.

---

38) An equivalent condition is that $1 - f(z)$ vanishes in $|z| \leqslant 1$ only at $z = 1$.

Suppose that $\mathfrak{p}_0 > 0$ and $m > 1$. Under these assumptions, the only solution in (0,1) of the equation $f(x) = x$ is $x = q$. The value $c = f'(q)$ plays an important role in subsequent considerations. It is easily seen that $0 < c < 1$.

Let us introduce the Hilbert space H consisting of all sequences $\mathbf{x} = (x_k)_{k \in N}$ of complex numbers such that

$$\| \mathbf{x} \|^2 = \sum_{k \in N} | x_k |^2 q^k < \infty.$$

The inner product in H is defined as

$$(\mathbf{x}, \mathbf{y}) = \sum_{k \in N} x_k \bar{y}_k q^k.$$

Let P be the linear operator defined on H to itself by

$$(P \mathbf{x})_j = \sum_{i \in N} x_i p_{ij}, \; j \in N.$$

It is easily verified that the adjoint operator P^* of P is defined by

$$(P^* \mathbf{y})_i = \frac{1}{q^i} \sum_{j \in N} p_{ij} y_j q^j, \quad i \in N.$$

It may be shown that the operator P is completely continuous. The spectral properties of a completely continuous operator are described in considerable detail in the classical Fredholm theory. The nonzero eigenvalues of P form at most a denumerable set (λ_n). The adjoint operator P^* is also completely continuous and has the same sequence of nonzero eigenvalues. Moreover, for each λ_n, the corresponding eigenspaces of P and P^* have the same (finite) dimension.

1.2.2.5.2. We begin by presenting a recursive procedure for determining eigenvalues and eigenvectors of the adjoint operator P^*. By differentiating the identity

$$\sum_{j \in N} p_{ij} q^{j-i} z^j = \left[\frac{f(qz)}{q} \right]^i$$

and by setting $z = 1$, we obtain

$$\sum_{j \in N} p_{ij} q^{j-i} (j)_r = c^r (i)_r + \pi_{r-1}(i), \quad r \in N^*, i \in N,$$

where we use the notation

$$(a)_r = a(a-1)\ldots(a-r+1), \quad r \in N^*,$$

and where π_{r-1} is a polynomial of degree $r-1$ which vanishes at 0. Let us define recurrently the polynomials $(Q_r)_{r \in N}$ by

$$Q_0(x) \equiv 1, \quad Q_1(x) = x,$$

$$Q_r(x) = (x)_r - \sum_{k=1}^{r-1} \frac{a_k}{c^k - c^n} Q_k(x), \quad r \geqslant 2,$$

where the constants a_k, $1 \leqslant k \leqslant r-1$, are (uniquely) determined from

$$\pi_{r-1}(x) = \sum_{k=1}^{r-1} a_k Q_k(x).$$

Theorem 1.2.19 (i) *The only nonzero eigenvalues of P, and of P^*, are $\lambda_n = c^n$, $n \in N$, and each eigenspace of P and P^* is one-dimensional.*

(ii) *The functional equation*

$$A(f(z)) = cA(z), \; A(q) = 0, \; A'(q) = 1 \tag{1.2.20}$$

as a unique analytic solution $A(z)$ which is regular at $z = q$. The solution is regular in $|z| < 1$ and satisfies $\lim_{n \to \infty} c^{-n}(f_n(z) - q) = A(z)$ uniformly in every closed disk $|z| \leqslant \rho < 1$.

(iii) *The eigenvectors of P^* are*

$$\mathbf{b}_n = \{Q_n(i)\}_{i \in N}.$$

The eigenvector $\mathbf{a}_n = \{a_n(i)\}_{i \in N}$ of P belonging to the eigenvalue λ_n has a generating function

$$a_n(z) = \sum_{i \in N} a_n(i) z^i$$

given by

$$a_n(z) = [A(z)]^n, \; n \in N.$$

It is easy to verify that

$$\mathbf{b}_0' = (1, 1, 1, \ldots)$$

$$\mathbf{b}_1' = (0, 1, 2, \ldots)$$

$$\mathbf{a}_0' = (1, 0, 0, \ldots).$$

Since $A(q) = 0$ and $A'(q) = 1$, the mapping

$$z \to w = A(z)$$

sends some neighborhood of $z = q$ conformally onto a neighborhood of $w = 0$. Therefore there exists an inverse $z = B(w)$ defined by

$$B(A(z)) = z \tag{1.2.21}$$

near $z = q$. The function $B(w)$ is regular near $w = 0$ and maps a neighborhood of $w = 0$ onto a neighborhood of $z = q$. From (1.2.21) and (1.2.20) we obtain

$$f(B(w)) = B(cw),\ B(0) = q,\ B'(0) = 1.$$

Let

$$[B(w)]^j = \sum_{r \in N} \psi_j(r)\, w^r,\ j \in N,$$

$$[A(z)]^r = \sum_{j \in N} \theta_j(r)\, z^j,\ r \in N.$$

In terms of the notation of Theorem 1.2.19 we have

$$\mathbf{a}_n = \{\theta_i(n)\}_{i \in N},$$

$$\mathbf{b}_n = \{q^{-i}\, \psi_i(n)\}_{i \in N}.$$

1.2.2.5.3. We can now state the general spectral representation theorem for P.

Theorem 1.2.20 j) *The n-step transition probabilities* $p_{ij}^{(n)}$ *admit the spectral representation*

$$p_{ij}^{(n)} = \sum_{r \in N} (c^r)^n\, \theta_j(r)\, \psi_i(r) \tag{1.2.22}$$

for all sufficiently large n and all $i, j \in N$.[39]
jj) *If the coefficients* $\{\psi_1(r)\}_{r \in N^*}$ *in*

$$B(w) = q + \sum_{r \in N^*} \psi_1(r)\, w^r$$

have alternating signs, that is $(-1)^{r-1}\, \psi_1(r) \geq 0$ *from some r on, then* (1.2.22) *is valid for all* $n \in N^*$, *the series being absolutely convergent. If we also have* $\mathfrak{p}_1 > 0$, *then* (1.2.22) *is valid for* $n \in N$.

[39] Theorem 1.2.10 is a special instance of (1.2.22).

A general class of probability generating functions for which the conditions in jj) are satisfied are of the form

$$f(z) = K e^{\gamma z} \frac{\prod_{i \in N^*} (1 + \alpha_i z)}{\prod_{i \in N^*} (1 - \beta_i z)},$$

where the parameters satisfy

$$\gamma \geqslant 0, \alpha_i \geqslant 0, 0 \leqslant \beta < 1,$$

$$\sum_{i \in N^*} (\alpha_i + \beta_i) < \infty, \quad K = e^{-\gamma} \frac{\prod_{i \in N^*} (1 - \beta_i)}{\prod_{i \in N^*} (1 + \alpha_i)}.$$

This class includes the Poisson, the binomial, the negative binomial and numerous other important distributions.

1.2.2.5.4. In order to obtain analogous results in the case $m < 1$, the authors make the essential simplifying assumption that the analytic function $f(z)$ is regular at $z = 1$. Without such a restriction, a continuous spectrum is present and the nature of the spectral representation appears to be quite intricate.

When $m = 1$, P ceases to have eigenvectors and has in fact a "continuous spectrum". If f satisfies certain regularity conditions, then we have a spectral representation of the form

$$p_{ij}^{(n)} = \int_0^{\infty} e^{-nx} P_j(x) d\psi_i(x), \quad i, j \in N, \; n \in N^*,$$

where ψ_i is of bounded variation on $[0, \infty)$ and P_j is a polynomial of degree j, which vanishes at zero if $j \geqslant 1$.

In closing we note that it is difficult to obtain explicit expressions for the basic functions A and B in concrete examples. The task of determining A is almost synonymous with that of finding in closed form the iterates f_n. Examples for which A and B can be displayed in closed form are to be found in the cited works by S. Karlin and J. McGregor.

For further considerations concerning the critical case $m = 1$ see Karlin and McGregor (1968).

1.2.2.6. Asymptotic properties

1.2.2.6.1. We shall first deal with the sequence $(W_n)_{n \in N}$ where

$$W_n = \frac{\xi_n}{m^n}.$$

From the very definition of Galton-Watson chains, we have

$$\mathsf{E}\,(\xi_{n+1} \mid \xi_n) = m\,\xi_n.$$

Dividing both sides by m^{n+1}, we obtain

$$\mathsf{E}\,(W_{n+1} \mid W_n) = \mathsf{E}\,(W_{n+1} \mid W_n, \ldots, W_0) = W_n,$$

the first equality holding by the Markov property. It follows that for arbitrary $m < \infty$ the sequence $(W_n)_{n \in N}$ is a martingale, and since $\mathsf{E}\,|W_n| = 1$, W_n converges a.s. to a random variable W with mean less than or equal to one (see e.g. DOOB (1953, p. 319)). This random variable is obviously identically 0 if $m \leqslant 1$. Thus we need only consider the case $m > 1$.

Theorem 1.2.21 (H. KESTEN, B. P. STIGUM)[40], *Let* $m > 1$. *One has either* $\mathsf{E}\,(W) = 1$ *or* $W = 0$ a.s. *The first alternative holds iff* $\mathsf{E}\,(\xi_1 \log \xi_1) < \infty$. *Further, if* $\mathfrak{p}_i < 1$ *for all* $i \in N$, *the equations* $\mathsf{E}(W) = 1$ *and* $\mathsf{P}\,(W = 0) = q$ *are equivalent.*

Proof. See KESTEN and STIGUM (1966 a). ◇

It follows from Theorem 1.2.21 and the corollary to Theorem 1.2.9 that if $\mathsf{E}\,(\xi_1 \log \xi_1) < \infty$, then $(W = 0) = (\xi_n \to 0) \cup$ (a set of probability zero), that is, $\xi_n \sim W m^n$ a.s. on $(\xi_n \to \infty)$. Therefore, if the population survives, it increases at a geometric rate, in accordance with the Malthusian law of growth.

Under a strong assumption we may obtain the mean square convergence of W_n to W.

Theorem 1.2.22 (T. E. HARRIS). *If* $m > 1$ *and* $\mathsf{E}\,\xi_1^2 < \infty$ *then* W_n *converges in mean square to* W, *and*

$$\mathsf{E}\,(W) = 1, \; \mathsf{D}\,(W) = \frac{\sigma^2}{m^2 - m}.$$

Proof. See HARRIS (1963, p. 13). ◇

In order to study the distribution of W we shall derive a functional equation for its characteristic function $\varphi(t) = \mathsf{E}\exp(itW)$, $t \in R$. Let $\varphi_n(t)$ be the characteristic function of W_n, $n \in N$, that is

$$\varphi_n(t) = f_n\left(\exp \frac{it}{m^n}\right), \quad t \in R.$$

Thus, by the very definition of f_n,

$$\varphi_{n+1}(mt) = f\,[\varphi_n(t)], \; n \in N, \; t \in R.$$

---

[40] Proved in part by LEVINSON (1959).

Since W_n converges a.s. to W as $n \to \infty$, we can assert that $\varphi_n(t) \to \varphi(t)$. Since $f(z)$, is continuous for $|z| \leqslant 1$, we obtain the functional equation

$$\varphi(mt) = f[\varphi(t)]. \tag{1.2.23}$$

By Theorem 1.2.21 the only interesting case is that for which $\mathsf{E}(W) = 1$, which holds iff $\mathsf{E}(\xi_1 \log \xi_1) < \infty$. Therefore in this case $\varphi'(0) = \mathrm{i}$.

Theorem 1.2.23 (T. E. Harris). *There is a unique characteristic function satisfying* (1.1.23) *with* $\varphi'(0) = \mathrm{i}$.

Proof. See Harris (1963, p. 15). ◇

Theorem 1.2.24 (H. Kesten, B. P. Stigum). *Let* $m > 1$. *If* $\mathsf{E}(\xi_1 \log \xi_1) < \infty$ *then the distribution of* W *has a jump of magnitude* q *at the origin and a continuous density function on the set of positive real numbers.*

Proof. See Stigum (1966) and Stigum and Kesten (1966 a). ◇

1.2.2.6.2. Nontrivial asymptotic results for the case $m \leqslant 1$ can be obtained if we consider conditional distributions given that $\xi_n \neq 0$. Thus we have

Theorem 1.2.25 (A. M. Jaglom)[41]. *Let* $m < 1$. *Then for each* $j \in N^*$,

$$\lim_{n \to \infty} \mathsf{P}(\xi_n = j \mid \xi_n \neq 0) = g_j$$

exists and $\sum_{j \in N^*} g_j = 1$. *The generating function* $g(z) = \sum_{j \in N^*} g_j z^j$ *is the unique probability generating function with* $g(0) = 0$ *satisfying the functional equation*

$$1 - g[f(z)] = m(1 - g(z)).$$

Moreover, $g'(1)$ *equals the constant* K *appearing in Theorem 1.2.11.*

Proof. See Heathcote, Seneta and Vere-Jones (1967), Joffe (1966) and A. V. Nagaev and Badalbaev (1967). ◇

Theorem 1.2.26 (A. M. Jaglom)[42]. *Let* $m = 1$, $\mathfrak{p}_1 \neq 1$, *and* $\sigma^2 < \infty$. *Then*

$$\lim_{n \to \infty} \mathsf{P}\left(\frac{2\xi_n}{n\sigma^2} \geqslant x \mid \xi_n \neq 0\right) = \mathrm{e}^{-x}, \; x \geqslant 0.$$

---

[41] Jaglom proved this theorem under the unnecessary condition $\sigma^2 < \infty$.

[42] Jaglom proved this theorem under the unnecessary condition $f'''(1) < \infty$. The Poissonian case $f(z) = \exp(z - 1)$ was previously considered by R. A. Fisher.

Proof. See JOFFE and SPITZER (1967). ◇

For an extension of Theorem 1.2.26 see SLACK (1968).

For other types of limiting distributions see LAMPERTI (1967 a, b)

1.2.2.7. Multitype Galton-Watson chains

1.2.2.7.1. Soon after the Second World War various concrete situations in physics and biology led to the study of growth processes involving several types of "particle". A simple model of such a process is the multitype Galton-Watson chain, a natural generalization of classical Galton-Watson chains[43]. The pioneers were M. S. BARTLETT, N. A. DMITRIEV, C. EVERETT, A. N. KOLMOGOROV, B. A. SEVAST'JANOV, and S. ULAM.

Let $\mathbf{e}_l$, $1 \leqslant l \leqslant s$ denote the vector whose lth component is 1 and the rest all 0. For each $1 \leqslant l \leqslant s$, let $(\boldsymbol{\eta}_n^l)_{n \in N^*}$ be a sequence of independent and identically distributed s-dimensional random vectors with generating function

$$f^l(z_1, \ldots, z_s) = \sum_{r_1, \ldots, r_s = 0}^{\infty} p^l(r_1, \ldots, r_s) z_1^{r_1} \ldots z_s^{r_s},$$

$$|z_1| \leqslant 1, \ldots, |z_s| \leqslant 1.$$

The quantity $p^l(r_1, \ldots, r_s)$ is to be interpreted as the probability that a particle of type l has r_1 offspring of type 1, ..., and r_s offspring of type s[44].

An *s-type Galton-Watson chain* is a Markov chain $(\boldsymbol{\xi}_n)_{n \in N} \equiv \equiv ((\xi_n^l)_{1 \leqslant l \leqslant s})_{n \in N}$ with state space $I = N^s$ and transition probabilities

$$p_{\mathbf{ij}} = \mathsf{P}(\boldsymbol{\xi}_{n+1} = (j_l)_{1 \leqslant l \leqslant s} \mid \boldsymbol{\xi}_n = (i_l)_{1 \leqslant l \leqslant s}) =$$

$$= Pr\left(\sum_{l=1}^{s} \sum_{k=1}^{i_l} \boldsymbol{\eta}_k^l = (j_l)_{1 \leqslant l \leqslant s}\right)$$

for $\mathbf{i} \neq \mathbf{0}$, and $p_{\mathbf{00}} = 1$.

It follows that the conditional generating function of $\boldsymbol{\xi}_1$ given that $\boldsymbol{\xi}_0 = \mathbf{e}_l$ is $f^l(z_1, \ldots, z_s)$, $1 \leqslant l \leqslant s$. Further, the conditional distribution of $\boldsymbol{\xi}_{n+1}$ given that $\boldsymbol{\xi}_n = (i_l)_{1 \leqslant l \leqslant s}$ coincides with the distribution

---

[43] We mention a generalization of Galton-Watson chains in which there exists a single type of particles but the probability distribution of offspring is generation-dependent (see ATHREYA and KARLIN (1971 a, b), GOODMAN (1968), SMITH (1968), SMITH and WILKINSON (1969)). Galton-Watson chains are also a special case of the so-called *percolation processes*, which may be considered as models for the penetration of a fluid into a porous medium through channels subject to random blocking or damming (see FRISCH and HAMMERSLEY (1963), HAMMERSLEY and WALTERS(1963)).

[44] For generalizations to infinitely (countable or otherwise) many types see BARTOSZYŃKI (1967), JIRINA (1958, 1964, 1966), and MOY (1967 a, b).

of the sum of $i_1 + \ldots + i_s$ independent random vectors, i_1 having the generating function $f^1, \ldots,$ and i_s having the generating function f^s.

1.2.2.7.2. Let us set

$$\mathbf{z} = (z_1, \ldots, z_s),$$

$$\mathbf{f}(\mathbf{z}) = (f^1(\mathbf{z}), \ldots, f^s(\mathbf{z})),$$

$$\mathbf{f}_0(\mathbf{z}) = \mathbf{z},\ \mathbf{f}_1(\mathbf{z}) = f(\mathbf{z}),$$

$$\mathbf{f}_{n+1}(\mathbf{z}) = (f^1_{n+1}(\mathbf{z}), \ldots, f^s_{n+1}(\mathbf{z})) = \mathbf{f}(\mathbf{f}_n(\mathbf{z})),\quad n \in N^*.$$

By a straightforward generalization of Theorem 1.2.6. we deduce

Theorem 1.2.27 (A. N. KOLMOGOROV, N. A. DMITRIEV). *The conditional distribution of* ξ_n *given that* $\xi_0 = \mathbf{e}_l$ *is* $f^l_n(\mathbf{z})$, $n \in N$, $1 \leqslant l \leqslant s$.

This theorem allows us to compute moments of ξ_n. Put

$$m_{lk} = \mathsf{E}(\xi^k_1 \mid \xi_0 = l) = \frac{\partial f^l(1, \ldots, 1)}{\partial z_k},\quad 1 \leqslant l, k \leqslant s,$$

and

$$\mathbf{M} = (m_{lk})_{1 \leqslant l, k \leqslant s}.$$

Throughout the remainder of this paragraph we assume that all the first moments m_{lk} are finite and that not all are zero.

It is easily seen that

$$\mathsf{E}(\xi_{m+n} \mid \xi_m) = \xi_m \mathbf{M}^n,\quad m, n \in N.$$

In particular

$$\mathsf{E}(\xi_n \mid \xi_0) = \xi_0 \mathbf{M}^n,\quad n \in N.$$

Let $\mathbf{V}_l$ denote the conditional covariance matrix of ξ_1 given that $\xi_0 = \mathbf{e}_l$, and let

$$\mathbf{C}_n = (\mathsf{E}\, \xi^l_n \xi^k_n)_{1 \leqslant l, k \leqslant s},\quad n \in N.$$

If the entries of the matrices $\mathbf{V}_l$, $1 \leqslant l \leqslant s$, are finite, then it is esily seen that

$$\mathbf{C}_n = \mathbf{M}'^n \mathbf{C}_0 \mathbf{M}^n + \sum_{j=1}^{n} \mathbf{M}'^{n-j} \left(\sum_{l=1}^{s} \mathbf{V}_l\, \mathsf{E}\, \xi^l_{j-1} \right) \mathbf{M}^{n-j},\quad n \in N^*.$$

The above relations emphasize the importance of the matrix **M** and its powers. An s-type Galton-Watson chain is said to be *positively regular* if $\mathbf{M}^{n_0}$ is positive (all its entries are positive) for some $n_0 \in N^*$. It follows from Theorem 1.1.54 that in this case $\mathbf{M}$ has a positive eigenvalue ρ which is simple and greater in absolute value than any other eigenvalue.

1.2.2.7.3. The analogue of Proposition 1.2.8 is

Proposition 1.2.28. *In positively regular s-type Galton-Watson chains such that*

$$\sum_{k=1}^{s} \frac{\partial f^l(0, \ldots, 0)}{\partial z_k} \neq 1,$$

for some $1 \leqslant l \leqslant s$ *the states* $\mathbf{i} \neq \mathbf{0}$ *are nonrecurrent.*

Proof. See HARRIS (1963, p. 38). ◇

In s-type Galton-Watson chains which are not positively regular recurrent states others than 0 may exist. Thus for $f^1(z_1, z_2) = z_1 z_2$. $f^2(z_1, z_2) = 1$, the state (1,1) is recurrent since it is absorbing.

1.2.2.7.4. Let q_l denote the *extinction probability* of an s-type Galton-Watson chain if $\boldsymbol{\xi}_0 = \mathbf{e}_l$, $1 \leqslant l \leqslant s$, that is

$$q_l = \mathsf{P}\,(\bigcup_{n \in N^*} (\boldsymbol{\xi}_n = \mathbf{0}) \mid \boldsymbol{\xi}_0 = \mathbf{e}_l).$$

Set

$$\mathbf{q} = (q_l)_{1 \leqslant l \leqslant s}.$$

The eigenvalue ρ plays a part similar to that of the expectation m for Galton-Watson chains in determining whether extinction occurs or not.

Theorem 1.2.29 (C. EVERETT, B. A. SEVAST'JANOV, S. ULAM). *Under the conditions of Proposition 1.2.28, one has* $\mathbf{q} = (1, \ldots, 1)$ *if* $\rho \leqslant 1$. *If* $\rho > 1$, *then* $\mathbf{0} \leqslant \mathbf{q} < (1, \ldots, 1)$ *and* $\mathbf{q}$ *is the only solution* $\neq (1, \ldots, 1)$ *of the equation* $\mathbf{q} = \mathbf{f}(\mathbf{q})$ *in the unit s-dimensional cube. In either case*

$$\mathbf{q} = \lim_{n \to \infty} \mathbf{f}_n(\mathbf{q}_0),$$

for arbitrary $\mathbf{q}_0 \neq (1, \ldots, 1)$ *in the unit s-dimensional cube.*

Proof. See HARRIS (1963, p. 41). ◇

In case

$$\sum_{k=1}^{s} \frac{\partial f^l(0, \ldots, 0)}{\partial z_k} = 1$$

for all $1 \leqslant l \leqslant s$, it is clear that $\mathbf{q} = \mathbf{0}$. A detailed treatment of the non-positively regular case can be found in SEVAST'JANOV (1951).

1.2.2.7.5. Under suitable conditions asymptotic properties for multitype Galton-Watson chains turn out to be analogous to those for Galton-Watson chains. We shall give such a result in

Theorem 1.2.30. *Suppose the s-type Galton-Watson chain is positively regular and* $\rho > 1$. *Let* $\mathbf{u}' = (u_1, \ldots, u_s)$ and $\mathbf{v}' = (v_1, \ldots, v_s)$ *be positive left and right eigenvectors of* $\mathbf{M}$ *associated with* ρ *and normalized such that* $\sum_{l=1}^{s} v_l = 1$, $\sum_{l=1}^{s} u_l v_l = 1$. *Then there exist a random vector* $\mathbf{W}$ *and a random variable* W *such that*

$$\lim_{n \to \infty} \frac{\boldsymbol{\xi}_n}{\rho^n} = \mathbf{W} \qquad \text{a.s.}$$

and

$$\mathbf{W} = W \mathbf{u} \qquad \text{a.s.}$$

Also one has either

$$\mathsf{E}\,(W \mid \boldsymbol{\xi}_0 = \mathbf{e}_l) = v_l, \qquad 1 \leqslant l \leqslant s, \tag{1.2.24}$$

or

$$W = 0 \qquad \text{a.s.}$$

Moreover (1.2.24) *holds iff*

$$\mathsf{E}\,(\xi_1^k \log \xi_1^k \mid \boldsymbol{\xi}_0 = \mathbf{e}_l) < \infty \tag{1.2.25}$$

for all $1 \leqslant k, l \leqslant s$. *Finally, if* (1.2.25) *holds and if there is an* l_0, $1 \leqslant l_0 \leqslant s$, *such that*

$$\mathsf{P}\left(\sum_{k=1}^{s} \xi_1^k v_k = \text{const} \mid \boldsymbol{\xi}_0 = \mathbf{e}_{l_0}\right) = 0 \tag{1.2.26}$$

then the distribution of W *conditional on* $\boldsymbol{\xi}_0 = \mathbf{e}_l$ *has a jump of magnitude* q_l *at the origin and a continuous density function on the set of positive real numbers; if* (1.2.26) *fails to hold for all* $1 \leqslant l_0 \leqslant s$, *then the distribution of* W *is concentrated on one point.*

Proof. See Athreya (1970), and Kesten and Stigum (1966 a). ◇

For further results concerning chains that are not positively regular see Kesten and Stigum (1966 b, 1967).

Asymptotic result in the positively regular case with $\rho \leqslant 1$ analoguous to those for Galton-Watson chains with $m \leqslant 1$ are to be found in Joffe and Spitzer (1967). For a history of asymptotic results prior to these see Harris (1963, p. 44).

1.2.3. Markov chains occurring in queueing theory

1.2.3.1. Preliminaries

1.2.3.1.1. Queueing theory (stochastic service theory or mass service theory) is an important chapter of operations research which studies flows of *customers* requiring *service*. The term customer is a generic one. It may refer to *bona fide* customers demanding service at a counter, ships entering a port, messages coming into an office, broken machines awaiting repair, aircraft awaiting to take-off, patients arriving at an out-patients clinic to see a doctor, etc. Usually there is some restriction on the service that can be provided e.g. that only a limited number of customers can be served at a time or that service is only available during limited periods.

1.2.3.1.2. Problems leading to queueing theory arose first in the telephone business. The earliest systematic mathematical work on queueing problems is that of A. K. ERLANG, an employee of the Copenhagen Telephone Company. His first paper on congestion in telephone exchanges was written in 1909. Other important early contributors were F. POLLACZEK, A. I. HINČIN and C. PALM. The reader may find in KENDALL (1951 b) a detailed account of the history of the subject.

Much recent investigations stem from KENDALL's (1951 b, 1953) papers which showed how certain problems in queueing theory can be solved using so-called imbedded Markov chains.

In the next paragraphs we give, following KENDALL (1953) a brief description of queueing systems and then study two imbedded Markov chains. The reader interested in the details of queueing theory is referred to BENEŠ (1963), COHEN (1970), COX and SMITH (1961), KENDALL (1964), MIHOC and CIUCU (1967), PRABHU (1965 a), RIORDAN (1962), SAATY (1961), SYSKI (1960) and TAKÁCS (1962).

1.2.3.2. Queueing systems

1.2.3.2.1. A *queueing system* is specified by (i) the input, (ii) the queue-discipline, (iii) the service-mechanism.

We suppose that if successive customers demand service at the instants $\ldots, t_{n-1}, t_n, t_{n+1}, \ldots$, then $(t_n)_{n\in Z}$ forms a chain with non-negative independent and identically distributed increments[45]. Let A be the distribution function of the increments $t_n - t_{n-1}$, $n \in Z$. The input is said to be

— *general independent* (*GI*) if no assumptions are made about A, except for the existence of a finite mean value $a > 0$ and $A(0+) = 0$ (there is zero probability of two customers arriving simultaneously) ;

— *deterministic* or *regular* (*D*) if

$$A(u) = \begin{cases} 0, & 0 \leqslant u < a \\ 1, & u \geqslant a\,; \end{cases}$$

— *random* or *Poisson* (*M*) if

$$A(u) = 1 - \exp\left(-\frac{u}{a}\right),\ u \geqslant 0;$$

---

[45] In some important applications the independence supposition will not be admissible. For example, it cannot be made when the customers (which may be ships, or aircraft) are scheduled to arrive at specified times but in fact arrive late or early according to random independent time-errors.

— *Erlang* (E_k) if

$$A'(u) = \frac{(k/a)^k}{\Gamma(k)} u^{k-1} \exp\left(-\frac{ku}{a}\right), \quad u \geqslant 0.$$

With a D-input the customers arrive at regular intervals of time (the interval-arrival time being fixed and equal to a) while with an M-input, the customers arrive "at random" (i.e. according to a Poisson process — see Paragraph 2.2.2.1). The input E_k coincides with M when $k = 1$ and approaches D as $k \to \infty$.

Concerning the queue-discipline it will be supposed that the customers form up into a single queue in the usual way and that the customer at the head of the queue is served as soon as a server is free to attend to him ("first come — first served")[46]. In general there will be s servers ($s = 1, 2, 3, \ldots$).

The service mechanism will be defined by the assertions that the service-times are non-negative random variables independent of one another and of the input (thus the presence of a long queue is supposed to have no effect on the speed of service), and that for all customers (irrespective of the identity of the server) the service-time has the (arbitrary) distribution function B with finite mean value $b > 0$ with $B(0+) = 0$ (there is zero probability of a customer receiving instantaneous service). The service mechanism will be classified with the aid of the symbols G, D, M and E_k. The symbol G implies that no special assumption is made about B. D means that it takes exactly the same length of time b to serve each customer. With an M-distribution the service-times follow a negative-exponential law (which has been found to hold for unrestricted telephone conversations). Once again the E_k-distribution is an intermediate form.

A standardized shorthand will be used for identifying queueing systems by giving them labels such as $M/G/2$ (Poisson input; no special assumption about the service-time distribution; two servers), etc.

1.2.3.2.2. The theory of queues is primarily concerned with the continuous parameter stochastic process $\xi(t)$ = the number of people waiting or being served at time $t \in R$. It does not constitute a Markov process (except in the special case $M/M/1$)[47]. In many cases its study can be greatly simplified by restricting attention to a so-called imbedded Markov chain. Let Θ_t denote a subset of the set of functions $N^{(-\infty, t]}$ and let Π be the (random) set of those values of $t \in R$ for which Θ_t contains the element $(\xi(\tau))_{-\infty<\tau\leqslant t}$. Let

$$\eta(t) = f_t\{(\xi(\tau))_{-\infty<\tau\leqslant t}\}, \quad t \in \Pi,$$

[46] One can imagine other types of queue-discipline, e.g. "last come, first served" (for example, articles for service are taken from the top of a stack).

[47] An expository account of the Markovian case is contained in Syski 1965).

where f_t is some specified functional with domain Θ_t. Now suppose that $\{\Theta_t, f_t; t \in R\}$ have been chosen in such a way that

(i) Π almost surely has no finite point of accumulation. (We then write its members in increasing order as $\ldots \tau_{n-1}, \tau_n, \tau_{n+1}, \ldots$).

(ii) If $\eta_n = \eta(\tau_n)$, then distribution $(\eta_{n+1} | \eta_n, \eta_{n-1}, \ldots) \equiv$ distribution $(\eta_{n+1} | \eta_n)$ for all $n \in Z$.

The variables $\ldots \eta_{n-1}, \eta_n, \eta_{n+1}, \ldots$ are said to constitute an *imbedded* Markov chain.

Such an imbedded chain can always be constructed, although perhaps in a trivial way. Thus we could choose $\Theta_t = N^{(-\infty, t]}$ for each $t \in Z$ and $\Theta_t = \emptyset$ for $t \notin Z$. Π would then be the set of integers, and by taking $f_t \equiv 1$ we would obtain an imbedded Markov chain. In practice, however, for the procedure to be of any value the functional f_t must be sufficiently and suitably sensitive to variations in its argument and the stochastic mechanism governing the transition from one instant in Π to the next must be simple enough to permit the calculation of the transition probabilities associated with distribution $(\eta_{n+1} | \eta_n)$.

1.2.3.3. An imbedded Markov chain in the M/G/1 queue

1.2.3.3.1. Let us define $\xi(.)$ to be continuous to the right at its points of discontinuity. We take Θ_t to be the set of N-valued functions g such that $g(t) = g(t-) - 1$; thus the set $\Pi = \{t : \xi(t) = \xi(t-) - 1\}$ consists of the *epochs of departure*. If we set $f_t(g) = g(t)$, $g \in \Theta_t$, then $\eta(t)$, $t \in \Pi$, is the number of customers left behind by a departing customer (including the customer, if any, whose service is just starting). KENDALL (1951 b) showed that this imbedded chain is Markov and that its transition matrix is given by

$$\mathbf{P} = \begin{matrix} & \begin{matrix} 0 & 1 & 2 & 3 & \cdot & \cdot & \cdot \end{matrix} \\ \begin{matrix} 0 \\ 1 \\ 2 \\ 3 \\ \vdots \end{matrix} & \begin{pmatrix} a_0 & a_1 & a_2 & a_3 & \cdot & \cdot & \cdot \\ a_0 & a_1 & a_2 & a_3 & \cdot & \cdot & \cdot \\ 0 & a_0 & a_1 & a_2 & \cdot & \cdot & \cdot \\ 0 & 0 & a_0 & a_1 & \cdot & \cdot & \cdot \\ \vdots & \vdots & \vdots & \vdots & \vdots & \vdots & \vdots \end{pmatrix} \end{matrix} \qquad (1.2.27)$$

where

$$a_k = \frac{1}{k!} \int_0^\infty e^{-\frac{v}{a}} \left(\frac{v}{a}\right)^k dB(v), \quad k \in N. \qquad (1.2.28)$$

1.2.3.2.2. In the following we shall not suppose that the a_k are given by (1.2.28), but only that $a_k > 0$, $k \in N$, $\sum_{k \in N} a_k = 1$, $\sum_{k \in N} k a_k = m <$

$< \infty$[48]. The first condition implies that our Markov chain is irreducible and aperiodic.

Proposition 1.2.31. *The Markov chain with the transition matrix* (1.2.27) *is positive, null-recurrent or nonrecurrent according as* $m < 1$, $m = 1$ *or* $m > 1$.

Proof. A first passage from $i > 0$ to 0 can occur only through $i-1$, $i-2, \ldots, 1$; moreover, since p_{ij} depends only on the difference $j-i$, the first passage time $\tau_{i,\,i-1}$ is independent of i. It follows that τ_{i0} is the sum of i independent random variables, each having the same distribution as τ_{10}. If we consider the generating functions

$$F_{i0}(z) = \sum_{n \in N^*} f_{i0}^{(n)} z^n, \quad i \in N^*,$$

we can write

$$F_{i0}(z) = [F_{10}(z)]^i, \quad i \in N^*.$$

Now notice that since in the matrix (1.2.27) the first two rows are identical we have $F_{10}(z) = F_{00}(z)$, therefore

$$F_{i0}(z) = [F_{00}(z)]^i, \quad i \in N^*.$$

Since

$$f_{00}^{(1)} = a_0, \quad f_{00}^{(n+1)} = \sum_{j \in N^*} a_j f_{j0}^{(n)}$$

we see that $F_{00}(z)$ satisfies the functional equation

$$F_{00}(z) = za[F_{00}(z)], \tag{1.2.29}$$

where

$$a(z) = \sum_{k \in N} a_k z^k.$$

Equation (1.2.29) is of the type appearing in Theorem 1.2.14[49]. Using this theorem we obtain

$$f_{00}^* = F_{00}(1) = \begin{cases} 1 \text{ if} & m \leqslant 1 \\ q \text{ if} & m > 1, \end{cases}$$

---

[48] For the imbedded Markov chain we have $m = b/a$. This quantity is known as the *relative traffic intensity* and is expressed in erlangs, the erlang being the international unit of telephone traffic.

[49] There is an interesting relationship between Galton-Watson chains and the $M/G/1$ queue. If we define the *busy period* as the time interval during which the server is continuously busy, then its evolution is equivalent to the history of a Galton-Watson chain with generating function a. This was pointed out by I. J. Good. See e.g. Kendall (1966 b, p. 393).

where q is the unique solution of $u = a(u)$ in (0,1). Differentiation of (1.2.29) yields

$$\frac{1}{\nu_{00}} = F'_{00}(1) = \begin{cases} (1-m)^{-1} & \text{if } m < 1 \\ \infty & \text{if } m = 1. \diamond \end{cases}$$

For classification of states when $m = \infty$ see PRABHU (1965 b, p. 61).

Proposition 1.2.32. *If $m < 1$, then the generating function of the stationary distribution of the Markov chain with the transition matrix* (1.2.27) *is*

$$\pi(z) = \frac{(1-m)\; a(z)(1-z)}{a(z) - z}.$$

Proof. The equations determining the stationary distribution are

$$\pi_j = \sum_{k \in N} \pi_k\, p_{kj} = \pi_0\, a_j + \sum_{l=1}^{j+1} \pi_l\, a_{j-l+1}\,, \quad j \in N.$$

Multiplying them by z^j and summing over $j \in N$ yields

$$\pi(z) = \pi_0\, a(z) + \frac{a(z)}{z}\,(\pi(z) - \pi_0),$$

whence

$$\pi(z) = \frac{\pi_0 a(z)\,(1-z)}{a(z) - z}.$$

It remains to note that $\pi_0 = f^*_{00}\, \nu_{00} = 1 - m. \diamond$

Proposition 1.2.33 (KENDALL (1960)). *In case $m > 1$ the Markov chain with the transition matrix* (1.2.27) *is geometrically ergodic. In case $m < 1$ it is geometrically ergodic iff the radius of convergence of $a(z)$ is greater than unity* [50].

---

[50] According to the corollary to Theorem 1.1.17, in case $m = 1$ our Markov chain is not geometrically ergodic.

Proof. First we shall prove that for $0 \leqslant x < \infty$ the equation $y = xa(y)$ has $y = F_{00}(x)$ as its smallest root in the interval $0 \leqslant y \leqslant \infty$. The fact that $y = F_{00}(x)$ is a root follows from (1.2.29). We now have to prove that y is the smallest non-negative root (note that it could be infinite). To this end we put

$$f_n(x) = \sum_{k=1}^{n} f_{00}^{(k)} x^k, \quad n \in N^*,$$

and observe that

$$f_{n+1}(x) = a_0 x + x \sum_{j \in N^*} a_j \left(\sum_{k=1}^{n} f_{j0}^{(k)} x^k \right) \leqslant$$

$$\leqslant a_0 x + x \sum_{j \in N^*} a_j [f_n(x)]^j = xa[f_n(x)], \tag{1.2.30}$$

and that

$$0 \leqslant f_n(x) \uparrow F_{00}(x) \leqslant \infty .$$

Now if $0 \leqslant y \leqslant \infty$ and if $y = x\, a(y)$, where $0 \leqslant x < \infty$, then $f_1(x) = a_0 x \leqslant y$, and so (by induction, using (1.2.30)) $f_n(x) \leqslant y$, and on letting $n \to \infty$ we obtain $F_{00}(x) \leqslant y$, as required.

The result proved permits us to assert that $F_{00}(x)$ will be finite iff equation $y = x\, a(y)$ has a root in the interval $0 \leqslant y < \infty$. Thus the chain will be geometrically ergodic iff for some real number $c > 1$ the equation

$$a(y) = c^{-1} y \tag{1.2.31}$$

has a finite positive solution (it cannot have the solution $y = 0$ because $a(0) = a_0 > 0$).

Now suppose that $m > 1$, so that $a'(1 -) > 1$. Then for a sufficiently small positive δ we shall have $[a(1) - a(1 - \delta)] / \delta > 1$, and thus $0 < a(1 - \delta) < 1 - \delta$. But this inequality tells us that (1.2.31) has the finite positive root $y = 1 - \delta$ when $c = (1 - \delta) / a(1 - \delta)$, and thus the chain is geometrically ergodic.

If $m < 1$ there are two possibilities: either $a(x) = \infty$ for all $x > 1$ or $a(d) < \infty$ for some $d > 1$. In the first case, (1.2.31) obviously cannot have a solution in $1 \leqslant y < \infty$, and it cannot have a solution in (0,1) because throughout this open interval $a(y) > y$; were this last statement not so, $a'(y)$ would be equal to unity somewhere in the open interval and yet

$$a'(y) = \sum_{j \in N^*} a_j\, j y^{j-1} \leqslant \sum_{j \in N^*} j a_j = m < 1.$$

Thus the chain cannot be geometrically ergodic. In the second case, $a(x)$ will be finite for $0 \leqslant x \leqslant d$ and continuously differentiable for $0 \leqslant x < d$. But $a'(1) = m < 1$, and so for sufficiently small positive δ we shall have $[a(1+\delta) - a(1)]/\delta < 1$ and thus $0 < a(1+\delta) < 1 + \delta$. But this inequality tells us that (1.2.31) has the finite positive root $y = 1 + \delta$ when $c = (1+\delta)/a(1+\delta)$, and thus the chain is geometrically ergodic. ◇

We note that if the a_k are given by (1.2.28), then the fact that the radius of convergence of $a(z)$ is greater than unity is equivalent to the existence of a real number $\varepsilon > 0$ such that $\int_0^\infty e^{\varepsilon v}\, dB(v) < \infty$.

1.2.3.4. An imbedded Markov chain in the GI/M/1 queue

1.2.3.4.1. Let us define $\xi(.)$ to be continuous to the right at its points of discontinuity. We take Θ_t to be the set of N-valued functions g such that $g(t) = g(t-) + 1$; thus the set $\Pi = \{t\colon \xi(t) = \xi(t-) + 1\}$ consists of the *epochs of arrival.* If we set $f_t(g) = g(t-)$, $g \in \Theta_t$, then $\eta(t)$, $t \in \Pi$, is the number of customers (waiting or being served) which a newly arrived customer has ahead of him. KENDALL (1953) showed that this imbedded chain is Markov and that its transition matrix is

$$\mathbf{P} = \begin{array}{c} \\ 0 \\ 1 \\ 2 \\ \vdots \end{array} \begin{array}{c} \begin{array}{ccccccc} 0 & 1 & 2 & 3 & 4 & \cdots \end{array} \\ \left(\begin{array}{ccccccc} \sum_{k\geqslant 1} b_k & b_0 & 0 & 0 & 0 & \cdots \\ \sum_{2\geqslant k} b_k & b_1 & b_0 & 0 & 0 & \cdots \\ \sum_{k\ 3} b_k & b_2 & b_1 & b_0 & 0 & \cdots \\ \vdots & \vdots & \vdots & \vdots & \vdots & \ddots \end{array}\right) \end{array} \tag{1.2.32}$$

where

$$b_k = \frac{1}{k!}\int_0^\infty e^{-\frac{u}{b}} \left(\frac{u}{b}\right)^k dA(u), \quad k \in N. \tag{1.2.33}$$

1.2.3.4.2. In the following we shall not suppose that the b_k are given by (1.2.33), but only that $b_k > 0$, $k \in N$, $\sum_{k\in N} b_k = 1$, $\sum_{k\in N} kb_k = m < \infty$. The first condition implies that the Markov chain is irreducible and aperiodic.

Proposition 1.2.34. *The Markov chain with transition matrix* (1.2.32) *is positive, null-recurrent or nonrecurrent according as* $m > 1$, $m = 1$ *or* $m < 1$.

Proof. It is easy to see that the taboo probabilities ${}_0p_{0j}^{(n)}$, $j \in N^*$, in the present case coincide with the $f_{j0}^{(n)}$ of the preceding case with the a_k replaced by the b_k. Let $f(z)$ denote the generating function $F_{00}(z)$ associated with (1.2.27) where a_k is replaced by b_k, $k \in N$. Then by (1.2.29) we shall have

$$f(z) = zb\,[f(z)], \tag{1.2.34}$$

with $b(z) = \sum_{k \in N} b_k z^k$.

In the present case we have

$$f_{00}^{(1)} = p_{00} = \sum_{k \geqslant 1} b_k,$$

$$f_{00}^{(n+1)} = \sum_{j \in N^*} {}_0p_{0j}^{(n)} p_{j0} = \sum_{j \in N^*} \left(\sum_{k \geqslant j+1} b_k \right) {}_0p_{0j}^{(n)}, \quad n \in N^*,$$

and using (1.2.34)

$$F_{00}(z) = \sum_{n \in N^*} f_{00}^{(n)} z^n = z \left\{ \sum_{k \geqslant 1} b_k + \sum_{j \in N^*} \left(\sum_{k \geqslant j+1} b_k \right) [f(z)]^j \right\} =$$

$$= z\, \frac{1 - b[f(z)]}{1 - f(z)} = \frac{z - f(z)}{1 - f(z)}.$$

It follows that

$$f_{00}^* = F_{00}(1) = \begin{cases} 1 & \text{if} \quad m \geqslant 1 \\ m & \text{if} \quad m < 1 \end{cases}$$

$$\frac{1}{\nu_{00}} = F'_{00}(1) = \begin{cases} (1-q)^{-1} & \text{if } m > 1 \\ \infty & \text{if } m = 1, \end{cases}$$

where q is the unique solution of $v = b(v)$ in $(0, 1)$. ◇

Proposition 1.2.35. *In case* $m > 1$ *the stationary distribution is given by* $\pi_j = (1 - q)\, q^j$, $j \in N$.

Proof. The values $\pi_j = (1 - q)\, q^j$, $j \in N$, satisfy equations (1.1.34). ◇

We quote also

Proposition 1.2.36. (Kendall (1960)). *In case* $m < 1$ *the Markov chain with the transition matrix* (1.2.32) *is geometrically ergodic. In* *case $m > 1$ it is geometrically ergodic iff the radius of convergence of $b(z)$ is greater than unity*[51].

1.2.3.4.3. For other imbedded Markov chains the reader is referred to AFANAS'EVA and MARTYNOV (1966), NEUTS (1967), VERE-JONES (1964) and WISHART (1956). A class of denumerable Markov chains which contains imbedded Markov chains as special cases was studied by MILLER (1965).

1.3. Markov chains with arbitrary state space

1.3.1. Preliminaries

In this subsection we deal with Markov chains whose state space is an arbitrary set X. It is possible to develop for this case a theory paralleling that of denumerable Markov chains. The interested reader is referred to CHUNG (1964); see also JAIN and JAMISON (1967), MOY (1965), ŠIDÁK (1967). The theory we develop in what follows is closely akin to that for finite state Markov chains.

1.3.1.1. Definition and existence theorem

1.3.1.1.1. Let $\mathcal{X}$ be a σ-algebra of subsets of X and denote by $(X^n, \mathcal{X}^n)$ the measurable product space of n copies of $(X, \mathcal{X})$. For $l \in N^*$ and $y^{(l)} = (y_1, \ldots, y_l) \in X^l$ let us consider the mapping $u(\,.\,; y^{(l)})$ of $X^{(k)}$ into itself defined by

$$u(x^{(k)}; y^{(l)}) = \begin{cases} (x_{l+1}, \ldots, x_k, y_1, \ldots, y_l) & \text{if } k > l \\ (y_{l-k+1}, \ldots, y_l) & \text{if } k \leqslant l. \end{cases}$$

Let $({}^nP)_{n \in N}$ be a sequence of *transition functions* from $(X^k, \mathcal{X}^k)$ to $(X, \mathcal{X})$. This means that for every $n \in N$, ${}^nP(x^{(k)}, .)$ is a probability on $\mathcal{X}$ for arbitrary $x^{(k)} \in X^k$, and ${}^nP(., A)$ is a $\mathcal{X}^k$-measurable function for arbitrary $A \in \mathcal{X}$.

Definition 1.3.1. An X-valued sequence $(\xi_{n-k+1})_{n \in N}$ of random variables on the probability space $(\Omega, \mathcal{K}, \mathsf{P})$ is said to be a *(nonhomo-*

---

[51] If the b_k are given by (1.2.33), then the fact that the radius of convergence of $b(z)$ is greater than unity is equivalent to the existence of a real number $\varepsilon > 0$ such that $\int_0^\infty e^{\varepsilon u}\, dA(u) < \infty$.

geneous) kth order (*simple* if $k = 1$) *Markov chain* with state space $(X, \mathscr{X})$ and transition functions ${}^nP$, $n \in N$, if

$$\mathsf{P}(\xi_{n+1} \in A \mid \xi_{n-j},\ 0 \leqslant j \leqslant n + k - 1) =$$

$$= \mathsf{P}(\xi_{n+1} \in A \mid \xi_n, \ldots, \xi_{n-k+1}) = {}^nP((\xi_{n-k+1}, \ldots, \xi_n), A)$$

P — a.s. for all $n \in N$, $A \in \mathscr{X}$. A kth order Markov chain is said to be *homogeneous* if the transition functions ${}^nP$, $n \in N$, do not depend on n.

1.3.1.1.2. The existence of kth order Markov chains follows from

Theorem 1.3.2. *Given the measurable space $(X, \mathscr{X})$ and the sequence of transition functions $({}^nP)_{n \in N}$, there exist a probability space $(\Omega, \mathscr{K}, \mathsf{P})$ and a kth order Markov chain $(\xi_{n-k+1})_{n \in N}$ defined on that probability space with state space $(X, \mathscr{X})$ and transition functions ${}^nP$, $n \in N$.*

Proof. We proceed in a manner completely analogous to the proof of Theorem 1.1.3. Thus, we take for Ω the set of all sequences $(x_{-k+1}, \ldots, x_1, \ldots, x_n, \ldots)$ of elements from X and for $\mathscr{K}$ the minimal σ-algebra containing all cylindrical sets

$$A_{-k+1} \times \ldots \times A_0 \times \ldots \times A_l \times \prod_{j=l+1}^{\infty} X_j$$

with $A_\nu \in \mathscr{X}$, $-k+1 \leqslant \nu \leqslant l$, $l \in N$, $X_j = X$, $j \geqslant l+1$. We consider a probability p on $\mathscr{X}^k$ and define the measure P on cylindrical sets by the equalities

$$\mathsf{P}\left(A_{-k+1} \times \ldots \times A_0 \times \prod_{j=1}^{\infty} X_j\right) = p(A_{-k+1} \times \ldots \times A_0)$$

$$\mathsf{P}\left(A_{-k+1} \times \ldots \times A_0 \times \ldots \times A_l \times \prod_{j=l+1}^{\infty} X_j\right) =$$

$$\int\limits_{A_{-k+1} \times \ldots \times A_0} p(\mathrm{d}y^{(k)}) \int\limits_{A_1} {}^0P(y^{(k)}, \mathrm{d}x_1) \ldots$$

$$\ldots \int\limits_{A_l} {}^{l-1}P(u(y^{(k)};\ x^{(l-1)}), \mathrm{d}x_l), \quad l \in N^*.$$

According to Ionescu Tulcea's theorem this measure may be extended to the whole $\mathcal{X}$. We then set

$$\xi_j(\omega) = x_j, \quad j \geqslant -k+1,$$

if

$$\omega = (x_{-k+1}, \ldots, x_0, \ldots, x_n, \ldots).$$

The reader will be able to verify that $(\xi_{n-k+1})_{n \in N}$ is a kth order Markov chain. ◇

It is obvious that

$$\mathsf{P}((\xi_{-k+1}, \ldots, \xi_0) \in A^{(k)}) = p(A^{(k)}), \quad A^{(k)} \in \mathcal{X}^k,$$

hence the probability p is called the *initial distribution* of the Markov chain. Sometimes to emphasize the dependence on p of the probability P, we shall write P_p. In particular, if $p = \delta_x$, where, for every $A \in \mathcal{X}$

$$\delta_x(A) = \begin{cases} 1 & \text{if } x \in A \\ 0 & \text{if } x \notin A, \end{cases}$$

the corresponding P_p will be denoted by P_x, and the corresponding Markov chain will be said to start in $x \in X$.

We note that the considerations in Paragraph 1.1.1.4. concerning the strong Markov property for the case of denumerable state space apply mutatis mutandis to the present case.

1.3.1.2. Generalized n-step transition functions

1.3.1.2.1. Now let us define the (generalized) *n-step transition functions* ${}^mP_l^n$.

For all $l \in N^*$, $m \in N$ let ${}^mP_l$ be the function defined on $X^k \times \mathcal{X}^l$ by

$${}^mP_l = {}^mP \quad \text{if } l = 1,$$

$${}^mP_l(x^{(k)}, A^{(l)}) =$$

$$= \int_X {}^mP(x^{(k)}, dy_1) \ldots \int_X {}^{m+l-1}P(u(x^{(k)}; y^{(l-1)}), dy_l)\, \chi_{A^{(l)}}(dy^{(l)})$$

if $l \geqslant 2$, where $\chi_{A^{(l)}}$ is the indicator of the set $A^{(l)} \in \mathcal{X}^{(l)}$.

For all $l, n \in N^*$, $m \in N$ let ${}^mP_l^n$ be the function defined on $X^k \times \mathscr{X}^l$ by

$${}^mP_l^n = {}^mP_l \quad \text{if } n = 1$$

$${}^mP_l^n(x^{(k)}, A^{(l)}) = \int_X {}^mP(x^{(k)}, \mathrm{d}y)\, {}^{m+1}P_l^{n-1}(u(x^{(k)}; y), A^{(l)}) \quad \text{if } n \geqslant 2.$$

The probabilistic interpretation of ${}^mP_l^n$ is as follows. We have

$$\mathsf{P}((\xi_{m+n}, \ldots, \xi_{m+n+l-1}) \in A^{(l)} \mid \xi_{m-j},\ 0 \leqslant j \leqslant m+k-1) =$$

$$= {}^mP_l^n((\xi_{m-k+1}, \ldots, \xi_m), A^{(l)})$$

P — a.s. for all $l, n \in N^*$, $m \in N$, $A^{(l)} \in \mathscr{X}^l$.

It is easy to see that the following generalization of the Chapman-Kolmogorov equation holds:

$${}^mP_l^n(x^{(k)}, A^{(l)}) = \int_{X^k} {}^mP_k^{s-k+1}(x^{(k)}, \mathrm{d}y^{(k)})\, {}^{m+s}P_l^{n-s}(y^{(k)}, A^{(l)}) \tag{1.3.1}$$

for $k \leqslant s < n$.

1.3.1.2.2. We shall now show how one can associate a simple Markov chain in a natural manner with a kth order Markov chain $(\xi_{n-k+1})_{n \in N}$. Consider the random vectors

$$\boldsymbol{\xi}_n = (\xi_{n-k+1}, \ldots, \xi_n),\ n \in N,$$

(of course, we assume that $k \geqslant 2$). The reader will be able to prove that $(\boldsymbol{\xi}_n)_{n \in N}$ is a simple Markov chain with space state $(X^k, \mathscr{X}^k)$ and transition functions

$${}^m\Pi(x^{(k)}, A^{(k)}) = {}^mP(x^{(k)}, \{x : u(x^{(k)}; x) \in A^{(k)})$$

for all $m \in N$, $x^{(k)} \in X^k$, $A^{(k)} \in \mathscr{X}^k$, so that

$${}^m\Pi^n(x^{(k)}, A^{(k)}) = {}^mP_k^{n-k+1}(x^{(k)}, A^{(k)})$$

for $n \geqslant k$.

On account of the considerations above it is sometimes asserted that the study of kth order Markov chains can always be reduced to that of simple Markov chains. Although this is true for some problems it is for others something of an oversimplification.

1.3.2. Uniform ergodicity

In this subsection we prove two theorems generalizing Theorem 1.1.48.

1.3.2.1. Uniform strong ergodicity

1.3.2.1.1. We start with

Definition 1.3.3. A kth order Markov chain is said to be *uniformly strong ergodic* iff for every $l \in N^*$ there is a probability P_l^∞ on $\mathcal{X}^l$ such that

$$\lim_{n\to\infty} {}^mP_l^n(x^{(k)}, A^{(l)}) = P_l^\infty(A^{(l)})$$

uniformly with respect to $x^{(k)} \in X^k$, $l, n \in N^*$, $m \in N$, $A^{(l)} \in \mathcal{X}^l$.

Condition $F_k(n_0)$ below is suggested by Theorem 1.1.48.

Condition $F_k(n_0)$. *Let* $n_0 \in N^*$; *there exists* $\delta > 0$ *such that for every partition* $A_1^{(k)} \cup A_2^{(k)} = X^k$, $A_i^{(k)} \in \mathcal{X}^k$, $i = 1, 2$, *we have either*

$${}^mP_k^{n_0}(x^{(k)}, A_1^{(k)}) \geqslant \delta \quad \textit{for all } m \in N,\ x^{(k)} \in X^k,$$

or

$${}^mP_k^{n_0}(x^{(k)}, A_2^{(k)}) \geqslant \delta \quad \textit{for all } m \in N,\ x^{(k)} \in X^k.$$

It is useful to consider an equivalent condition, namely

Condition $M_k(n_0)$. *Let* $n_0 \in N^*$; *there exists* $\delta > 0$ *such that*

$$|{}^{m'}P_k^{n_0}(x^{(k)}, A^{(k)}) - {}^{m''}P_k^{n_0}(y^{(k)}, A^{(k)})| \leqslant 1 - \delta$$

for every $m', m'' \in N$, $x^{(k)}, y^{(k)} \in X^k$, $A^{(k)} \in \mathcal{X}^k$.

The equivalence of Conditions $F_k(n_0)$ and $M_k(n_0)$ can be proved as follows. The implication $F_k(n_0) \Rightarrow M_k(n_0)$, follows immediately from the relation

$$|{}^{m'}P_k^{n_0}(x^{(k)}, A_1^{(k)}) - {}^{m''}P_k^{n_0}(y^{(k)}, A_1^{(k)})| =$$
$$= |{}^{m'}P_k^{n_0}(x^{(k)}, A_2^{(k)}) - {}^{m''}P_k^{n_0}(y^{(k)}, A_2^{(k)})|,$$

valid for every partition $A_1^{(k)} \cup A_2^{(k)} = X^k$, $A_i^{(k)} \in \mathcal{X}^k$, $i = 1, 2$. Now suppose that $M_k(n_0)$ holds. Put

$$f(A^{(k)}) = \inf_{m \in N,\, x^{(k)} \in X^k} {}^mP_k^{n_0}(x^{(k)}, A^{(k)}).$$

We have

$$\sup_{m \in N, x^{(k)} \in X^k} {}^{m}P_k^{n_0}(x^{(k)}, A^{(k)}) = \sup_{m \in N, x^{(k)} \in X^k} [1 - {}^{m}P_k^{n_0}(x^{(k)}, X^k - A^{(k)})] =$$

$$= 1 - f(X^k - A^{(k)}).$$

Condition $M_k(n_0)$ then implies

$$1 - f(X^k - A^{(k)}) - f(A^{(k)}) \leqslant 1 - \delta,$$

that is

$$f(A^{(k)}) + f(X^k - A^{(k)}) \geqslant \delta.$$

Therefore Condition $F(n_0)$ is valid, δ being replaced by $\delta/2$.

1.3.2.1.2. Let us set

$$a_n = \sup | {}^{m'+n}P(x^{(k)}, A) - {}^{m''+n}P(x^{(k)}, A)|, \quad n \in N^*,$$

the supremum being taken over all $x^{(k)} \in X^k$, $A \in \mathscr{X}$, $m', m'' \in N$. It is easy to prove that

$$| {}^{m'+n}P_l(x^{(k)}, A^{(l)}) - {}^{m''+n}P_l(x^{(k)}, A^{(l)})| \leqslant \sum_{j=n}^{n+l-1} a_j$$

for all $x^{(k)} \in X^k$, $A^{(l)} \in \mathscr{X}^l$, $m', m'' \in N$, $l, n \in N^*$ and this implies

$$| {}^{m'+n}P_l^h(x^{(k)}, A^{(l)}) - {}^{m''+n}P_l^h(x^{(k)}, A^{(l)})| \leqslant \sum_{j=n}^{n+h+l-2} a_j \quad (1.3.2)$$

for all $x^{(k)} \in X^k$, $A^{(l)} \in \mathscr{X}^l$, $m', m'' \in N$, $l, h, n \in N^*$.

Theorem 1.3.4 (Iosifescu (1965)). *Condition $F_k(n_0)$ is necessary for the uniform strong ergodicity of a kth order Markov chain. If*

$$\sum_{n \in N^*} a_n < \infty, \text{[52]}$$

---

[52] Notice that this condition implies the existence of the limit $\lim_{m \to \infty} {}^{m}P(x^{(k)}, A)$ uniformly with respect to $x^{(k)} \in X^k$ and $A \in \mathscr{X}$.

then Condition $F_k(n_0)$ is also sufficient for uniform strongergodicity and we have

$$|{}^m P_l^n(x^{(k)}, A^{(l)}) - P_l^\infty(A^{(l)})| \leqslant \inf_{k+n_0-1\leqslant s\leqslant n-1} \left(\frac{\sum\limits_{j\geqslant s} a_j}{\delta} + (1-\delta)^{\frac{n}{s}-1} \right)$$

for all $l \in N^*$, $m \in N$, $A^{(l)} \in \mathscr{X}^l$, $x^{(k)} \in X^k$.

Proof. The first part follows easily from Definition 1.3.3. To prove the second part, let us first note that on account of (1.3.1) we have

$${}^m P_k^{n_0+k'}(x^{(k)}, A^{(k)}) = \int_{X^k} {}^m P_k^{k'-k+1}(x^{(k)}, \mathrm{d}y^{(k)})\, {}^{m+k'}P_k^{n_0}(y^{(k)}, A^{(k)}) \quad (1.3.3)$$

for $k' \geqslant k$. It follows without difficulty that Condition $F_k(n_0)$ implies

$$|{}^{m'}P_k^n(x^{(k)}, A^{(k)}) - {}^{m''}P_k^n(y^{(k)}, A^{(k)})| \leqslant 1 - \delta \quad (1.3.4)$$

for all $x^{(k)}, y^{(k)} \in X^k$, $A^{(k)} \in \mathscr{X}^k$, $m', m'' \in N$, $n \geqslant n_0$.

Using (1.3.1) again we obtain

$$\begin{aligned} &{}^{m'}P_l^n(x^{(k)}, A^{(l)}) - {}^{m''}P_l^n(y^{(k)}, A^{(l)}) = \\ &= \int_{X^k} [{}^{m'}P_k^{s-k+1}(x^{(k)}, \mathrm{d}z^{(k)}) - \\ &- {}^{m''}P_k^{s-k+1}(y^{(k)}, \mathrm{d}z^{(k)})]\, {}^{m'+s}P_l^{n-s}(z^{(k)}, A^{(l)}) + \\ &+ \int_{X^k} [{}^{m'+s}P_l^{n-s}(z^{(k)}, A^{(l)}) - \\ &- {}^{m''+s}P_l^{n-s}(z^{(k)}, A^{(l)})]\, {}^{m''}P_k^{s-k+1}(y^{(k)}, \mathrm{d}z^{(k)}) \end{aligned} \quad (1.3.5)$$

for $k \leqslant s < n$.

Set

$$\underline{P}_l^n(A^{(l)}) = \inf_{m\in N,\, x^{(k)}\in X^k} {}^m P_l^n(x^{(k)}, A^{(l)})$$

$$\overline{P}_l^n(A^{(l)}) = \sup_{m\in N,\, x^{(k)}\in X^k} {}^m P_l^n(x^{(k)}, A^{(l)}).$$

From (1.3.5), taking into account (1.3.2) and (1.3.4) and using Lemma 1.3.8, we get

$$\overline{P}_l^n(A^{(l)}) - \underline{P}_l^n(A^{(l)}) \leqslant (1-\delta)\,[\overline{P}_l^{n-s}(A^{(l)}) - \underline{P}_l^{n-s}(A^{(l)})] + \sum_{j\geqslant s} a_j$$

for $k + n_0 - 1 \leqslant s < n$. It follows that

$$\overline{P}_l^n(A^{(l)}) - \underline{P}_l^n(A^{(l)}) \leqslant \sum_{j\geqslant s} a_j \sum_{m=0}^{p-1} (1-\delta)^m +$$

$$+ (1-\delta)^p\,[\overline{P}_l^{n-ps}(A^{(l)}) - \underline{P}_l^{n-ps}(A^{(l)})]$$

for all $p \in N^*$ such that $ps < n$. Since the sequences $(\overline{P}_l^n(A^{(l)}))_{n\in N^*}$ and $(\underline{P}_l^n(A^{(l)}))_{n\in N^*}$ are respectively nonincreasing and nondecreasing, we deduce without any difficulty the existence of the probabilities P_l^∞ and the claimed domination. ◇

1.3.2.1.3. It is easy to verify that for a simple homogeneous Markov chain the limiting probability P_1^∞ is also the corresponding stationary one, that is

$$P_1^\infty(A) = \int_X P_1^\infty(\mathrm{d}x)\, P(x, A)$$

for all $A \in \mathcal{X}$. For $p = P_1^\infty$, the Markov chain $(\xi_n)_{n\in N}$ on $(\Omega, \mathcal{K}, \mathsf{P}_p)$ is strictly stationary.

1.3.2.2. Uniform weak ergodicity

1.3.2.2.1. Now we pass to the consideration of another kind of ergodicity.

Definition 1.3.5. A kth order Markov chain is said to be *uniformly weak ergodic* iff

$$\lim_{n\to\infty} |{}^mP_l^n(x^{(k)}, A^{(l)}) - {}^mP_l^n(y^{(k)}, A^{(l)})| = 0$$

uniformly with respect to $x^{(k)}, y^{(k)} \in X^{(k)}$, $l \in N^*$, $m \in N$, $A^{(l)} \in \mathcal{X}^l$.

We shall use

Condition $F_k'(n_0)$. *Let $n_0 \in N^*$; there exists $\delta > 0$ such that for every partition $A_1^{(k)} \cup A_2^{(k)} = X^k$, $A_i^{(k)} \in \mathcal{X}^k$, $i = 1, 2$, and every $m \in N$, we have either*

$${}^mP_k^{n_0}(x^{(k)}, A_1^{(k)}) \geqslant \delta \quad \textit{for all } x^{(k)} \in X^k,$$

or

$$ {}^{m}P_{k}^{n_0}(x^{(k)}, A_2^{(k)}) \geqslant \delta \quad \textit{for all } x^{(k)} \in X^{k}. $$

It is obvious that Condition $F_k'(n_0)$ is weaker than Condition $F_k(n_0)$. An equivalent condition is

Condition $M_k'(n_0)$. *Let* $n_0 \in N^*$; *there exists* $\delta > 0$ *such that*

$$ |{}^{m}P_{k}^{n_0}(x^{(k)}, A^{(k)}) - {}^{m}P_{k}^{n_0}(y^{(k)}, A^{(k)})| \leqslant 1 - \delta $$

for all $m \in N$, $x^{(k)}, y^{(k)} \in X^k$, $A^{(k)} \in \mathscr{X}^k$.

The equivalence of Conditions $F_k'(n_0)$ and $M_k'(n_0)$ may be proved as was done for Conditions $F_k(n_0)$ and $M_k(n_0)$.

Theorem 1.3.6. *Condition* $F_k'(n_0)$ *is necessary and sufficient for the uniform weak ergodicity of a kth order Markov chain. Moreover,*

$$ |{}^{m}P_{l}^{n}(x^{(k)}, A^{(l)}) - {}^{m}P_{l}^{n}(y^{(k)}, A^{(l)})| \leqslant (1 - \delta)^{\frac{n}{n_0 + 2k - 1} - 1} $$

for all $l \in N^*$, $m \in N$, $x^{(k)}, y^{(k)} \in X^k$, $A^{(l)} \in \mathscr{X}^l$.

Proof. The necessity follows from Definition 1.3.5. To prove the sufficiency, we first note that Condition $M_k'(n_0)$ and (1.3.3) imply

$$ |{}^{m}P_{k}^{n}(x^{(k)}, A^{(k)}) - {}^{m}P_{k}^{n}(y^{(k)}, A^{(k)})| \leqslant 1 - \delta \tag{1.3.7} $$

for all $x^{(k)}, y^{(k)} \in X^k$, $A^{(k)} \in \mathscr{X}^k$, $m \in N$, $n \geqslant n_0 + k$. Using (1.3.1) we get

$$ {}^{m}P_{l}^{n}(x^{(k)}, A^{(l)}) - {}^{m}P_{l}^{n}(y^{(k)}, A^{(l)}) = $$

$$ = \int_{X^k} [{}^{m}P_{k}^{s-k+1}(x^{(k)}, dz^{(k)}) - $$

$$ - {}^{m}P_{k}^{s-k+1}(y^{(k)}, dz^{(k)})]\, {}^{m+s}P_{l}^{n-s}(z^{(k)}, A^{(l)}) \tag{1.3.8} $$

for $k \leqslant s < n$. Put

$$ {}^{m}\underline{P}_{l}^{n}(A^{(l)}) = \inf_{x^{(k)} \in X^k} {}^{m}P_{l}^{n}(x^{(k)}, A^{(l)}), $$

$$ {}^{m}\overline{P}_{l}^{n}(A^{(l)}) = \sup_{x^{(k)} \in X^k} {}^{m}P_{l}^{n}(x^{(k)}, A^{(l)}). $$

From (1.3.8) on account of Lemma 1.3.8 it follows that

$$ {}^{m}\overline{P}_l^n(A^{(l)}) - {}^{m}\underline{P}_l^n(A^{(l)}) \leqslant (1-\aleph)[{}^{m+s}\overline{P}_l^{n-s}(A^{(l)}) - {}^{m+s}\underline{P}_l^{n-s}(A^{(l)})] $$

for $2k + n_0 - 1 \leqslant s < n$, and hence the domination in the statement follows immediately. ◇

1.3.2.2.2. It is obvious that uniform strong ergodicity implies uniform weak ergodicity but the converse is generally not true. In the homogeneous case, however, the two types of ergodicity coincide.

For further developments concerning uniform ergodicity of nonhomogeneous Markov chains see Bui-Trong-Lieu and Dorel (1967) and Deberghes and Bui-Trong-Lieu (1968).

1.3.3. The coefficient of ergodicity

We now introduce a "measure" of chain independence, namely the coefficient of ergodicity of a transition function. This coefficient was used by Markov himself in some special cases. It was rediscovered by Dynkin (1954) and applied systematically to the investigation of Markov chains by Dobrusin (1956) and Ueno (1957) (see also Hajnal (1956, 1958)). Details can be found in Iosifescu and Theodorescu (1969).

In Paragraph 1.3.3.3 we give some limit theorems expressed in terms of the coefficient of ergodicity. For various other limit theorems the reader is referred to Chung (1967, §§ 15 and 16), Doob (1953, Ch. 5). In the case of a finite number of states, material on the central limit theorem can be found in Fréchet (1938), Hanen (1963), Onicescu and Mihoc (1943), and Romanovski (1949).

1.3.3.1. Definition and properties

1.3.3.1.1. Let $(X, \mathscr{X})$ and $(Y, \mathscr{Y})$ be two measurable spaces. Consider a transition function from the first to the second measurable space, that is, a real-valued function defined on $X \times \mathscr{Y}$ such that $P(x,.)$ is a probability on $\mathscr{Y}$ for any $x \in X$, and $P(., B)$ is an $\mathscr{X}$-measurable function for any $B \in \mathscr{Y}$.

Definition 1.3.7. The real number

$$ \alpha(P) = 1 - \sup_{x', x'' \in X \; B \in \mathscr{Y}} |P(x', B) - P(x'', B)| $$

is called the *coefficient of ergodicity* of the transition function P.

Obviously, $0 \leqslant \alpha(P) \leqslant 1$. It is easy to see that $\alpha(P) = 1$ iff $P(x, B) = P(B)$, i.e. iff P does not depend on $x \in X$. Furthermore, $\alpha(P) = 0$

iff for any $\varepsilon > 0$ there exist $x', x'' \in X$ so that the probabilities $P(x', .)$ and $P(x'',.)$ are concentrated on two sets $B_{x'}$ and $B_{x''}$ for which $P(x', B_{x'}, \cap B_{x''}) < \varepsilon$, $P(x'', B_{x'} \cap B_{x''}) < \varepsilon$.

1.3.3.1.2. Given a measurable space $(V, \mathcal{V})$, denote by $\mathfrak{L}(\mathcal{V})$ the linear space of all finite, completely additive signed measures λ defined on $\mathcal{V}$ such that $\lambda(V) = 0$. With the norm

$$||\lambda|| = \sup_{A \in \mathcal{V}} |\lambda(A)|,$$

$\mathfrak{L}(\mathcal{V})$ is a Banach space.

Thus, we may write

$$\alpha(P) = 1 - \sup_{x', x'' \in X} || P(x',.) - P(x'',.)||. \tag{1.3.8}$$

1.3.3.1.3. In the following we shall use

Lemma 1.3.8. *Consider a measure space $(E, \mathcal{E}, \lambda)$, such that $\lambda \in \mathfrak{L}(\mathcal{E})$. If f is a bounded real-valued and $\mathcal{E}$-measurable function defined on E, then*

$$\left| \int_E f \, d\lambda \right| \leqslant ||\lambda|| \text{ ess osc } f. \text{ [53]}$$

Proof. Let $E^+ \cup E^- = E$ be the Hahn decomposition of E with respect to λ. The relation $\lambda(E) = 0$ implies that $\lambda(E^+) = -\lambda(E^-) = ||\lambda||$. Consequently,

$$\int_E f d\lambda = \int_{E^+} f d\lambda + \int_{E^-} f d\lambda \leqslant \text{ess sup} f . \lambda(E^+) + \text{ess inf} f . \lambda(E^-) =$$

$$= ||\lambda|| \text{ ess osc} f.$$

On the other hand, taking into account the fact that

$$\text{ess osc } (-f) = \text{ess osc } f,$$

we also have

$$-\int_E f \, d\lambda \leqslant ||\lambda|| \text{ ess osc} f. \diamond$$

---

[53] ess osc f = ess sup f − ess inf f.

1.3.3.1.4. An operator $\mathfrak{P}$ sending $\mathfrak{L}(\mathcal{X})$ into $\mathfrak{L}(\mathcal{Y})$ is associated with the transition function P by setting

$$(\mathfrak{P}\lambda)(B) = \int_X P(x, B)\, \lambda\,(\mathrm{d}x), \qquad B \in \mathcal{Y}.$$

The norm of the operator $\mathfrak{P}$

$$\|\mathfrak{P}\| = \sup_{\lambda \in \mathfrak{L}(\mathcal{X})} \frac{\|\mathfrak{P}\lambda\|}{\|\lambda\|}$$

and the coefficient of ergodicity of P are connected by a very simple relation.

Proposition 1.3.9. *We have*

$$\alpha(P) = 1 - \|\mathfrak{P}\|.$$

Proof. The inequality

$$\alpha(P) \leqslant 1 - \|\mathfrak{P}\|$$

follows immediately from Lemma 1.3.8. Thus, it remains to show that

$$\alpha(P) \geqslant 1 - \|\mathfrak{P}\|. \tag{1.3.9}$$

If $x', x'' \in X$, then [54] $\delta_{x'} - \delta_{x''} \in \mathfrak{L}(\mathcal{X})$ and

$$(\mathfrak{P}(\delta_{x'} - \delta_{x''}))(B) = P(x', B) - P(x'', B), \quad B \in \mathcal{Y}.$$

We distinguish two possibilities. If there exists an $A \in \mathcal{X}$ such that $x' \in A$ and $x'' \notin A$, then $\|\delta_{x'} - \delta_{x''}\| = 1$ and (1.3.9) holds. If for every $A \in \mathcal{X}$ which contains x' we have $x'' \in A$, then the $\mathcal{X}$-measurability of the function $P(., B)$ implies $P\,(x', B) = P\,(x'', B)$. Therefore, if the second possibility occurs for all pairs of points $x', x'' \in X$, then we have $\alpha\,(P) = 1$, that is, (1.3.9) also holds. ◇

1.3.3.1.5. Consider the Banach space $O\,(V, \mathcal{V})$ of all real-valued bounded and $\mathcal{V}$-measurable functions ψ defined on V with the seminorm

$$\|\psi\| = \operatorname{osc} \psi = \sup_{v \in V} \psi\,(v) - \inf_{v \in V} \psi\,(v).$$

---

[54] δ_x, $x \in X$, was defined on p. 122.

It is easy to prove that if with the transition function P from $(X, \mathscr{X})$ to $(Y, \mathscr{Y})$ we associate an operator $\mathfrak{P}^*$ from $\mathcal{O}(Y, \mathscr{Y})$ into $\mathcal{O}(X, \mathscr{X})$ by setting

$$(\mathfrak{P}^*\psi)(x) = \int_Y P(x, dy)\, \psi(y),$$

then we have also

$$\alpha(P) = 1 - \|\mathfrak{P}^*\|.$$

1.3.3.1.6. We shall now calculate the coefficient of ergodicity of the transition function determined by a stochastic matrix $\mathbf{P}=(p_{ij})_{i,j \in I}$, where I is at most denumerable. The meaning of $\alpha(\mathbf{P})$ is as follows: it is the coefficient of ergodicity of the transition function P defined by

$$P(i, A) = \sum_{j \in A} p_{ij}, \quad i \in I, \quad A \subset I.$$

Thus, in this special case, X and Y coincide with I and $\mathscr{X}$ and $\mathscr{Y}$ with $\mathcal{P}(I)$, the set of all subsets of I. The space $\mathfrak{L}(\mathcal{P}(I))$ comprises the families of real numbers $\mathbf{a} = (a_i)_{i \in I}$ such that $\sum_{i \in I} |a_i| < \infty$ and $\sum_{i \in I} a_i = 0$. Further

$$\|\mathbf{a}\| = \sum_{i \in I} (a_i)^+ = \sum_{i \in I} (a_i)^-.$$ [55]

The part of the operator $\mathfrak{P}$ is played here by the matrix $\mathbf{P}$ itself ($\mathbf{a}\mathbf{P} \in \mathfrak{L}(\mathcal{P}(I))$ corresponds to $\mathbf{a} \in \mathfrak{L}(\mathcal{P}(I))$.

Proposition 1.3.10. *We have*

$$\alpha(\mathbf{P}) = \inf_{i,j \in I} \sum_{k \in I} \min(p_{ik}, p_{jk}) = 1 - \frac{1}{2} \sup_{i,j \in I} \sum_{k \in I} |p_{ik} - p_{jk}|.$$

Proof. Consider an element $\mathbf{a} = \mathfrak{L}(\mathcal{P}(I))$ with only two non-null components: $a_i = a > 0$ and $a_j = -a$. Therefore $\|\mathbf{a}\| = a$. We have

$$\begin{aligned}\|\mathbf{a}\mathbf{P}\| &= \sum_{k \in I} (a_i p_{ik} + a_j p_{jk})^+ = \|\mathbf{a}\| \sum_{k \in I} (p_{ik} - p_{jk})^+ = \\ &= \|\mathbf{a}\| \sum_{k \in I} [p_{ik} - \min(p_{ik}, p_{jk})] = \|\mathbf{a}\| \left(1 - \sum_{k \in I} \min(p_{ik}, p_{jk})\right).\end{aligned} \tag{1.3.10}$$

---

[55] $(a)^+ = \max(a, 0)$, $(a)^- = \min(a, 0)$, $a \in R$.

Proposition 1.3.9 implies

$$\alpha(\mathbf{P}) \leqslant \sum_{k \in I} \min(p_{ik}, p_{jk})$$

for every $i \neq j \in I$, and hence

$$\alpha(\mathbf{P}) \leqslant \inf_{i,j \in I} \sum_{k \in I} \min(p_{ik}, p_{jk}).$$

Now note that any $\mathbf{a} \in \mathfrak{L}(\mathcal{P}(I))$ may be represented as an absolutely convergent series of elements $\mathbf{a}^{(l)} \in \mathfrak{L}(\mathcal{P}(I))$, $l \in I$, having only two non-null components and such that $\|\mathbf{a}\| = \sum_{l \in I} \|\mathbf{a}^{(l)}\|$. According to (1.3.10) we may write

$$\|\mathbf{aP}\| \leqslant \sum_{l \in I} \|\mathbf{a}^{(l)}\mathbf{P}\| \leqslant \sum_{l \in I} \|\mathbf{a}^{(l)}\| \left(1 - \inf_{i,j \in I} \sum_{k \in I} \min(p_{ik}, p_{jk})\right) =$$

$$= \|\mathbf{a}\| \left(1 - \inf_{i,j \in I} \sum_{k \in I} \min(p_{ik}, p_{jk})\right).$$

Then, Proposition 1.3.9 implies

$$\alpha(\mathbf{P}) \geqslant \inf_{i,j \in I} \sum_{k \in I} \min(p_{ik}, p_{jk}).$$

Thus, we have proved that

$$\alpha(\mathbf{P}) = \inf_{i,j \in I} \sum_{k \in I} \min(p_{ik}, p_{jk}).$$

Taking into account that

$$\min(a, b) = \frac{a + b - |a - b|}{2},$$

we obtain

$$\alpha(\mathbf{P}) = 1 - \frac{1}{2} \sup_{i,j \in I} \sum_{k \in I} |p_{ik} - p_{jk}|. \quad \diamond$$

1.3.3.2. Application to uniform ergodicity

1.3.3.2.1. Now we turn to study the coefficients of ergodicity of the transition functions ${}^mP_l^n$, $m \in N$, $l, n \in N^*$.

Proposition 1.3.11. *We have*

$$\alpha({}^mP_l^n) = \alpha({}^mP_k^n)$$

for all $m \in N$, $n \in N^*$, $l > k$.

Proof. It is easy to verify that for $l > k$

$$^{m}P_l^n(x^{(k)}, A^{(l)}) =$$

$$= \int_{X^k} {}^{m}P_k^n(x^{(k)}, dy^{(k)}) \int_{X^{l-k}} {}^{m+n+k-1}P_{l-k}(y^{(k)}, dz^{(l-k)})\, \chi_{A^{(l)}}(y^{(k)}, z^{(l-k)}).$$

This relation and Lemma 1.3.8 imply

$$\|{}^{m}P_l^n(x^k, \cdot) - {}^{m}P_l^n(y^{(k)}, \cdot)\| \leqslant \|{}^{m}P_k^n(x^{(k)}, \cdot) - {}^{m}P_k^n(y^{(k)}, \cdot)\|$$

for all $m \in N$, $n \in N^*$, $l > k$, $x^{(k)}$, $y^{(k)} \in X^k$. Therefore

$$\alpha({}^{m}P_l^n) \geqslant \alpha({}^{m}P_k^n).$$

The converse inequality is immediate on account of the relation

$$^{m}P_l^n(x^{(k)}, A^{(k)} \times X^{l-k}) = {}^{m}P_k^n(x^{(k)}, A^{(k)}). \diamond$$

Proposition 1.3.12. *We have*

$$1 - \alpha({}^{m}P_k^n) \leqslant (1 - \alpha({}^{m}P_k^{s-k+1}))\,(1 - \alpha({}^{m+s}P_k^{n-s}))$$

for all $m \in N$, $n \in N^*$, $k \leqslant s < n$.

Proof. Consider the operators ${}^{m}\mathfrak{P}_k^n$, ${}^{m}\mathfrak{P}_k^{s-k+1}$, ${}^{m+s}\mathfrak{P}_k^{n-s}$ associated respectively with the transition functions ${}^{m}P_k^n$, ${}^{m}P_k^{s-k+1}$, ${}^{m+s}P_k^{n-s}$. They act on $\mathfrak{L}(\mathcal{X}^k)$ and by virtue of (1.3.1)

$$^{m}\mathfrak{P}_k^n = {}^{m}\mathfrak{P}_k^{s-k+1}\, {}^{m+s}\mathfrak{P}_k^{n-s},$$

whence

$$\|{}^{m}\mathfrak{P}_k^n\| \leqslant \|{}^{m}\mathfrak{P}_k^{s-k+1}\|\, \|{}^{m+s}\mathfrak{P}_k^{n-s}\|.$$

Now apply Proposition 1.3.9. $\diamond$

1.3.3.2.3. Notice that Condition $M_k'(n_0)$ characterizing the uniform weak ergodicity of a kth order Markov chain amounts to the existence of a $\delta > 0$ such that

$$\alpha({}^{m}P_k^{n_0}) \geqslant \delta$$

for all $m \in N$.

The theorem below weakens this condition with a corresponding loss of the uniformity.

Theorem 1.3.13. (DOBRUŠIN (1956)). *In order that*

$$\lim_{n\to\infty} |{}^mP_l^n(x^{(k)}, A^{(l)}) - {}^mP_l^n(y^{(k)}, A^{(l)})| = 0 \tag{1.3.11}$$

uniformly with respect to $x^{(k)}, y^{(k)} \in X^k$, $l \in N^*$, $A^{(l)} \in \mathcal{X}^l$, *it suffices that the series* $\sum_{t\in N} \alpha({}^{r+tk}P_k)$, $0 \leqslant r < k$, *be divergent.*

Proof. On account of Proposition 1.3.11 relation (1.3.11) amounts to

$$\lim_{n\to\infty} \alpha({}^mP_k^n) = 1$$

for all $m \in N$. Since the sequence $\{\alpha({}^mP_k^n)\}_{n\in N^*}$ is nondecreasing, the last relation is equivalent to

$$\lim_{n\to\infty} \alpha({}^mP_k^{nk+1}) = 1 \tag{1.3.12}$$

for all $m \in N$. Proposition 1.3.12 yields

$$1 - \alpha({}^mP_k^{nk+1}) \leqslant \prod_{t=0}^{n} (1 - \alpha({}^{m+tk}P_k)),$$

and therefore the divergence of the series $\sum_{t\in N} \alpha({}^{r+tk}P_k)$, $0 \leqslant r < k$, is sufficient for (1.3.12) to hold. ◇

1.3.3.3. Some limit theorems

1.3.3.3.1. Let $X = I$ be at most a denumerable set. Consider a simple homogeneous Markov chain $(\xi_n)_{n\in N}$ with state space I and transition matrix $(p_{ij})_{i,j\in I}$. Suppose that I is a positive class and the transition function P defined by

$$P(i, A) = \sum_{j\in A} p_{ij}, \quad i \in I, \quad A \subset I,$$

has a positive coefficient of ergodicity. On account of Proposition 1.3.10 this amounts to

$$\inf_{i,j\in I} \sum_{k\in I} \min(p_{ik}, p_{jk}) > 0.$$

Consider the real-valued function f defined on I by

$$f(i) = a + k_i h, \quad i \in I,$$

where a is a given real number, h is a given positive number and k_i, $i \in I$ are given integers.

Theorem 1.3.14 (S. V. NAGAEV). *Suppose that*

$$\sum_{i \in I} \pi_i f^2(i) < \infty.$$

Then the finite limit

$$\sigma^2 = \lim_{n\to\infty} \frac{1}{n} \mathsf{E}_\pi \left[\sum_{k=1}^{n} f(\xi_k) - n \sum_{i \in I} \pi_i f(i) \right]^2$$

exists. If $\sigma > 0$ and the greatest common divisor of numbers k_i, $i \in I$, is 1, then, for every $i \in I$ we have

$$\lim_{n\to\infty} \left[\frac{\sigma\sqrt{n}}{h} \mathsf{P}_i \left(\sum_{k=1}^{n} f(\xi_k) = na + sh \right) - \frac{1}{\sqrt{2\pi}} \exp\left(- \frac{x_{ns}^2}{2} \right) \right] = 0$$

uniformly with respect to $s \in Z$ where

$$x_{ns} = \frac{1}{\sigma\sqrt{n}} [(n+1)\, a + sh - (n+1) \sum_{i \in I} \pi_i f(i)].$$

Proof. See S. V. NAGAEV (1957). ◇

1.3.3.3.2. Let $(X, \mathscr{X})$ be an arbitrary measurable space. Consider a simple homogeneous Markov chain $(\xi_n)_{n \in N}$ with state space $(X, \mathscr{X})$ and transition function P for which uniform (strong) ergodicity holds. We know that this amounts to the existence of an $n_0 \in N^*$ such that $\alpha(P^{n_0}) > 0$.

Consider a real-valued function f defined on X and $\mathscr{X}$-measurable.

Theorem 1.3.15. (S. V. NAGAEV). *Suppose that*

$$\int_X f^2(x)\, P_1^\infty(\mathrm{d}x) < \infty.$$

Then the finite limit

$$\sigma^2 = \lim_{n\to\infty} \frac{1}{n} \mathsf{E} \left[\sum_{k=1}^{n} f(\xi_k) - n \int_X f(x)\, P_1^\infty(\mathrm{d}x) \right]^2$$

exists. If $\sigma > 0$, *then for every* $x \in X$

$$\lim_{n\to\infty} \mathsf{P}_x \left(\frac{\sum_{k=1}^{n} f(\xi_k) - n \int_X f(x) P_1^\infty(\mathrm{d}x)}{\sigma \sqrt{n}} < a \right) = \frac{1}{\sqrt{2\pi}} \int_{-\infty}^{a} \mathrm{e}^{-\frac{t^2}{2}} \mathrm{d}t$$

uniformly with respect to $a \in R$.

Proof. See S. V. NAGAEV (1957). ◇

1.3.3.3.3. Consider a "Markov triangular array"

$$\begin{array}{l} \xi_1^{(1)} \\ \xi_1^{(2)},\ \xi_2^{(2)} \\ \xi_1^{(3)},\ \xi_2^{(3)},\ \xi_3^{(3)} \\ \cdot\ \cdot\ \cdot\ \cdot\ \cdot\ \cdot\ \cdot\ \cdot \\ \xi_1^{(n)},\ \cdot\ \cdot\ \cdot\ \cdot,\ \xi_n^{(n)} \\ \cdot\ \cdot\ \cdot\ \cdot\ \cdot\ \cdot\ \cdot\ \cdot \end{array}$$

By this we mean that the random variables $\xi_i^{(n)}$, $1 \leqslant i \leqslant n$, on a probability space $(\Omega_n, \mathcal{K}_n, \mathsf{P}^n)$ form for every $n \in N^*$ a nonhomogeneous Markov chain with state space X_n (endowed with a σ-algebra $\mathscr{X}_n$) and transition functions ${}^jP_{(n)}$, $1 \leqslant j < n$.

Consider for every $n \in N^*$ a set of n real-valued functions $f_i^{(n)}$, $1 \leqslant i \leqslant n$, defined on X_n and $\mathscr{X}_n$-measurable. Put

$$g_i^{(n)} = f_i^{(n)}(\xi_i^{(n)}), \quad 1 \leqslant i \leqslant n,$$

$$S_n = \sum_{i=1}^{n} g_i^{(n)}, \qquad n \in N^*.$$

Theorem 1.3.16 (R. L. DOBRUŠIN). *Suppose that the functions* $f_i^{(n)}$ *are uniformly bounded and*

$$\mathsf{D}^{(n)} g_i^{(n)} \geqslant c > 0 \tag{1.3.13}$$

for $1 \leqslant i \leqslant n$, $n \in N^*$. *Let*

$$\alpha^{(n)} = \min_{1 \leqslant j < n} \alpha({}^jP_{(n)})$$

If

$$\lim_{n\to\infty} \alpha^{(n)} n^{\frac{1}{3}} = \infty, \tag{1.3.14}$$

then

$$\lim_{n\to\infty} \mathsf{P}^{(n)}\left(\frac{S_n - \mathsf{E}^{(n)} S_n}{\mathsf{D}^{(n)} S_n} < a\right) = \frac{1}{\sqrt{2\pi}}\int_{-\infty}^{a} e^{-\frac{t^2}{2}}\, dt.$$

uniformly with respect to $a \in R$. *Condition* (1.3.14) *cannot be replaced by* $\lim_{n\to\infty} \alpha^{(n)} n^{\frac{1}{3}} = k < \infty$.[56)]

Proof. See DOBRUŠIN (1956). ◇

It may be shown that conditions (1.3.13) and (1.3.14) may be replaced by the single less restrictive condition

$$\lim_{n\to\infty} n^{-\frac{2}{3}} \alpha^{(n)}\left(\sum_{i=1}^{n} \mathsf{D}^{(n)} g_i^{(n)}\right) = \infty.$$

A complete treatment of the problem has been given by STATULEVICIUS (1969, 1970).

1.3.3.3.4. For local limit theorems the reader is referred to KOLMOGOROV (1949), STATULEVICIUS (1961) and RAUDELIUNAS (1961).

1.3.4. Compact Markov chains

In this subsection we first describe following NORMAN (1968 c) a class of simple homogeneous Markov chains whose theory is completely analogous to that of finite state Markov chains. Then, since such Markov chains occur in the theory of random systems with complete connections, we give the first elements of that theory. For a full treatment of random systems with complete connections and mathematical learning theory see IOSIFESCU and THEODORESCU (1969).

1.3.4.1 Definition and properties

1.3.4.1.1. Let X be a compact metric space with respect to a metric d, $\mathcal{X}$ the σ-algebra of Borel sets in X, $B(X)$ the set of all complex-valued bounded $\mathcal{X}$-measurable functions on X, $C(X)$ the set of all continuous complex-valued functions on X, and $L(X)$ the set of all Lipschitz functions on X (i.e. functions such that $\|f\| = m(f) + |f| < \infty$ where

$$|f| = \sup_{x\in X} |f(x)| \text{ and } m(f) = \sup_{x'\neq x''} |f(x') - f(x'')|/d(x', x'')).$$

56) It is interesting to note that condition (1.3.14) which is optimal for Markov triangular arrays, is no longer so for arbitrary (infinite) Markov chains.

It is well known that $B(X)$ ($C(X)$) and $L(X)$ are Banach spaces respectively under the norms $|\cdot|$ and $\|\cdot\|$.

Let U be the linear operator on $B(X)$ defined by

$$(Uf)(x) = \int_X P(x, \mathrm{d}y) f(y), \qquad x \in X.$$

If $(\xi_n)_{n \in N}$ is a Markov chain associated with $(X, \mathcal{X})$ and P, then it is easily seen that

$$(Uf)(x) = \mathsf{E}\,(f(\xi_{n+1}) \mid \xi_n = x), \quad n \in N.$$

Note that $|U| = 1$ and for $k > 1$,

$$(U^k f)(x) = \int_X P^k(x, \mathrm{d}y) f(y) = \mathsf{E}(f(\xi_{n+k}) \mid \xi_n = x), \qquad n \in N.$$

A Markov chain will be called *compact* iff the following conditions are satisfied:

i) U maps $L(X)$ into itself and is bounded with respect to $\|.\|$, i.e.

$$\|U\| = \sup_{f \in L(X),\, f \not\equiv 0} \frac{\|Uf\|}{\|f\|} < \infty,$$

and

ii) there are $k \in N^*$ and real numbers $0 \leqslant q < 1$ and $Q \geqslant 0$ such that $m(U^k f) \leqslant q\, m(f) + Q|f|$ for all $f \in L(X)$.

If X is a finite set, and d is any metric on X, then X is compact with respect to d, all complex-valued functions on X are Lipschitz, and $m(f) \leqslant a|f|$ where $a = 2(\min_{x' \neq x''} d(x', x''))^{-1}$. Thus $m(Uf) \leqslant a|Uf| \leqslant a|f|$, and $\|Uf\| \leqslant (a+1)|f| \leqslant (a+1)\|f\|$. Therefore i) and ii) are satisfied so that all finite Markov chains are compact.

Note also that if U satisfies i) and ii), then the sequence $(U^n f)_{n \in N}$ is equicontinuous for any $f \in C(X)$. The theory of operators U having this property was developed by JAMISON (1964, 1965) and ROSENBLATT (1964 a, b, 1967).

1.3.4.1.2. The following basic theorem is a specialization of a uniform ergodic theorem of IONESCU TULCEA and MARINESCU (1950).

Theorem 1.3.17. *For a compact Markov chain*

(a) *there are at most a finite number of eigenvalues* λ_i, $1 \leqslant i \leqslant p$ *of* U *for which* $|\lambda_i| = 1$;

(b) *for all* $n \in N^*$

$$U^n = \sum_{i=1}^{p} \lambda_i^n U_i + V^n,$$

where V *and* U_i *are linear operators on* $L(X)$ *bounded with respect to* $\|\cdot\|$;

(c) $U_i^2 = U_i$, $U_i U_j = 0$ *for* $i \neq j$, $U_i V = VU_i = 0$;

(d) $\mathfrak{D}_i = \{f \in L(X)\colon Uf = \lambda_i f\}$ *is finite dimensional and* $\mathfrak{D}_i = U_i(L(X))$, $1 \leqslant i \leqslant p$;

(e) *For some* $\alpha < 1$ *and* $M > 0$, $\|V^n\| \leqslant M\alpha^n$ *for all* $n \in N^*$.

Clearly, 1 is an eigenvalue of U and all nonzero complex constants are corresponding eigenfunctions. Without loss of generality, we suppose that $\lambda_1 = 1$ and $\lambda_i \neq 1$ for $i \neq 1$. Theorem 1.3.17 implies that

$$\lim_{n\to\infty} |\bar{U}_n f - U_1 f| = 0$$

for any $f \in C(X)$, where $\bar{U}_n = \dfrac{1}{n} \sum_{j=0}^{n-1} U^j$ and U_1 has been extended (uniquely) to a bounded linear operator on $C(X)$. It follows that $\dfrac{1}{n} \sum_{j=0}^{n-1} P^j(x, \cdot)$ [57] converges weakly to a probability $P^\infty(x,.)$ for every $x \in X$ and U_1 has the representation

$$(U_1 f)(x) = \int_X P^\infty(x, \mathrm{d}y) f(y).$$

(If 1 is the only eigenvalue of modulus 1 of U, the Cesaro averaging in the above statements is not necessary and $P^\infty(x, A)$ does not depend on $x \in X$).

1.3.4.1.3. Let $\mathfrak{M}(\mathcal{X})$ be the space of complex valued Borel measures on $\mathcal{X}$, and let V be the operator on $\mathfrak{M}(\mathcal{X})$ defined by

$$(V\mu)(A) = \int_X P(x, A)\, \mu(\mathrm{d}x), \quad A \in \mathcal{X}.$$

Then V is the adjoint of the operator U on $C(X)$, and, more intuitively, V takes the distribution of ξ_n into that of ξ_{n+1}. Thus the stationary distributions are those that are fixed under V.

1.3.4.1.4. A Borel set $A \subset X$ is said to be *stochastically closed* iff it is non-empty and $P(x, A) = 1$ for all $x \in A$. A Borel set $A \subset X$ is

[57] $P^0(x, \cdot) = \delta_x(\cdot)$.

said to be an *essential set* if it is stochastically and topologically closed and it has no proper subset with these properties. It is easy to show that distinct essential set are disjoint.

1.3.4.1.5. For arbitrary initial state, the chain is attracted to its essential sets, as the following theorem shows.

Theorem 1.3.18 (M. F. NORMAN). *A compact Markov chain* $(\xi_n)_{n\in N}$ *has l essential sets, where l is the dimension of* $\mathfrak{D}_1$. *Denote these* $E_1, \ldots, E_l$ *and let*

$$\gamma_i(x) = \mathrm{P}_x(\lim_{n\to\infty} d(\xi_n, E_i) = 0).$$

Then $\sum_{i=1}^{l} \gamma_i(x) = 1$ *for every* $x \in X$ *and* $\{\gamma_1, \ldots, \gamma_l\}$ *is a basis for* $\mathfrak{D}_1$. *There is a unique stationary distribution* μ_i *with support* E_i, *and* $\{\mu_1, \ldots, \mu_l\}$ *is a basis for* $\{\mu \in \mathfrak{M}(\mathfrak{X}) : V\mu = \mu\}$. *Finally*

$$P^\infty(x, A) = \sum_{i=1}^{l} \gamma_i(x)\,\mu_i(A).$$

Proof. See NORMAN (1968 c). ◇

It follows that there is a unique stationary distribution iff there is only one essential set.

1.3.4.1.6. Any subchain of a compact Markov chain obtained by restricting the chain to a stochastically and topologically closed subset of the state space is a compact Markov chain. Just as in the theory of finite state Markov chains, subchains on the essential sets may have a cyclic character. The periods of these cycles determine the eigenvalues of modulus 1 of the original chain.

Theorem 1.3.19 (M. F. NORMAN). *A complex number of modulus* 1 *is an eigenvalue of a compact Markov chain iff it is an eigenvalue of the subchain on some essential set. For any essential set E there is a positive integer d (called the period of E) such that the eigenvalues of modulus* 1 *of the corresponding subchain are* $\exp(2\pi \mathrm{i}\, j/d)$, $1 \leqslant j \leqslant d$. *There are d topologically closed pairwise disjoint sets* $E^1, \ldots, E^d$ *with union E, such that* $P(x, E^{j+1}) = 1$ *for all* $x \in E^j$, $1 \leqslant j \leqslant d$ $(E^{d+1} = E^1)$.

Proof. See NORMAN (1968 c). ◇

1.3.4.2. Random systems with complete connections

1.3.4.2.1. Let $(X, \mathfrak{X})$ and $(W, \mathcal{W})$ be two measurable spaces, $P(.;.)$ a transition function from $(W, \mathcal{W})$ to $(X, \mathfrak{X})$, and $u(.;.)$ a mapping of $W \times X$ into W such that

$$\{(w, x^{(n)}) : u(w; x^{(n)}) \in B\} \in \mathcal{W} \times \mathfrak{X}^n$$

for any $n \in N^*$ and $B \in \mathcal{W}$, where

$$u(\cdot\,; x^{(n)}) = u(\cdot\,; x_n) \circ \ldots \circ u(\cdot\,; x_1)$$

for $(x_1, \ldots, x_n) \in X^n$.

Definition 1.3.20. A system $\{(W, \mathcal{W}), (X, \mathcal{X}), u, P\}$ is called *a* (homogeneous) *random system with complete connections.*

1.3.4.2.2. We have the following existence theorem

Theorem 1.3.21. *For a given random system with complete connections and a given $w \in W$, there exists a probability space $(\Omega, \mathcal{K}, \mathsf{P}_w)$ and a sequence of X-valued random variables $(\xi_n)_{n \in N^*}$ defined on Ω such that*

$$\mathsf{P}_w(\xi_1 \in A) = P(w; A),$$

$$\mathsf{P}_w(\xi_{n+1} \in A \mid \xi_j,\ 1 \leqslant j \leqslant n) = P(u(w; \xi^{(n)}); A) \quad \mathsf{P}_w - \text{a.s.}$$

for any $n \in N^*$, $A \in \mathcal{X}$, where $\xi^n = (\xi_1, \ldots, \xi_n)$.

Proof. We proceed analogously as in the proof of Theorem 1.3.2. We take $\Omega = X^{N^*}$ and we set

$$\xi_n(\omega) = x_n,\ n \in N^*,$$

if

$$\omega = (x_n)_{n \in N^*}.$$

Now, for $n > 1$ we put

$$\mathsf{P}_w(\xi_1 \in A_1, \ldots, \xi_n \in A_n)$$

$$= \int_{A_1} P(w; dx_1) \int_{A_2} P(u(w; x_1); dx_2) \ldots \int_{A_n} P(u(w; x^{(n-1)}); dx_n)$$

and

$$\mathsf{P}_w(\xi_1 \in A_1) = P(w; A_1),$$

where $A_1, \ldots, A_n \in \mathcal{X}$. According to Ionescu Tulcea's theorem (see Loève (1963), p. 137) there exists a probability P_w on $\mathcal{X}^{N^*}$ satisfying the above relations and therefore those of the theorem. ◇

1.3.4.2.3. We now consider the W-valued random variables

$$\zeta_n = u(w; \xi^{(n)}),\quad n \in N^*,$$

on $(\Omega, \mathcal{K}, \mathsf{P}_w)$. We have

$$\mathsf{P}_w(\zeta_1 \in B) = P(w; B_w)$$

$$\mathsf{P}_w(\zeta_{n+1} \in B \mid \zeta_j,\ 1 \leqslant j \leqslant n) = \mathsf{P}_w(\zeta_{n+1} \in B \mid \zeta_n) = P(\zeta_n; B_{\zeta_n})$$

for $w \in W$, $B \in \mathcal{W}$, $n \in N^*$, where

$$B_w = \{x: u(w; x) \in B\}.$$

We set

$$Q(w; B) = P(w; B_w).$$

It is obvious that Q is a transition function from $(W, \mathcal{W})$ to itself. Therefore, the sequence $(\zeta_n)_{n \in N}$ with $\zeta_0 \equiv w$ is a simple Markov chain with state space $(W, \mathcal{W})$ and transition function Q. It is easy to see that the sequence $(\xi_n, \zeta_{n-1})_{n \in N^*}$ is also a Markov chain. Under suitable conditions these two Markov chains are compact.

1.3.4.2.4. We may imagine the following scheme representing the random variables associated with a random system with complete connections.

$$\begin{array}{lcl}
\zeta_0 = w & \xrightarrow{\ P(w;\ \cdot)\ } & \xi_1 \\
 & \swarrow & \\
\zeta_1 = u(w; \xi^{(1)}) & \xrightarrow{\ P(\zeta_1;\ \cdot)\ } & \xi_2 \\
 & \swarrow & \\
\zeta_2 = u(w; \xi^{(2)}) & \xrightarrow{\ P(\zeta_2;\ \cdot)\ } & \xi_3 \\
 & \swarrow & \\
\vdots & & \vdots
\end{array}$$

In what follows we briefly describe some random systems with complete connections.

1.3.4.2.5. *kth order (homogeneous) Markov chains.* A kth order Markov chain may be considered as the random system with complete connections for which

$$W = X^k, \quad \mathcal{W} = \mathcal{X}^k,$$

$$u(w; x) = \begin{cases} (x_2, \ldots, x_k, x) & \text{if } k \geqslant 2 \\ x & \text{if } k = 1, \end{cases}$$

for $w = x^{(k)} = (x_1, \ldots, x_k) \in W$, $x \in X$.

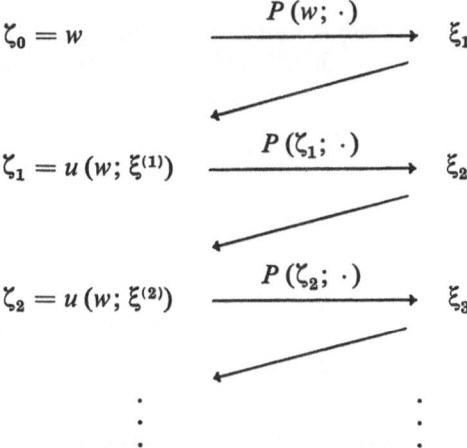

1.3.4.2.6. *OM-chains (chains with complete connections).*
Let us set

$$\Delta = \left\{ \mathbf{p} = (p_1, \ldots, p_{m+1}) : 0 \leqslant p_i \leqslant 1,\ 1 \leqslant i \leqslant m+1,\ \sum_{i=1}^{m+1} p_i = 1 \right\}$$

for a given $m \in N^*$ and let φ_i, $1 \leqslant i \leqslant m+1$, be a family of measurable mappings of Δ into itself.

An *OM*-chain $(\Delta, (\varphi_i)_{1 \leqslant i \leqslant m+1})$ is a random system with complete connections for which

$$(W, \mathcal{W}) = (\Delta, \mathcal{B}_\Delta)\,^{58)}, \quad X = \{1, \ldots, m+1\},$$
$$u(\mathbf{p}; i) = \varphi_i(\mathbf{p}),$$
$$P(\mathbf{p}; i) = p_i .$$

Historically *OM*-chains are the first example of a random system with complete connections (ONICESCU and MIHOC (1935)).

1.3.4.2.7. *Chains of infinite order.* Formally, a random system with complete connections $((W, \mathcal{W}), (X, \mathcal{X}), u, P)$ is a chain of infinite order if

$$(W, \mathcal{W}) = (\prod_{j \in -N} X_j, \prod_{j \in -N} \mathcal{X}_j)$$

where $X_j = X$, $\mathcal{X}_j = \mathcal{X}$ for every $j \in -N$. Furthermore, if $w = (\ldots, x_{-n}, \ldots, x_{-1}, x_0)$, then

$$u(w; x) = (\ldots, x'_{-n}, \ldots, x'_{-1}, x'_0),$$

with

$$x'_{-n} = x_{-n+1},\ n \in N^*,\ x'_0 = x.$$

Chains of infinite order have been considered by DOEBLIN and FORTET (1937).

1.3.4.2.8. *Learning models.* In describing a learning process we follow NORMAN (1968 a, b). Suppose that a subject is repeatedly exposed to an experimental situation in which various responses are possible, and suppose that each such exposure or trial can alter the subject's response tendencies in the learning situation. It is assumed that the subject's response tendencies on trial n are determined by its "*state*" S_n at that time. The set of possible states is denoted by S and is called the *state space*. The effect of the nth trial is represented by the occurrence

58) $\mathcal{B}_\Delta$ is the Borel σ-algebra in Δ.

of a certain *event* E_n. The set of possible events is denoted by E and referred to as the *event space*. We assume that E is a finite set. The quantities S_n and E_n are to be considered random variables. To represent the fact that the occurrence of an event effects a change of state, we associate with each event e a mapping $f_e(.)$ of S into itself such that, if $E_n = e$, then $S_{n+1} = f_e(S_n)$, $n \in N$. The mappings f_e, $e \in E$ are assumed to be $\mathcal{S}$-measurable, where $\mathcal{S}$ is a σ-algebra of subsets of S. It is further supposed that the learning situation is memory-less and temporally homogeneous in the sense that the probabilities of the various possible events on trial n depend only on the state on trial n, and not on earlier states or events, or on the trial number. That is, there is a real valued function $\varphi_.(.)$ on $E \times S$ such that

$$\mathsf{P}_s(E_1 = e_1) = \varphi_{e_1}(s),$$

and

$$\mathsf{P}_s(E_{n+1} = e_{n+1} \mid E_j = e_j,\ 1 \leqslant j \leqslant n) = \varphi_{e_{n+1}}(f_{e_1 \dots e_n}(s))$$

for $n \in N^*$, where

$$f_{e_1 \dots e_n}(s) = f_{e_n}(f_{e_{n-1}}(\dots(f_{e_1}(s)))).$$

Subscripts on probabilities denote initial states, that is, values of S_0. The system $((S, \mathcal{S}), E, (f_e)_{e \in E}, \varphi)$ is said to be a *learning model*. It is obvious that it is a random system with complete connections for which

$$X = E, \quad (W, \mathcal{W}) = (S, \mathcal{S})$$

$$u(s; e) = f_e(s),\ P(s; e) = \varphi_e(s).$$

References

Afanas'eva, L. G., and A. V. Martynov

(1966) An ergodicity condition for certain queueing systems. Kibernetika (Kiev), no. 4, 57–63. (Russian)

Albert, E.

(1964) Generalized symmetric random walks. Scripta Math. **27**, 185–187.

Aleškjavičene, A.

(1965) A local limit theorem for recurrent events. Litovsk. Mat. Sb. **5**, 373–380. (Russian)

(1967) Large deviations for the number of appearances of a recurrent event. Litovsk. Mat. Sb. **7**, 185–193 and 363. (Russian)

Athreya, K. B.

(1970) A simple proof of a result of Kesten and Stigum on supercritical multitype Galton-Watson branching process. Ann. Math. Statist. **41**, 195–202.

(1971) A note on a functional equation arising in Galton-Watson branching processes. J. Appl. Probability **8**, 589–598.

Athreya, K. B., and S. Karlin

(1971 a) Branching processes with random environments, I: Extinction probabilities. Ann. Math. Statist. **42**, 1499–1520.

(1971 b) Branching processes with random environments, II: Limit theorems. Ann. Math. Statist. **42**, 1843–1858.

Atkinson, R. C., and W. K. Estes

(1963) Stimulus sampling theory, pp. 121–268 in R. D. Luce, R. R. Bush, E. Galanter (Eds.) *Handbook of mathematical psychology*, Vol. II. New York-London: Wiley.

Barnett, V. D.

(1963) Some explicit results for an asymmetric two-dimensional random walk. Proc. Cambridge Philos. Soc. **59**, 451–462.

(1964) A three-player extension of the gambler's ruin problem. J. Appl. Probability **1**, 321–334.

(1965) Wald's identity and absorption probabilities for two-dimensional random walks. Proc. Cambridge Philos. Soc. **61**, 747–762.

Bartlett, M. S.

(1957) On theoretical models for competitive and predatory biological systems. Biometrika **44**, 27–42.

(1966) An introduction to stochastic processes with specia reference to methods and applications. 2nd Edition. London—New York: Cambridge Univ. Press.

BARTOSZYŃSKI, R.

(1967) Some limit properties of generalized branching processes Bull. Acad. Polon. Sci. Sér. Sci. Math. Astronom. Phys. **15**, 157—160.

BELLMAN, R., and T. HARRIS

(1951) Recurrence times for the Ehrenfest model. Pacific J. Math. **1**, 179—193.

BENEŠ, V. E.

(1963) General stochastic processes in the theory of queues. Reading, Massachussets: Addison-Wesley.

BOROVKOV, A. A.

(1972) Stochastic processes in queueing theory. Moscow: Nauka. (Russian)

BRENY, H.

(1962 a) Sur un point de la théorie des files d'attente. Ann. Soc. Sci. Bruxelles **76**, 5—12.

(1962 b) Cheminements conditionnels de chaînes de Markov absorbantes. Ann. Soc. Sci. Bruxelles **76**, 81—87.

BRETAGNOLLE, J., and D. DACUNHA-CASTELLE

(1967) Sur une classe de marches aléatoires. Ann. Inst. H. Poincaré Sect. B **3**, 403—431.

BROOKNER, E.

(1966) Recurrent events in a Markov chain. Information and Control **9**, 215—229.

BÜHLER, W. J.

(1966) Probability generating functions and their iterates. Proc. Nat. Acad. Sc. USA **56**, 1912—1916.

(1967) Slowly branching processes. Ann. Math. Statist. **38**, 919—921.

(1968) Some results on the behaviour of branching processes. Teor. Verojatn. i Primenen. **13**, 51—62.

BUI-TRONG-LIEU, and M. DOREL

(1967) Sur le comportement asymptotique de processus de Markov non homogènes. Studia Math. **28**, 253—274.

CARTIER, P.

(1962/1963) Fluctuations dans les suites de variables aléatoires indépendantes. Séminaire Bourbaki, 15e année, no. 241.

CHUNG, K. L.

(1964) The general theory of Markov processes according to Doeblin. Z. Wahrscheinlichkeitstheorie **2**, 230—254.

(1967) Markov chains with stationary transition probabilities 2nd Edition. Berlin-Heidelberg—New York: Springer.

CHUNG, K. L., and W. H. J. FUCHS

(1951) On the distribution of values of sums of independent random variables. Mem. Amer. Math. Soc. **6**, 1—12.

COHEN, J. W.
(1969) The single server queue. New York : Wiley – Interscience.

COX, D. R., and H. D. MILLER
(1965) The theory of stochastic processes. London: Methuen.

COX, D. R., and W. L. SMITH
(1961) Queues. London: Methuen.

CRAMÉR, H., and M. R. LEADBETTER
(1967) Stationary and related stochastic processes. Sample function properties and their applications. New York – London – Sydney: Wiley.

DALEY, D. J.
(1968) Stochastically monotone Markov chains. Z. Wahrscheinlichkeitstheorie **10**, 305 – 317.
(1969) Quasi-stationary behaviour of a left-continuous random walk. Ann. Math. Statist. **40**, 532 – 539.

DARLING, D. A.
(1970) The Galton-Watson process with infinite mean. J. Appl. Probability **7**, 455 – 456.

DARROCH, J. N., and K. W. MORRIS
(1967) Some passage-time generating functions for discrete-time and continuous-time finite Markov chains. J. Appl. Probability **4**, 496 – 507.

DARROCH, J. N., and E. SENETA
(1965) On quasi-stationary distributions in absorbing discrete-time finite Markov chains. J. Appl. Probability **2**, 88 – 100.

DAYANITHY, K.
(1971) Spectralities of branching processes. Z. Wahrscheinlichkeitstheorie **20**, 279 – 307.

DEBERGHES, H., and BUI-TRỌNG-LIEU
(1968) On the ergodicity for nonstationary multiple Markov processes. Ann. Math. Statist. **39**, 1448 – 1462.

DECELL JR., H. P., and P. L. ODELL
(1967 a) On the fixed point probability vector of regular or ergodic transition matrices. J. Amer. Statist. Assoc. **62**, 600 – 602.
(1967 b) On computing the fixed-point probability vector of ergodic transition matrices. J. Assoc. Comput. Mach. **14**, 765 – 768.

DEPAIX, M.
(1964/1965) Étude des distributions de certains processus de marche au hasard dans l'espace Z^n limité par des barrières. Ann. Inst. H. Poincaré Sect. B **1**, 311 – 439.

DERMAN, C.
(1954) A solution to a set of fundamental equations in Markov chains. Proc. Amer. Math. Soc. **5**, 332 – 334.
(1955) Some contributions to the theory of denumerable Markov chains. Trans. Amer. Math. Soc. **79**, 541 – 555.

DOBRUŠIN, R. L.
(1956) Central limit theorem for nonstationary Markov chains. Teor. Verojatn. i Primenen. **1**, 72 – 89, 365 – 425. (Russian)

DOEBLIN, W., and R. FORTET
(1937) Sur des chaînes à liaisons complètes. Bull. Soc. Math. France **65**, 132–148.

DONEY, R. A.
(1965) Recurrent and transient sets for 3-dimensional random walks. Z. Wahrscheinlichkeitstheorie **4**, 253–259.

DOOB, J. L.
(1953) Stochastic processes. New York: Wiley.

DYNKIN, E. B.
(1954) On some limit theorems for Markov chains. Ukrain. Mat. Ž. **6**, 21–27. (Russian)
(1969) The boundary theory of Markov processes (discrete case). Uspehi Mat. Nauk **24**, no. 2 (1946), 3–42. (Russian)

DWASS, M.
(1967) Simple random walk and rank order statistics. Ann. Math. Statist. **38**, 1042–1054.
(1969) The total progeny in a branching process and a related random walk. J. Appl. Probability **6**, 682–686.

ENGELBERG, O.
(1965) On some problems concerning a restricted random walk. J. Appl. Probability **2**, 394–404.

ERDÖS P., H. POLLARD, and W. FELLER
(1949) A property of power series with positive coefficients. Bull. Amer. Math. Soc. **55**, 201–204.

ERDÖS, P., and S. J. TAYLOR
(1960) Some problems concerning the structure of random walk paths. Acta Math. Acad. Sci. Hungar. **11**, 137–162.

ERICKSON, R. V.
(1971) On the existence of absolute moments for the extinction time of a Galton-Watson process. Ann. Math. Statist. **42**, 1124–1128.

EWENS, W. J.
(1964) The pseudo-transient distribution and its uses in genetics. J. Appl. Probability **1**, 141–156.

FAGIN, R.
(1968) Representation theory for a class of denumerable Markov chains. J. Math. Anal. Appl. **23**, 500–530.

FATOU, P.
(1919) Sur les équations fonctionnelles. Bull. Soc. Math. France, **47**, 161–271.
(1920) Sur les équations fonctionnelles (Troisième mémoire). Bull. Soc. Math. France **48**, 208–314.

FELLER, W.
(1949) Fluctuation theory of recurrent events. Trans. Amer. Math. Soc. **67**, 98–119.
(1966 a) On the Fourier representation for Markov chains and the strong ratio theorem. J. Math. Mech. **15**, 273–283.
(1966 b) An introduction to probability theory and its applications, Vol. 2. New York–London: Wiley.

(1968) An introduction to probability theory and its applications. Vol. 1, 3rd Edition. New York—London: Wiley.

FINCH, P. D.

(1963) A limit theorem for Markov chains with continuous state space. J. Austral. Math. Soc. **3**, 351–358.

FISHER, M. E.

(1966) Shape of a self-avoiding walk or polymer chain. J. Chem. Phys. **44**, 616–622.

FOGUEL, S. R.

(1966) Limit theorems for Markov processes. Trans. Amer. Math. Soc. **121**, 200–209.

FOLKMAN, J. H., and S. C. PORT

(1966) On Markov chains with the strong ratio limit property. J. Math. and Mech. **15**, 113–121.

FOSTER, F. G.

(1951) Markov chains with an enumerable number of states and a class of cascade processes. Proc. Cambridge Philos. Soc. **47**, 633–634.

(1952 a) On Markov chains with an enumerable infinity of states. Proc. Cambridge Philos. Soc. **48**, 587–591.

(1952 b) A Markov chain derivation of discrete distributions. Ann. Math. Statist. **23**, 624–627.

(1953) On the stochastic matrices associated with certain queueing processes. Ann. Math. Statist. **24**, 355–360.

FRÉCHET, M.

(1938) Recherches théoriques modernes sur le calcul des probabilités, Vol. II. Méthode des fonctions arbitraires. Théorie des événements en chaîne dans le cas d'un nombre fini d'états possibles. Paris: Hermann.

FREEDMAN, D. A.

(1965) Bernard Friedman's urn. Ann. Math. Statist. **36**, 956–970.

(1967) Some invariance principles for functionals of a Markov chain. Ann. Math. Statist. **38**, 1–7.

FRIEDMAN, B.

(1949) A simple urn model. Comm. Pure Appl. Math. **2**, 59–70.

FRISCH, H. L., and J. M. HAMMERSLEY

(1963) Percolation processes and related topics. J. Soc. Indust. Appl. Math. **11**, 894–918.

GANTMACHER, F. R.

(1959) Applications of the theory of matrices. New York—London: Interscience.

GARSIA, A.

(1963) Some Tauberian theorems and the asymptotic behaviour of probabilities of recurrent events. J. Math. Anal. Appl. **7**, 146–162.

GARSIA, A., S. OREY, and E. RODEMICH

(1962) Asymptotic behaviour of the successive coefficients of some power series. Illinois J. Math. **6**, 620–629.

GATUN, V. P. and A. V. SKOROHOD
(1969) Difference equations and Markov chains. Ukrain. Mat. Ž. **21**, 305–315 (Russian).

GEL'FOND, A. O.
(1964) An estimate of the remainder term in a limit theorem for recurrent events. Teor. Verojatn. i Primenen. **9**, 327–331. (Russian)

GEORGE, C.
(1961) Les urnes d'Ehrenfest et les processus stochastiques du premier ordre. Acad. Roy. Belg. Bull. Cl. Sci. (5) **47**, 92–110.

GILLIS, J.
(1955) Correlated random walk. Proc. Cambridge Philos. Soc. **51**, 639–651.
(1960) A random walk problem. Proc. Cambridge Philos. Soc. **56**, 390–392.

GITTINS, J. C.
(1971) Stochastic monotonicity and queues subject to tidal interruptions. Proc. Cambridge Philos. Soc. **70**, 61–75.

GOODMAN, L. A.
(1968) How to minimize or maximize the probabilities of extinction in a Galton-Watson process and in some related multiplicative population processes. Ann. Math. Statist. **39**, 1700–1710.

GUPTA, H. C.
(1966) Random walk in the presence of a multiple-function barrier. J. Math. Sci. **1**, 18–29.

HAJNAL, J.
(1956) The ergodic properties of nonhomogeneous finite Markov chains. Proc. Cambridge Philos. Soc. **52**, 67–77.
(1958) Weak ergodicity in nonhomogeneous Markov chains. Proc. Cambridge Philos. Soc. **54**, 233–246.

HAMMERSLEY, J. M. and R. S. WALTERS
(1963) Percolation and fractional branching processes. J. Soc Indust. Appl. Math. **11**, 831–839.

HANEN, A.
(1963) Théorèmes limites pour une suite de chaînes de Markov. Ann. Inst. H. Poincaré **18**, 197–301.

HARDIN, J. C. and A. L. SWEET
(1969) A note on absorption probabilities for a random walk between a reflecting and an absorbing barrier. J. Appl. Probability **6**, 224–226.
(1970) Moments of the time to absorption in the random walk between a reflecting and an absorbing barrier. SIAM Rev. **12**, 140–142.

HARDY, G. H. and W. W. ROGOSINSKI
(1956) Fourier series. 3rd Edition. Cambridge Tracts in Mathematics and Mathematical Physics no. 38.

HARRIS, T. E.
(1952) First passage and recurrence distributions. Trans. Amer. Math. Soc. **73**, 471–486.

(1956) The existence of stationary measures for certain Markov processes. Proc. 3rd Berkeley Symp. Math. Statist. Prob., Vol. 2, pp. 113–124. Berkeley: Univ. California Press.

(1957) Transient Markov chains with stationary measures. Proc. Amer. Math. Soc. **8**, 937–942.

(1963) The theory of branching processes. Berlin – Göttingen – Heidelberg: Springer.

HEATHCOTE, C. R.

(1966) Corrections and comments on the paper "A branching process allowing immigration". J. Roy. Statist. Soc. Ser. B **28**, 213–217.

HEATHCOTE, C. R., and E. SENETA

(1966–67) Inequalities for branching processes. J. Appl. Probability **3**, 261–267. Correction, ibid. **4**, 215.

HEATHCOTE C. R., E. SENETA, and D. VERE-JONES

(1967) A refinement of two theorems in the theory of branching processes. Teor. Verojatn. i Primenen. **12**, 341–346.

HEYDE, C. C.

(1967) A limit theorem for random walks with drift. J. Appl. Probability **4**, 144–150.

(1970 a) A rate of convergence result for the supercritical Galton-Watson process. J. Appl. Probability **7**, 451–454.

(1970 b) Extension of a result of Seneta for the supercritical Galton-Watson process. Ann. Math. Statist. **41**, 739–742.

HEYDE, C. C. and E. SENETA

(1971) Analogues of classical limit theorems for the supercritical Galton-Watson process with immigration. Math. Biosci. **11**, 249–259.

HIGGINSON, J. A.

(1969) A matrix representation of a class of denumerable homogeneous Markov chains. J. Math. Anal. Appl. **28**, 26–31.

HOLLAND, P. W.

(1968) Some properties of an algebraic representation of stochastic processes. Ann. Math. Statist. **39**, 164–170.

HOLMES, P. T.

(1966–67) On some spectral properties of a transition matrix. Sankhyā Ser. A. **28**, 205–214. Corrigendum, ibid. **29**, 106.

(1967) On non-dissipative Markov chains. Sankhyā Ser. A **29**, 383–390.

HOSTINSKY, B.

(1931) Méthodes générales du calcul de probabilités. Paris: Gauthier-Villars.

IONESCU TULCEA, C. T., and G. MARINESCU

(1950) Théorie ergodique pour des classes d'opérations non complètement continues. Ann. Math. **52**, 140–147.

IOSIFESCU, M.

(1963) Random systems with complete connections with an arbitrary set of states. Rev. Math. Pures Appl. **8**, 611–645. (Russian)

(1965) Sur l'ergodicité uniforme des chaînes de Markoff variables et multiples. C. R. Acad. Sci. Paris **261**, 875–877.

Iosifescu, M., and R. Theodorescu
(1969) Random processes and learning. Berlin—Heidelberg—New York: Springer.

Isaac, R.
(1967) On the ratio-limit theorem for Markov processes recurrent in the sense of Harris. Illinois J. Math. **11**, 608—615.
(1968) Some topics in the theory of recurrent Markov processes. Duke Math J. **35**, 641—652.

Jain, G. C.
(1966) On cumulative sums in one-dimensional symmetric random walk. J. Indian Statist. Assoc. **4**, 73—85.

Jain, N. C.
(1966 a) A note on invariant measures. Ann. Math. Statist. **37**, 729—732.
(1966 b) Some limit theorems for a general Markov process. Z. Wahrscheinlichkeitstheorie **6**, 206—223.

Jain, N., and B. Jamison
(1967) Contributions to Doeblin's theory of Markov processes. Z. Wahrscheinlichkeitstheorie **8**, 19—40.

Jamison, B.
(1964) Asymptotic behaviour of successive iterates of continuous functions under a Markov operator. J. Math. Anal. Appl. **9**, 203—214.
(1965) Ergodic decompositions induced by certain Markov operators. Trans. Amer. Math. Soc. **117**, 451—468.

Jiřina, M.
(1958) Stochastic branching processes with continuous state space. Czechoslovak Math. J. **8**, 292—313.
(1964) Branching processes with measure-valued states. Trans. 3rd Prague Conf. Information Theory, Statist. Decision Functions, Random Processes, pp. 333—357. Prague: Academia.
(1966) Asymptotic behaviour of measure-valued branching processes. Rozpravy Československé Akad. Věd. 76, no. 3, 55 pp.
(1970) A simplified proof of the Sevastyanov theorem on branching processes. Ann. Inst. Poincaré Sect. B **6**, 1—7.

Joffe, A.
(1967) On the Galton-Watson branching process with mean less than one. Ann. Math. Statist. **38**, 264—266.

Joffe, A., and F. Spitzer
(1967) On multitype branching processes with $\rho \leqslant 1$. J. Math. Anal. Appl. **19**, 409—430.

Kac, M.
(1947) Random walk and the theory of Brownian motion. Amer. Math. Monthly **54**, 369—391.

Karlin, S.
(1955) On the renewal equation. Pacific J. Math. **5**, 229—257.
(1966) A first course in stochastic processes. New York—London: Academic Press.

KARLIN, S., and J. MCGREGOR
(1959) Random walks. Illinois J. Math. **3**, 66–81.
(1965) Ehrenfest urn models. J. Appl. Probability **2**, 352–376.
(1966 a) Spectral theory of branching processes I. The case of discrete spectrum. Z. Wahrscheinlichkeitstheorie **5**, 6–33.
(1966 b) Spectral representation of branching processes II. Case of continuous spectrum. Z. Wahrscheinlichkeitstheorie **5**, 34–54.
(1967 a) Uniqueness of stationary measures for branching processes and applications. Proc. 5th Berkeley Symp. Math. Statist. Prob. Vol. II, Part 2, pp. 243–254. Berkeley: Univ. California Press.
(1967 b) Properties of the stationary measure of the critical case simple branching process. Ann. Math. Statist. **38**, 977–991.
(1968) On the spectral representation of branching processes with mean one. J. Math. Anal. Appl. **21**, 485–495.

KEILSON, J.
(1965) Green's function methods in probability theory. London: Griffin.

KEMENY, J. G.
(1966 a) Slowly spreading chain of the first kind. J. Math. Anal. Appl. **15**, 295–310.
(1966 b) Representation theory for denumerable Markov chains. Trans. Amer. Math. Soc. **125**, 47–62.

KEMENY, J. G., and J. L. SNELL
(1960) Finite Markov chains. Princeton–Toronto–London–New York: Van Nostrand.

KEMENY, J. G., J. L. SNELL, and A. W. KNAPP
(1966) Denumerable Markov chains. Princeton–Toronto–New York–London: Van Nostrand.

KEMPERMAN, J. H. B.
(1961) The passage problem for a stationary Markov chain. Chicago: Univ. Chicago Press.

KENDALL, D. G.
(1951 a) On non-dissipative Markov chains with an enumerable number of states. Proc. Cambridge Philos. Soc. **47**, 633–634.
(1951 b) Some problems in the theory of queues. (With discussion). J. Roy. Statist. Soc. Ser. B **13**, 151–185.
(1953) Stochastic processes occurring in the theory of queues and their analysis by the method of the imbedded Markov chain. Ann. Math. Statist. **24**, 338–354.
(1958) Integral representations for Markov transition probabilities. Bull. Amer. Math. Soc. **64**, 358–362.
(1959) Unitary dilations of Markov transition operators and the corresponding integral representations for transition-probability matrices, pp. 139–161 in U. Grenander (Ed.) *Probability & Statistics* (The Harald Cramér Volume). Stockholm: Almqvist and Wiksell.
(1960) Geometric ergodicity and the theory of queues, pp. 176–195 in K. J. Arrow, S. Karlin, and P. Suppes (Eds.) *Mathematical methods in the social sciences, 1959*. Stanford: Stanford University Press.

(1961 a) The distribution of energy perturbations for Halley's and some other comets. Proc. 4th Berkeley Symp. Math. Statist. Prob. Vol. 3, pp. 87–98. Berkeley: Univ. California Press.
(1961 b) Some problems in the theory of comets I, II. Proc. 4th Berkeley Symp. Math. Statist. Prob. Vol. 3, pp. 99–147. Berkeley: Univ. California Press.
(1963) Information theory and the limit theorem for Markov chains and processes with a countable infinity of states. Ann. Inst. Statist. Math. **15**, 137–143.
(1964) Some recent work and further problems in the theory of queues. Teor. Verojatn. i Primenen. **9**, 3–15.
(1966 a) On super-critical branching processes with a positive chance of extinction, pp. 157–165 in F. N. David (Ed.) assisted by Evelyn Fix, *Research papers in statistics* (Festschrift for J. Neyman). London–New York–Sydney: Wiley.
(1966 b) Branching processes since 1873. J. London Math. Soc. **41**, 385–406.
(1966 c) Discussion on Seneta's (1966) paper. J. Roy. Statist. Soc. Ser. B. **28**, 266–269.
(1967) Renewal sequences and their arithmetic. Symposium on Probability Methods in Analysis (Loutraki, 1966), pp. 147–175. Berlin–Heidelberg–New York: Springer.

KENDALL D. G., and G. E. H. REUTER
(1957) The calculation of the ergodic projection for Markov chains and processes with a countable infinity of states. Acta Math. **97**, 103–144.

KESTEN, H.
(1963) On the number of self-avoiding walks. J. Mathematical Phys. **4**, 960–969.

KESTEN, H., and B. P. STIGUM
(1966 a) A limit theorem for multi-dimensional Galton-Watson processes. Ann. Math. Statist. **37**, 1211–1223.
(1966 b) Additional limit theorems for indecomposable multidimensional Galton-Watson processes. Ann. Math. Statist. **37**, 1463–1481.
(1967) Limit theorems for decomposable multidimensional Galton-Watson processes. J. Math. Anal. Appl. **17**, 309–338.

KESTEN, H., P. NEY, and F. SPITZER
(1966) The Galton-Watson process with mean one and finite variance. Teor. Verojatn. i Primenen. **11**, 579–611.

KINGMAN, J. F. C.
(1963) A continuous time analogue of the theory of recurrent events. Bull. Amer. Math. Soc. **69**, 268–272.
(1964) The stochastic theory of regenerative events. Z. Wahrscheinlichkeitstheorie **2**, 180–224.
(1965) Stationary measures for branching processes. Proc. Amer. Math. Soc. **16**, 245–247.
(1966 a) On the algebra of queues. J. Appl. Probability **3**, 285–326.
(1966 b) An approach to the study of Markov processes. (With discussion). J. Roy. Statist. Soc. Ser. B. **28**, 417–447.

KINGMAN, J. F. C., and S. OREY
(1964) Ratio limit theorems for Markov chains. Proc. Amer. Math. Soc. **15**, 907–910.

KLIMKO, L. A. and L. SUCHESTON
(1968) On probabilistic limit theorems for a class of positive matrices. J. Math. Anal. Appl., **24**, 191–201.

KLIMOV, G. P.
(1966) Stochastic service systems. Moscow: Nauka. (Russian)

KOLMOGOROV, A. N.
(1949) A local limit theorem for classical Markov chains. Izv. Akad. Nauk SSSR, Ser. Mat. **13**, 281–300. (Russian)

KOLMOGOROV, A. N., and N. A. DMITRIEV
(1947) Branching stochastic processes. Dokl. Akad. Nauk SSSR, **56**, 7–10. (Russian)

KRICKEBERG, K.
(1967) Strong mixing properties of Markov chains with infinite invariant measure. Proc. 5th Berkeley Symp. Math. Statist. Prob. Vol. II, Part 2, pp. 431–446. Berkeley: Univ. California Press.

KULLBACK, S.
(1967) An extension of an information-theoretic derivation of certain limit relations for a Markov chain. SIAM J. Control **5**, 51–53.

LAMPERTI, J.
(1967 a) Limiting distributions for branching processes. Proc. 5th Berkeley Symp. Math. Statist. Prob. Vol. II, Part 2, pp. 225–241. Berkeley: Univ. California Press.
(1967 b) The limit of a sequence of branching processes. Z. Wahrscheinlichkeitstheorie **7**, 271–288.
(1967 c) Continuous state branching processes. Bull. Amer. Math. Soc. **73**, 382–386.

LAMPERTI, J., and P. NEY
(1968) Conditioned branching processes and their limiting diffusions. Teor. Verojatn. i Primenen. **13**, 126–137.

LEHMAN, R. S., and WEISS G. H.
(1958) A study of the restricted random walk. J. Soc. Indust. Appl. Math. **6**, 257–278.

LEHNER, G.
(1963) One dimensional random walk with a partially reflecting barrier. Ann. Math. Statist. **34**, 405–412.

LESLIE, R. T.
(1967) Recurrent composite events. J. Appl. Probability **4**, 34–61.

LEVINSON, N.
(1959) Limiting theorems for Galton-Watson branching process. Illinois J. Math. **3**, 554–565.
(1965) An elementary proof of the stationary distribution for an irreducible Markov chain. Amer. Math. Monthly **72**, 366–369.

LOÈVE, M.
(1963) Probability theory. 3rd. Edition. Princeton–Toronto–New-York–London: Van Nostrand.

LOTKA, A. J.
(1931) The extinction of families I, II. J. Wash. Acad. Sci. **21**, 377–380 and 453–459.

LOYNES, R. M.
(1966) On certain applications of the spectral representation of stationary processes. Z. Wahrscheinlichkeitstheorie **9**, 20–25.

MALOŠEVSKI, S. G.
(1965) A limit theorem for some special random walks. Teor. Verojatn. i Primenen. **10**, 560–567. (Russian)

MANDL, P.
(1959) On the asymptotic behaviour of the probabilities within groups of states of a homogeneous Markov process. Časopis Pěst. Math. **85**, 448–456. (Russian)

MIHOC, G., and G. CIUCU
(1967) An introduction to queueing theory. Bucharest: Technical Publishing House. (Romanian)

MILLER, H. D.
(1961) A generalization of Wald's identity with applications to random walk. Ann. Math. Statist. **32**, 549–560.
(1966) Geometric ergodicity in a class of denumerable Markov chains. Z. Wahrscheinlichkeitstheorie **4**, 354–373.

MIRSKY, L.
(1962/63) Results and problems in the theory of doubly-stochastic matrices. Z. Wahrscheinlichkeitstheorie **1**, 319–334.

MOHANTY, S. G.
(1968) On some generalization of a restricted random walk. Studia Scient. Math. Hungarica **3**, 225–241.

MOLČANOV, S. A.
(1967) A limit theorem for the ratio of transition probabilities of Markov chains. Uspehi Mat. Nauk **22**, no. 2 (134), 124–125. (Russian)

MOY, S. C.
(1965) λ-continuous Markov chains I, II. Trans. Amer. Math. Soc. **117**, 68–91; **120**, 83–107.
(1967 a) Ergodic properties of expectation matrices of a branching process with countably many types. J. Math. Mech. **16**, 1207–1225.
(1967 b) Extensions of a limit theorem of Everett, Ulam and Harris on multitype branching processes to a branching process with countably many types. Ann. Math. Statist. **38**, 992–999.

NAGAEV, A. V.
(1967) Estimation of the mean number of direct descendants of a particle in a branching random process. Teor. Verojatn. i Primenen. **12**, 363–369. (Russian)

NAGAEV, A. V., and I. BADALBAEV
(1967) A refinement of certain theorems on branching random processes. Litovsk. Mat. Sb. **7**, 129–136. (Russian)

NAGAEV, S. V.
(1957) Some limit theorems for homogeneous Markov chains. Teor. Verojatn. Primenen. **2**, 389–416. (Russian)

(1965) Ergodic theorems for discrete-time Markov processes. Sibirsk. Mat. Z. **6** 413–432. (Russian)

Nagaev, S. V., and R. Muhamedhanova

(1966 a) Transition phenomena in branching random processes with discrete time, pp. 83–89 in *Limit theorems statist. inference*, Tashkent: Izdat. "Fan". (Russian)

(1966 b) Certain limit theorems in the theory of branching random processes, pp. 90–112 in *Limit theorems statist. inference*, Tashkent: Izdat. "Fan". (Russian)

Nash-Williams, C. St. J. A.

(1959) Random walk and electric currents in networks. Proc. Cambridge Philos. Soc. **55**, 181–194.

Neuts, M. F.

(1963) Absorption probabilities for a random walk between a reflecting and an absorbing barrier. Bull. Soc. Math. Belg. **15**, 253–258.

(1967) Two Markov chains, arising from examples of queues with state-dependent service times. Sankhyā Ser. A **29**, 259–264.

Norman, M. F.

(1968 a) Some convergence theorems for stochastic learning models with distance diminishing operators. J. Math. Psychology **5**, 61–101.

(1968 b) Mathematical learning theory, pp. 283–313 in G. V. Dantzig and A. F. Veinott (Eds.) *Mathematics of the decision sciences*. Part 2. Providence: Amer. Math. Soc.

(1968 c) Compact Markov processes. Technical rep. no. 2, University of Pennsylvania.

Onicescu, O.

(1964) Nombres et systèmes aléatoires. Bucarest et Paris: Publishing House of the Romanian Academy et Eyrolles.

Onicescu, O., and G. Mihoc

(1935) Sur les chaînes de variables statistiques. Bull. Sci. Math. **59**, 174–192.

(1943) Les chaînes de variables aléatoires. Problèmes asymptotiques. Bucarest: Académie Roumaine.

Orey, S.

(1961) Strong ratio limit property. Bull. Amer. Math. Soc. **67**, 571–574.

(1962) An ergodic theorem for Markov chains. Z. Wahrscheinlichkeitstheorie **1**, 174–176.

Ornstein, D. S.

(1969) Random walks. I, II. Trans. Amer. Math. Soc. **138**, 1–43; 45–60.

Pakes, A. G.

(1971 a) Some limit theorems for the total progeny of a branching process. Adv. Appl. Probability **3**, 176–192.

(1971 b) Branching processes with immigration. J. Appl. Probability **8**, 32–42.

Papangelou, F.

(1967) Strong ratio limits, R-recurrence and mixing properties of discrete parameter Markov processes. Z. Wahrscheinlichkeitstheorie **8**, 259–297.

(1968) A lemma on the Galton-Watson process and some of its consequences. Proc. Amer. Math. Soc. **19**, 1469–1479.

PENROSE, R.
(1955) A generalized inverse for matrices. Proc. Cambridge Philos. Soc. **51**, 406–413.

PHATARFOD, R. M.
(1964) Correlated random walk. J. Indian Statist. Assoc. **2**, 229–233.

PORT, S. C.
(1963) An elementary probability approach to fluctuation theory. J. Math. Anal. Appl. **6**, 109–151.
(1967) Limit theorems for transient Markov chains. J. Combinatorial Theory **2**, 107–128.

PRABHU, N. U.
(1965 a) Queues and inventories. New York: Wiley.
(1965 b) Stochastic processes. New York–London: Macmillan.

PRUITT, W. E.
(1964) Eigenvalues of non-negative matrices. Ann. Math. Statist. **35**, 1797–1800.
(1965) Strong ratio limit property for R-recurrent Markov chains. Proc. Amer. Math. Soc. **16**, 196–200.

PULLMAN, N.
(1965) The geometry of finite Markov chains. Canad. Math. Bull. **8**, 345 – 358.

QUINE, M. P.
(1970) The multitype Galton-Watson process with immigration. J. Appl. Probability **7**, 411–422.

RAUDELIUNAS, A.
(1961) Limit theorems for sums of random vectors forming a nonhomogeneous Markov chain. Litovsk. Mat. Sb. **1**, 203–230. (Russian)

RIORDAN, J.
(1962) Stochastic service systems. New York: Wiley.

ROMANOVSKI, V. I.
(1949) Discrete Markov chains. Moscow–Leningrad: Gostekhizdat. (Russian)

ROSENBLATT, M.
(1964 a) Equicontinuous Markov operators. Teor. Verojatn. i Primenen. **9**, 205 – 222.
(1964 b) Almost periodic transition operators acting on the continuous functions on a compact space. J. Math. Mech. **13**, 837–847.
(1967) Transition probability operators. Proc. 5th. Berkeley Symp. Math. Statist. Prob. Vol. II, Part 2, pp. 473–483. Berkeley: Univ. California Press.

ROZANOV, JU. A.
(1967) Stationary random processes. San Francisco–London–Amsterdam: Holden-Day.

RUBIN, H., and D. VERE-JONES
(1968) Domains of attraction for the subcritical Galton-Watson branching process. J. Appl. Probability **5**, 216–219.

RUNNEBURG, J. T.
(1964) Random walk and the triangle of Pascal. Math. Centrum Amsterdam Afd. Zuivere Wisk. ZW–004,10 pp. (Dutch)

SAATY, T. L.
(1961) Elements of queueing theory with applications. New York: Mc Graw-Hill
(1966) Seven more years of queues. A lament and a bibliography. Naval Res. Logist. Quart. **13**, 447–476.

SARYMSAKOV, T. A.
(1954) Elements of the theory of Markov processes. Moscow: Gostekhizdat. (Russian)

SELIVANOV, B. I.
(1969) Some explicit formulas for Galton-Watson chains. Teor. Verojatn. i Primenen. **14**, 348–354. (Russian)

SENETA, E.
(1966) Quasi-stationary distributions and time-reversion in genetics. (With discussion). J. Roy. Statist. Soc. Ser. B **28**, 253–277.
(1967 a) The Galton-Watson process with mean one. J. Appl. Probability, **4**, 489–495.
(1967 b) On the transient behaviour of a Poisson branching process. J. Austral. Math. Soc. **7**, 465–480.
(1967–68) Finite approximations to infinite non-negative matrices I, II. Proc. Cambridge Philos. Soc. **63**, 983–992; **64**, 465–470.
(1968 a) The stationary distribution of a branching process allowing immigration; A remark on the critical case. J. Roy. Statist. Soc. Ser. B **30**, 176–179.
(1968 b) On recent theorems concerning the supercritical Galton-Watson process. Ann. Math. Statist. **39**, 2098–2102.
(1969 a) Functional equations and the Galton-Watson process. Advances in Applied Probability **1**, 1–42.
(1969 b) Some second order properties of the Galton-Watson extinction time distribution. Sankhyā Ser. A, **31**, 75–78.
(1970 a) An explicit limit theorem for the critical Galton-Watson process with immigration. J. Roy. Statist. Soc. Ser. B, **32**, 149–152.
(1970 b) On the supercritical Galton-Watson process with immigration. Math. Biosci. **7**, 9–14.
(1971) On invariant measures for simple branching processes. J. Appl. Probability **8**, 43–51.

SENETA, E., and D. VERE-JONES
(1966) On quasi-stationary distributions in discrete-time Markov chains with a denumerable infinity of states. J. Appl. Probability **3**, 403–434.
(1968) On the asymptotic behaviour of subcritical branching processes with continuous state space. Z. Wahrscheinlichkeitstheorie **10**, 212–225.

SEVAST'JANOV, B. A.
(1951) The theory of branching random processes. Uspehi Mat. Nauk **6**, no. 6 (46), 47–99. (Russian)

SHENTON, L. R.
(1955) A semi-infinite random walk with discrete steps. Proc. Cambridge Philos. Soc. **51**, 442 – 448

ŠIDÁK, Z.
(1967) Classification of Markov chains with a general state space. Trans. 4th Prague Conf. Information Theory, Statist. Decision Functions, Random Processes, pp. 547–571. Prague: Academia.

SINGER, A.
(1965) A random walk in which absorption takes place at the second visit to a point. Bull. Soc. Math. Belg. **17**, 27–36.

SLACK, R. S.
(1968) A branching process with mean one and possibly infinite variance. Z. Wahrscheinlichkeitstheorie **9**, 139–145.

SMIRNOV, N. V., O. V. SARMANOV, and V. K. ZAHAROV
(1966) A local limit theorem for the number of transitions in a Markov chain and its applications. Dokl. Akad. Nauk SSSR **167**, 1238–1241. (Russian)

SMITH, W. L.
(1968) Necessary conditions for almost sure extinction of branching processes with random environment. Ann. Math. Statist. **39**, 2136–2140.

SMITH, W. L., and W. E. WILKINSON
(1969) On branching processes with random environments. Ann. Math. Statist. **40**, 814–827.

SPITZER, F.
(1964) Principles of random walk. Princeton–Toronto–New York–London: Van Nostrand.

STATULEVICIUS, V.
(1961) Local limit theorems and asymptotic expansions for nonhomogeneous Markov chains. Litovsk. Mat. Sb. **1**, 231–314. (Russian)
(1969–70) Limit theorems for sums of random variables related to a Markov chain. I–III. Litovsk. Mat. Sb. **9**, 345–362, 635–672; **10**, 161–169. (Russian)

STIGUM, B. P.
(1966) A theorem on the Galton-Watson process. Ann. Math. Statist. **37**, 695–698.

SYSKI, R.
(1960) Introduction to congestion theory in telephone systems. Edinburgh–London: Oliver & Boyd.
(1965) Markovian queues. (With discussion), pp. 170–222 in Proc. Symp. Congestion Theory. Chapel Hill: Univ. North Carolina Press.

TAKÁCS, L.
(1962) Introduction to the theory of queues. London–New York: Oxford Univ. Press.
(1964) A combinatorial method in the theory of Markov chains. J. Math. Anal. Appl. **9**, 153–161.
(1969) On the classical ruin problems. J. Amer. Statist. Assoc. **64**, 889–906.

UENO, T.
(1957) Some limit theorems for temporally discrete Markov processes. J. Fac. Sci. Univ. Tokyo **7**, 449–462, 557–565.

VEECH, W.
(1963) The necessity of Harris' condition for the existence of an invariant measure for transient Markov chains. Proc. Amer. Math. Soc. **14**, 856–860.

VERE-JONES, D.
(1962) Geometric ergodicity in denumerable Markov chains. Quart. J. Math. Oxford (2nd series) **13**, 7–28.

(1963) On the spectra of some linear operators associated with queueing systems. Z. Wahrscheinlichkeitstheorie **2**, 12–21.

(1964) A rate of convergence problem in the theory of queues. Teor. Verojatn. i Primenen. **9**, 104–112.

(1967–1968) Ergodic properties of non-negative matrices I, II. Pacific J. Math. **22**, 361–386; **26**, 601–620.

VERE-JONES, D., and D. G. KENDALL

(1959) A commutativity problem in the theory of Markov chains. Teor. Verojatn. i Primenen. **4**, 97–100.

VINCZE, I.

(1964) Über das Ehrenfestsche Modell der Wärmeübertragung. Arch. Math. **15**, 394–400.

WEESAKUL, B.

(1961) The random walk between a reflecting and an absorbing barrier. Ann. Math. Statist. **32**, 765–769.

WIELANDT, H.

(1950) Unzerlegbare nichtnegative Matrizen. Math. Z. **52**, 642–648.

WISHART, D. M. G.

(1956) A queueing system with χ^2 service-time distribution. Ann. Math. Statist. **27**, 768–779.

WOLFOWITZ, J.

(1963) Products of indecomposable, aperiodic, stochastic matrices. Proc. Amer. Math. Soc. **14**, 733–737.

YEH, R. Z.

(1970) A geometric proof of Markov ergodic theorem. Proc. Amer. Math. Soc. **26**, 335–340.

YOSIDA, K

(1965) Functional analysis. Berlin—Göttingen—Heidelberg: Springer.

ZAHAROV, V. K., and O. V. SARMANOV

(1968) The distribution law of the number of runs in a homogeneous Markov chain. Dokl. Akad. Nauk SSSR **179**, 526–528. (Russian)

CHAPTER 2

CONTINUOUS PARAMETER STOCHASTIC PROCESSES

Stochastic processes were defined in the introduction to Chapter 1 as families of random variables on the same probability space. In this chapter we shall study the case where the parameter range is a connex subset T of R (i.e. T is either a finite or an infinite interval). The parameter $t \in T$ may be thought of as time[1].

The first continuous parameter stochastic process was rigorously introduced in 1923 by N. WIENER who made precise heuristic ideas set forth by L. BACHELIER. This process is now called the Wiener process. It is interesting to note that WIENER constructed his process a decade before probabilists had made their subject respectable; he came into probability theory from analysis and made no concession to non-rigorous treatments. Since then, research in the theory of continuous parameter stochastic processes has developed along three main lines: a) the general theory including questions of existence, properties of trajectories, ergodic theorems, etc. (this line will be sketched in Section 2.1); b) the investigation of particular processes (or classes of processes) and the computation of various probability characteristics (this line will be developed in Sections 2.2 and 2.3); and c) linear problems including: linear transformations, spectral representation, linear prediction, and filtering. The last line of investigation deals with special problems in the theory of Hilbert spaces which will not be treated here.

2.1. Some general problems

This section is intended to give a very brief survey of basic notions and results in the general theory of stochastic processes. As a rule we will not give proofs but refer the reader to the appropriate sources.

[1] There are interesting classes of stochastic processes whose parameter does not run over a set of real numbers. (The interpretation of the parameter as time must then be dropped) Such a class is made up of the so-called stochastic point processes. See e.g. GOLDMAN (1967 a) and KARLIN (1966, Ch. 12).

2.1.1. Preliminaries

2.1.1.1. Definition of a stochastic process

2.1.1.1.1. Let $\mathcal{M} = \mathcal{M}[(\Omega, \mathcal{K}, \mathsf{P}), (X, \mathcal{X})]$ be the set of all random variables defined on the probability space $(\Omega, \mathcal{K}, \mathsf{P})$ with values in the measurable space $(X, \mathcal{X})$.

Definition 2.1.1 (N. WIENER). Any mapping $\xi = (\xi(t))_{t \in T}$ of T into $\mathcal{M}$ is said to be a *continuous parameter stochastic process* with state space $(X, \mathcal{X})$.

In what follows we shall simply refer to a process rather than to a continuous parameter stochastic process.

According to Definition 2.1.1, for any $t \in T$, $\xi(t)$ is an element of $\mathcal{M}$; we agree to denote by $\xi(t, \omega)$ the value of this element at $\omega \in \Omega$.

2.1.1.1.2. Let us consider the special case for which Ω is the Cartesian product X^T (in other words the set of all X-valued functions defined on T) and $\mathcal{K}$ equals $\mathcal{X}^T$, the smallest σ-algebra containing all the sets of the form

$$(\omega\colon \mathrm{pr}_u\, \omega \in A)^{2)}$$

for arbitrary $u \in T, A \in \mathcal{X}$. Define a process η by setting

$$\eta(t, \omega) = \mathrm{pr}_t\, \omega$$

and introducing a probability Q on $\mathcal{X}^T$. Thus in our present case the process reduces to a probability on the function space $(X^T, \mathcal{X}^T)$. Clearly, in order to obtain interesting processes, we must restrict our attention to probabilities on $(X^T, \mathcal{X}^T)$ supported by subsets of X^T into which it is possible to introduce a suitable topology.

This method of defining stochastic processes was first considered by J. L. DOOB. Notice that the general definition can be reduced to this special case. Indeed, let ξ be a process in the sense of Definition 2.1.1, and let us consider the mapping f of Ω in X^T defined by

$$f(\omega) = (\xi(t, \omega))_{t \in T}.$$

Define a probability Q on $\mathcal{X}^T$ by

$$\mathsf{Q}(B) = \mathsf{P}f^{-1}(B)), \quad B \in \mathcal{X}^T. \tag{2.1.1}$$

We have thus constructed a process η in the *Doob sense* such that

$$\eta(t, f(\omega)) = \mathrm{pr}_t f(\omega) = \xi(t, \omega), \tag{2.1.2}$$

and our assertion is justified.

---

2) ω belongs to X^T so that $\mathrm{pr}_u\, \omega$ is well defined.

2.1.1.2. Finite dimensional distributions

2.1.1.2.1. It is obvious that we can handle practically only the so-called *finite dimensional distributions* of the process. These are the multivariate distributions of the random variables $\xi(t_1), \ldots, \xi(t_n)$ for finite sets $t_1, \ldots, t_n$ of parameter values, i.e.

$$F_{t_1 \ldots t_n}(A_1, \ldots, A_n) = \mathsf{P}(\xi(t_1) \in A_1, \ldots, \xi(t_n) \in A_n),$$

with $A_1, \ldots, A_n \in \mathcal{X}$, $n \in N^*$.

It is easily seen from (2.1.2) that the finite dimensional distributions of the processes ξ and η coincide. This suggests

Definition 2.1.2. Two processes with the same state space and parameter set are said to be *equivalent* iff their finite dimensional distributions are identical.

Another somewhat more restrictive concept of equivalence is given in

Definition 2.1.3. Two processes $(\xi_1(t))_{t \in T}$ and $(\xi_2(t))_{t \in T}$ with the same state space and parameter set and defined on the same probability space are said to be *versions* of each other iff

$$\mathsf{P}(\xi_1(t) \neq \xi_2(t)) = 0$$

for all $t \in T$.

Clearly, if ξ_1 and ξ_2 are versions of each other, the corresponding countably infinite dimensional distributions are identical, that is

$$\mathsf{P}\{\bigcap_{i \in N^*}(\xi_1(t_i) \in A_i)\} = \mathsf{P}\{\bigcap_{i \in N^*}(\xi_2(t_i) \in A_i)\}$$

for $t_i \in T$, $A_i \in \mathcal{X}$, $i \in N^*$.

2.1.1.2.2. It is obvious that the finite dimensional distributions satisfy the following *Kolmogorov consistency conditions* [3]: whatever $t_1 < \ldots < t_n$, $A_i \in \mathcal{X}$, $1 \leqslant i \leqslant n$, $n \in N^*$, we have

$$F_{t_1 t_2 \ldots t_n}(X, A_2, \ldots, A_n) = F_{t_2 \ldots t_n}(A_2, \ldots, A_n),$$

[3] If T is an arbitrary set, then the consistency conditions are expressed as follows:

a) $$F_{t_1 \ldots t_{n-1} t_n}(A_1, \ldots, A_{n-1}, X) = F_{t_1 \ldots t_{n-1}}(A_1, \ldots, A_{n-1})$$

for all $t_i \in T$, $A_j \in \mathcal{X}$, $1 \leqslant i \leqslant n$, $1 \leqslant j < n$, $n \geqslant 2$.

b) for an arbitrary permutation π of $(1, \ldots, n)$ we have

$$F_{t_{\pi(1)} \ldots t_{\pi(n)}}(A_{\pi(1)}, \ldots, A_{\pi(n)}) = F_{t_1 \ldots t_n}(A_1, \ldots, A_n$$

for all $t_i \in T$, $A_i \in \mathcal{X}$, $1 \leqslant i \leqslant n$, $n \geqslant 2$.

$$F_{t_1 \ldots t_{n-1} t_n}(A_1, \ldots, A_{n-1}, X) = F_{t_1 \ldots t_{n-1}}(A_1, \ldots, A_{n-1}),$$
$$F_{t_1 \ldots t_{k-1} t_k t_{k+1} \ldots t_n}(A_1, \ldots, A_{k-1}, X, A_{k+1}, \ldots, A_n) =$$
$$= F_{t_1 \ldots t_{k-1} t_{k+1} \ldots t_n}(A_1, \ldots, A_{k-1}, A_{k+1}, \ldots, A_n), \ 1 < k < n.$$

Knowledge of the finite dimensional distributions is sufficient to determine the process up to an equivalence. We have, namely

Theorem 2.1.4. (P. J. Daniell, A. N. Kolmogorov). *Let X be a σ-compact topological space and $\mathcal{X}$ the σ-algebra of Borel sets in X. For any given family of finite dimensional distributions satisfying the consistency conditions, there exists a unique process in the Doob sense whose finite dimensional distributions coincide with the given ones.*

Proof. See Dynkin (1961, Ch. 1, § 5). ◇

2.1.1.2.3. A different method of specifying a process is to use the infinite-dimensional analogue of the concept of a characteristic function, first introduced by A. N. Kolmogorov in 1935. For details we refer the reader to Prohorov (1961) and Skorohod (1965 b).

2.1.2. Basic concepts

2.1.2.1. Separability

2.1.2.1.1. It follows from the properties of measurable functions that we cannot determine on the measurability of sets defined by means of a non-countable set of parameter values. For example, if X is a topological space, we cannot say whether or not the set

$$(\omega : \xi(\cdot, \omega) \text{ is continuous on } T)$$

is measurable (i.e. belong to $\mathcal{K}$).

The first satisfactory removal of this difficulty is due to J. L. Doob who introduced the notion of separability.

In this subsection we suppose that X is a metric space with metric ρ and that $\mathcal{X}$ is the σ-algebra of Borel sets in X.

Definition 2.1.5. A process $(\xi(t))_{t \in T}$ with state space $(X, \mathcal{X})$ is said to be *separable* iff there exist a countable set $(s_j)_{j \in N^*}$ everywhere dense in T and a set $\Lambda \in \mathcal{K}$ of probability zero such that whatever the open interval $\mathcal{J} \subset T$ and the closed set $F \subset X$ the difference of the sets

$$(\omega : \xi(s_j, \omega) \in F, \ s_j \in \mathcal{J})$$

and

$$(\omega : \xi(t, \omega) \in F, \ t \in \mathcal{J})$$

is contained in Λ. The set $(s_j)_{j \in N^*}$ is said to be a *separability set* of the process.

2.1.2.1.2. The utility of the concept of separability follows from

Proposition 2.1.6. *Let $X = R$. The following properties are equivalent to the separability of the process $(\xi(t))_{t \in T}$, the separability set being $(s_j)_{j \in N^*}$: for every open interval $\mathcal{J} \subset T$:*

i) $\inf_{s_j \in \mathcal{J}} \xi(s_j) = \inf_{t \in \mathcal{J}} \xi(t), \quad \sup_{s_j \in \mathcal{J}} \xi(s_j) = \sup_{t \in \mathcal{J}} \xi(t);$

ii) $\inf_{s_j \in \mathcal{J}} \xi(s_j) \leqslant \inf_{t \in \mathcal{J}} \xi(t), \quad \sup_{t \in \mathcal{J}} \xi(t) \leqslant \sup_{s_j \in \mathcal{J}} \xi(s_j);$

iii) $\inf_{s_j \in \mathcal{J}} \xi(s_j) \leqslant \xi(t) \leqslant \sup_{s_j \in \mathcal{J}} \xi(s_j), \ t \in \mathcal{J};$

for every $t \in \overline{T}$

iv) $\underline{\lim}_{s_j \to t} \xi(s_j) = \underline{\lim}_{t' \to t} \xi(t'), \quad \overline{\lim}_{t' \to t} \xi(t') = \overline{\lim}_{s_j \to t} \xi(s_j);$

v) $\underline{\lim}_{s_j \to t} \xi(s_j) \leqslant \underline{\lim}_{t' \to t} \xi(t'), \quad \overline{\lim}_{t' \to t} \xi(t') \leqslant \overline{\lim}_{s_j \to t} \xi(s_j);$

vi) $\underline{\lim}_{s_j \to t} \xi(s_j) \leqslant \xi(t) \leqslant \overline{\lim}_{s_j \to t} \xi(s_j), \ t \in T.$

Proof. See Loève (1963, p. 505). ◇

If $X = R^n$, then the properties from Proposition 2.1.6. hold for each of the components of a separable process.

Notice that if $X = R$, then on account of Proposition 2.1.6 for a separable process ξ the sets

$$(\omega : \xi(\cdot, \omega) \text{ is continuous on } T)$$

$$(\omega : \xi(\cdot, \omega) \text{ is monotone on } T)$$

$$(\omega : \xi(\cdot, \omega) \text{ is bounded on } T)$$

are all events (i.e. belong to $\mathcal{K}$).

2.1.2.1.3. Separability is not a serious restriction as it is shown by

Theorem 2.1.7 (J. L. Doob). *Let X be a compact metric space. Given a process ξ with state space $(X, \mathcal{X})$ there exists a separable version of ξ. Further, if X is a separated locally compact but not compact metric space, then, given a process ξ with state space $(X, \mathcal{X})$, there exists a separable version of ξ taking values in the one-point compactification of X.*

Proof. See Gihman and Skorohod (1965, p. 201). ◇

For more general results concerning separability of a process, the reader is referred to Borges (1966) and Meyer (1966, p. 92).

Note that Theorem 2.1.7 implies we need study only separable processes.

2.1.2.2. Stochastic continuity and measurability

2.1.2.2.1. In many cases it is useful to determine the separability set of a process more precisely. This can be done for the so-called stochastically continuous processes.

Definition 2.1.8. A process $(\xi(t))_{t\in T}$ is said to be *stochastically continuous at* $t_0 \in T$ iff for any $\varepsilon > 0$

$$\mathsf{P}(\rho(\xi(t), \xi(t_0)) > \varepsilon) \to 0$$

as $t \to t_0$. A process is said to be *stochastically continuous* iff it is stochastically continuous at any point of T.

We have

Theorem 2.1.9. *Let ξ be a separable and stochastically continuous process. Any countable, everywhere dense set in T is a separability set of ξ.*

Proof. See Gihman and Skorohod (1965, p. 207). ◇

2.1.2.2.2. For some purposes it is necessary to suppose that $\xi(\cdot, \cdot)$ is a measurable function of the pair of variables (t, ω). Let $\mathcal{L}_T$ be the σ-algebra of Lebesgue sets in T.

Definition 2.1.10. A process ξ is said to be *measurable* iff $\xi(\cdot, \cdot)$ is a $\mathcal{K}\times\mathcal{L}_T$-measurable function.

Stochastic continuity is sufficient to ensure the existence of a measurable and separable version of a given process. More precisely, we have

Theorem 2.1.11. *Let X be either a compact metric space or a separated locally compact but not compact metric space. Suppose that, given a process ξ with state space $(X, \mathcal{X})$, there is a set $T_1 \subset T$ of Lebesgue measure zero such that ξ is stochastically continuous on $T - T_1$. Under these conditions, there exists a measurable and separable version of ξ taking values either in X or in the one-point compactification of X.*

Proof. See Gihman and Skorohod (1965, p. 209). ◇

2.1.3. Trajectories

2.1.3.1. Generalities

2.1.3.1.1. A function $\xi(\cdot, \omega)$ on T obtained by fixing $\omega \in \Omega$ is said to be a *trajectory* [4)] of the process. The study of trajectories is very important because they shed light on the dynamics of the considered process.

In this subsection we suppose that $T = [a, b]$, X is a complete metric space and $\mathcal{X}$ is the σ-algebra of Borel sets in X.

2.1.3.1.2. We are interested in those cases for which the trajectories are either continuous or have only discontinuities of the first kind.

[4)] The nomenclatures *sample function* and *path* are also used.

Definition 2.1.12. A process is said to be *continuous* iff almost all[5] its trajectories are continuous.

Clearly, continuity implies stochastic continuity but the converse is not true in general.

Definition 2.1.13. A process is said *to have no discontinuities of the second kind* iff almost all its trajectories have only discontinuities of the first kind.

In particular, a trajectory has no discontinuities of the second kind if it is a *step function*, that is, a continuous function in the discrete topology in X except for a countable set of parameter values having no limit points in T.

2.1.3.1.3. It is important to distinguish carefully between an assertion concerning $\xi(t, \omega)$ for almost all $\omega \in \Omega$ at a fixed $t \in T$ and an assertion about $\xi(\cdot, \omega)$ for almost all $\omega \in \Omega$. In the first case the exceptional null set may depend on t — so that the second type of assertion may be false.

2.1.3.2. Continuous trajectories

2.1.3.2.1. A number of sufficient conditions for the continuity of trajectories are available.

Set

$$\alpha(\varepsilon, \delta) = \operatorname*{ess\,inf}_{\omega \in \Omega} \sup_{a \leqslant s \leqslant t \leqslant s+\delta \leqslant b} \mathsf{P}[\rho(\xi(s), \xi(t)) \geqslant \varepsilon \mid \mathcal{K}_s],$$

where $\mathcal{K}_s$ is the σ-algebra generated by the random variables $\xi(u)$, $u \leqslant s$.

Theorem 2.1.14 (E. B. Dynkin, J. R. Kinney). *Let ξ be a separable process. If for any $\varepsilon > 0$*

$$\lim_{\delta \to 0} \frac{\alpha(\varepsilon, \delta)}{\delta} = 0,$$

then ξ is continuous.

Proof. See Gihman and Skorohod (1965, p. 225).

2.1.3.2.2. We note also

Theorem 2.1.15 (A. N. Kolmogorov). *Let ξ be a separable process. If there are positive constants C, r, and β such that for any $\varepsilon > 0$, $t_1, t_2 \in [a, b]$*

$$\mathsf{P}[\rho(\xi(t_1), \xi(t_2)) > \varepsilon] \leqslant \frac{C|t_2 - t_1|^{1+r}}{\varepsilon^\beta}$$

then ξ is continuous.

Proof. See Gihman and Skorohod (1965, p. 223). ◇

[5] "Almost all" means "except for a set of points ω of probability zero."

2.1.3.3. Trajectories without discontinuities of the second kind

2.1.3.3.1. We have

Theorem 2.1.16 (J. R. KINNEY). *Let ξ be a separable process. If for any $\varepsilon > 0$*

$$\lim_{\delta \to 0} \alpha(\varepsilon, \delta) = 0,$$

then ξ has no discontinuities of the second kind.

Proof. See GIHMAN and SKOROHOD (1965, p. 221). ◇

Further sufficient conditions for a process to have no discontinuities of the second kind are to be found in ČENCOV (1956) and CRAMÉR (1966).

2.1.3.3.2. The structure of the trajectories of a separable stochastically continuous process having no discontinuities of the second kind can be made more precise as follows.

Proposition 2.1.17. *Let ξ be a separable stochastically continuous process having no discontinuities of the second kind. There exists a version $\tilde{\xi}$ of ξ almost all trajectories of which are right-continuous.*

Proof. See GIHMAN and SKOROHOD (1965, p. 222). ◇

2.1.4. Convergence of stochastic processes

2.1.4.1. Weak convergence of processes

2.1.4.1.1. We shall now briefly describe a concept of convergence for stochastic processes. Its importance lies in the fact that we can represent complicated processes as limits of more simple ones. The fundamental paper in this area is due to PROHOROV (1956).

In what follows we suppose that all processes we are concerned with have the same state space $(X, \mathscr{X})$ and parameter set T.

Definition 2.1.18. A sequence $(\xi_n)_{n \in N^*}$ of processes is said to be *weakly convergent* to a process ξ if the finite dimensional distributions of ξ_n converge as $n \to \infty$ to the corresponding ones of ξ.

Clearly, for the above definition to make sense, X must be endowed with a topology.

2.1.4.1.2. From now on we suppose that the trajectories of ξ_n, $n \in N^*$, and ξ belong to the same separable metric space $Y \subset X^T$ and that the Borel sets in Y belong to $\mathscr{X}^T$.

We note that Definition 2.1.18 has the following interpretation. Consider the probabilities Q_n and Q on $\mathscr{X}^T$ associated with ξ_n and ξ by means of (2.1.1)[6]. The weak convergence of the sequence $(\xi_n)_{n \in N^*}$ to ξ

[6] Actually Q_n and Q can be considered as defined only on $\mathscr{X}^T \cap Y$. If $B \cap Y = \emptyset$, $B \in \mathscr{X}^T$, then obviously $Q_n(B) = Q(B) = 0$.

is equivalent to the convergence of the sequence $(Q_n(C))_{n \in N^*}$ to $Q(C)$ for all cylinder sets $C \in \mathcal{X}^T$ which are continuity sets for Q[7].

2.1.4.2. The Prohorov theorem

The problem of determining the distributions of various functionals defined on the trajectories of a given process is particularly important. Examples of such functionals are (let $X = R$)

$$f(\xi(\cdot)) = \inf_{t \in T} \xi(t) \text{ or } f(\xi(\cdot)) = \sup_{t \in T} \xi(t).$$

A natural problem is the following: if the sequence $(\xi_n)_{n \in N^*}$ converges weakly to ξ, under which conditions does $f(\xi_n(\cdot))$ converge in distribution to $f(\xi(\cdot))$? A solution to this problem is given by

Theorem 2.1.19 (Ju.V. Prohorov). *Suppose $(\xi_n)_{n \in N^*}$ converges weakly to ξ and that for any $\varepsilon > 0$ there is a compact set $K_\varepsilon \subset Y$ such that*

$$\sup_{n \in N^*} Q_n(Y - K_\varepsilon) < \varepsilon.$$

Under these conditions, $f(\xi_n(\cdot))$ converges in distribution to $f(\xi(\cdot))$ for arbitrary $\mathcal{X}^T \cap Y$-measurable and Q—a.s. *continuous functionals f on Y.*

Proof. See e.g. Gihman and Skorohod (1965, Ch. 9, § 1). ◇

2.2. Processes with independent increments

Processes with independent increments are the continuous parameter analogue of chains with independent increments (see the introduction to Chapter 1). While the study of the latter leads to that of partial sums of sequences of independent random variables, the former have no analogue in classical probability theory. The systematic study of processes with independent increments was initiated by B. de Finetti and developed further by A. Ja. Hinčin, K. Itô, P. Lévy.

2.2.1. Preliminaries

2.2.1.1. Definition and existence theorem

2.2.1.1.1. In what follows, we shall suppose that $T = [0, a)$ with either finite or infinite a and that the state space X is R, although almost

[7] A set $A \in \mathcal{X}^T$ is said to be a continuity set for Q if $Q(\mathring{A}) = Q(\bar{A})$ where $\mathring{A}$ and $\bar{A}$ are, respectively, the interior and the closure of A.

all results can be extended without difficulty to the case $X = R^n$, $n > 1$, [see SKOROHOD (1964)] or, more generally, to the case where X is a topological group [see CUCULESCU (1968, Ch. 11)].

Definition 2.2.1. A process $(\xi(t))_{t\in T}$ is said to *have independent increments*[8] if $\xi(0) = 0$ and for all sequences $0 \leqslant t_0 < t_1 < \ldots < t_n < a$ the random variables $\xi(t_0), \xi(t_1) - \xi(t_0), \ldots, \xi(t_n) - \xi(t_{n-1})$ are independent.

Let F_{st}, $s < t$, be the distribution function of the random variable $\xi(t) - \xi(s)$.

If all distribution functions F_{st}, $s < t$, are of a specific type, say normal, Poisson, stable, infinitely divisible, we shall say that the process itself is of this type. Moreover, if F_{st} depends only on the difference $t-s$, the process is said to be *homogeneous*.

Since for $s < u < t$, the variable $\xi(t) - \xi(s)$ is the sum of the independent random variables $\xi(u) - \xi(s)$ and $\xi(t) - \xi(s)$, we must have

$$F_{st}(x) = \int_R F_{su}(x - y)\, \mathrm{d}F_{ut}(y). \tag{2.2.1}$$

If the process is stochastically continuous, then this condition together with (2.2.1) shows that F_{st} is infinitely divisible. According to the Hinčin-Lévy formula [see e.g. DOOB (1953, p. 130)], the logarithm of the characteristic function φ_{st} of F_{st} admits the representation

$$\log \varphi_{st}(\theta) = \mathrm{i}\theta\,[m(t) - m(s)] + \int_R \left(\mathrm{e}^{\mathrm{i}\theta x} - 1 - \frac{\mathrm{i}\theta x}{1+x^2}\right)\frac{1+x^2}{x^2}\,\mathrm{d}_x(G(t, x) - G(s, x)), \tag{2.2.2}$$

where m is a continuous function and G is a continuous function with respect to the first variable and a bounded and monotone function with respect to the second; moreover, for $s < t$, $G(t, .) - G(s, .)$ is also monotone.

In case the process is homogeneous, F_{st} is obviously infinitely divisible and in (2.2.2) $m(t) - m(s)$ and $G(t, \cdot) - G(s, \cdot)$ depend linearly on $t - s$.

2.2.1.1.2. The existence of processes with independent increments follows from

[8] The nomenclatures *differential process, additive process, decomposable process*, and *integral with independent random elements* have also been used.

Theorem 2.2.2. *Given a family* $(F_{st})_{s<t\in T}$ *of distribution functions satisfying* (2.2.1), *there exists a process* ξ *with independent increments such that the distribution function of the increment* $\xi(t)-\xi(s)$ *is* F_{st}, $s<t\in T$.

Proof. Assuming we have already demonstrated the existence of the process, we note that the characteristic functions of its finite dimensional distributions are uniquely determined by the characteristic functions φ_{st}. Indeed, for $0\leqslant t_1<\ldots<t_n<a$,

$$\varphi_{t_1\ldots t_n}(\theta_1,\ldots,\theta_n)=\mathsf{E}\exp\left(i\sum_{j=1}^{n}\theta_j\,\xi(t_j)\right)=$$

$$=\mathsf{E}\exp i\left(\left(\sum_{j=1}^{n}\theta_j\right)\xi(t_1)+\left(\sum_{\kappa=2}^{n}\theta_k\right)(\xi(t_2)-\xi(t_1))+\cdots\right.$$

$$\left.+\theta_n(\xi(t_n)-\xi(t_{n-1}))\right)=\varphi_{0t_1}\left(\sum_{j=1}^{n}\theta_j\right)\varphi_{t_1t_2}\left(\sum_{k=2}^{n}\theta_k\right)\cdots\varphi_{t_{n-1}t_n}(\theta_n).$$

Thus the proof results from the following considerations: the characteristic functions $\varphi_{t_1\ldots t_n}$ uniquely determine a consistent family of finite dimensional distributions. Indeed, the consistence follows from the fact that (2.2.1) implies $\varphi_{st}=\varphi_{su}\,\varphi_{ut}$ whence

$$\varphi_{t_1\ldots t_n}(0,\theta_2,\ldots,\theta_n)=\varphi_{t_2\ldots t_n}(\theta_2,\ldots,\theta_n),\text{ etc.}$$

To complete the proof apply Theorem 2.1.4. ◇

2.2.1.1.3. Anticipating some of the initial ideas of the next section, we emphasize the Markovian character of processes with independent increments. In this respect, the analogy with the discrete parameter case is complete.

Proposition 2.2.3. *Every process* ξ *with independent increments has the Markov property: for all* $x\in R$, $0\leqslant t_0<\ldots<t_n<a$ *we have*

$$\mathsf{P}(\xi(t_n)<x\mid\xi(t_k),0\leqslant k<n)=\mathsf{P}(\xi(t_n)<x\mid\xi(t_{n-1}))=$$

$$=F_{t_{n-1}t_n}(x-\xi(t_{n-1})).$$

Proof. Note that the σ-algebra generated by the random variables $\xi(t_k)$, $0\leqslant k<n$, coincides with that generated by the random variables $\xi(t_0)$, $\xi(t_1)-\xi(t_0)$, $\ldots$, $\xi(t_{n-1})-\xi(t_{n-2})$. Let Λ be a set belonging to

this σ-algebra. Starting from the above remark, one proves immediately the equality

$$\int_\Lambda F_{t_{n-1} t_n}(x - \xi(t_{n-1}, \omega))\, \mathsf{P}(d\omega) = \mathsf{P}(\Lambda \cap (\xi(t_n) < x))$$

which is equivalent to those stated. ◇

2.2.1.2. Stochastic continuity

2.2.1.2.1. The stochastic continuity of processes with independent increments can be characterized in terms of the characteristic functions φ_{0t}, $t \in T$.

Proposition 2.2.4. *A process with independent increments is stochastically continuous at t_0 iff φ_{0t} is continuous with respect to t at t_0.*

Proof. The "if" implication is a consequence of the very definition of stochastic continuity. Thus we need only prove the "only if" part. Let $t_n \to t_0+$. Then $\varphi_{0t_n}(\theta) = \varphi_{0t_0}(\theta)\, \varphi_{t_0 t_n}(\theta)$ and since $\varphi_{0t_n}(\theta) \to \varphi_{0t_0}(\theta)$, we have $\varphi_{t_0 t_n}(\theta) \to 1$ for all θ such that $\varphi_{0t_0}(\theta) \neq 0$. Choose $\delta > 0$ so that $\varphi_{0t_0}(\theta) \neq 0$ for $|\theta| < \delta$. It follows that $\int_R \cos \theta x \, dF_{t_0 t_n}(x) \to 1$, whence

$$\frac{1}{\delta} \int_0^\delta d\theta \int_R \cos \theta x \, dF_{t_0 t_n}(x) \to 1,$$

or

$$\int_R \frac{\sin \delta x}{\delta x} \, dF_{t_0 t_n}(x) \to 1.$$

Since

$$\sup_{|x| > \varepsilon} \frac{\sin \delta x}{\delta x} < 1,$$

we have

$$\mathsf{P}(|\xi(t_n) - \xi(t_0)| > \varepsilon) \to 0.$$

The case $t_n \to t-$ can be treated in an analogous manner. ◇

2.2.1.2.2. Stochastic continuity for processes with independent increments ensures some regularity of their trajectories. We have

Proposition 2.2.5. *Every separable and stochastically continuous process with independent increments has no discontinuities of the second kind.*

Proof. This result follows from Theorem 2.1.16 and the very definition of processes with independent increments. ◇

From now on, by virtue of Proposition 2.1.17, we can and do assume that the trajectories of separable and stochastically continuous processes with independent increments are right-continuous.

2.2.2. Basic processes with independent increments

Historically, the first processes with independent increments considered were the Poisson and the Wiener processes, both born from physical phenomena such as radioactive disintegrations or the motion of gas molecules. In a sense, every separable and stochastically continuous process with independent increments is the "sum" of a Wiener process and a family of Poisson processes (see Theorem 2.2.17).

2.2.2.1. The Poisson process

2.2.2.1.1. We begin with a formal definition.

Definition 2.2.6. A process ξ with independent increments is said to be a *Poisson process* $P(\lambda(\cdot))$ if the increments $\xi(t) - \xi(s)$ are distributed according to the Poisson law with parameter $\lambda(t) - \lambda(s)$, where $\lambda(\cdot)$ is a non-decreasing real-valued function defined on T. In case $\lambda(t) = \lambda t$, $\lambda > 0$, the process ξ is said to be a *homogeneous* Poisson process $P(\lambda)$.

Thus, for the process $P(\lambda(\cdot))$ we have

$$\mathsf{P}(\xi(t) - \xi(s) = k) = \frac{1}{k!}[\lambda(t) - \lambda(s)]^k \exp[\lambda(s) - \lambda(t)], \quad k \in N,$$

$$\mathsf{E}(\xi(t) - \xi(s)) = \mathsf{D}(\xi(t) - \xi(s)) = \lambda(t) - \lambda(s),$$

$$\varphi_{st}(\theta) = \exp\{[\lambda(t) - \lambda(s)](e^{i\theta} - 1)\}.$$

On account of Proposition 2.2.4, the process $P(\lambda(\cdot))$ is stochastically continuous iff $\lambda(\cdot)$ is a continuous function on T.

2.2.2.1.2. Stochastically continuous Poisson processes can be characterized by a property of their trajectories.

Theorem 2.2.7. *If the process $P(\lambda(\cdot))$ is separable and stochastically continuous and $\lim_{t\to a} \lambda(t) = \infty$, then with probability 1 all its trajectories are monotone non-decreasing step functions whose jumps equal 1.*

Proof. By definition $\mathsf{P}(\xi(t) - \xi(s) \geqslant 0) = 1$ for fixed s and t, $s < t$. It then follows that, with probability 1, $\xi(\cdot)$ is a N-valued non-decreasing function on every countable set $(t_j)_{j \in N} \subset T$. If we choose as $(t_j)_{j \in N}$ a separability set of the process, then by Proposition 2.1.6, with probability 1, every trajectory takes on the same upper and lower bounds on every open interval as it does on the t_j in the interval. Hence, with probability 1, all trajectories are monotone non-decreasing step functions with jumps of integral magnitude. We shall now show that the probability of jumps greater than 1 is zero. It is obviously sufficient to prove this for any given finite interval, say $(0, t)$. Note that if a trajectory has a jump of magnitude greater than 1 in $(0, t)$, then

$$\max_{1 \leqslant k \leqslant n} \left[\xi\left(\frac{kt}{n}\right) - \xi\left(\frac{(k-1)t}{n}\right)\right] \geqslant 2$$

for every $n \in N^*$. But

$$\mathsf{P}\left(\max_{1 \leqslant k \leqslant n} \left[\xi\left(\frac{kt}{n}\right) - \xi\left(\frac{(k-1)t}{n}\right)\right] \geqslant 2\right) \leqslant$$

$$\leqslant \sum_{k=1}^{n} \mathsf{P}\left(\left[\xi\left(\frac{kt}{n}\right) - \xi\left(\frac{(k-1)t}{n}\right)\right] \geqslant 2\right) \leqslant$$

$$\leqslant \max_{1 \leqslant k \leqslant n} \left(\lambda\left(\frac{kt}{n}\right) - \lambda\left(\frac{(k-1)t}{n}\right)\right) e^{\lambda(t) - \lambda(0)} \to 0 \quad (n \to \infty).$$

It remains to note that the trajectories of the process $P(\lambda(\cdot))$ are not continuous functions since the probability of continuity throughout an interval (s, t) is $\exp[\lambda(s) - \lambda(t)]$ which goes to 0 as $t \to a$ [9]. ◇

In some applications it is convenient to describe each jump of a trajectory as an "event". Thus, on account of Theorem 2.2.7, the number of events occurring within the interval (s, t) is just $\xi(t) - \xi(s)$ with probability 1. Furthermore, the expected number of these events is $\lambda(t) - \lambda(s)$. In particular, for the process $P(\lambda)$, λ is the (average) rate of occurrence of the events. If events occur according to a Poisson process as described here, they are sometimes referred to as "purely random". It is possible to describe them in a manner independent of the nomenclature of stochastic processes. For details and applications see e.g. COX and LEWIS (1966).

2.2.2.1.3. We now prove the converse to Theorem 2.2.7.

---

[9] Nevertheless there is continuity at each $t \in T$ with probability 1 since $\mathsf{P}(\xi(t + \varepsilon) - \xi(t - \varepsilon) > 0) = 1 - \exp[\lambda(t - \varepsilon) - \lambda(t + \varepsilon)] \to 0$ as $\varepsilon \to 0$.

Theorem 2.2.8. *Let ξ be a stochastically continuous process with independent increments. If with probability 1 its trajectories are monotone non-decreasing step functions whose jumps equal 1, then ξ is a separable Poisson process.*

Proof. [ITÔ (1960), p. 59]. Separability follows from the properties of the trajectories on account of Proposition 2.1.6. It remains to show that the distribution of $\xi(t) - \xi(s)$, $s < t$, is Poisson.

It follows from stochastic continuity that for all $u \in [s, t]$ and $\varepsilon > 0$ there exists $\delta = \delta(u, \varepsilon)$ such that

$$\mathsf{P}(|\xi(v) - \xi(u)| > \varepsilon) < \varepsilon \quad \text{for} \quad |v - u| < \delta.$$

The Heine-Borel theorem assures us that δ can be chosen independently of $u \in [s, t]$. Let us divide the closed interval $[s, t]$ into n equal subintervals and denote by $y_{n1}, \ldots, y_{nn}$. the increments of ξ on them. We have

$$y = \xi(t) - \xi(s) = y_{n1} + \ldots + y_{nn}.$$

Set

$$y'_{nk} = \begin{cases} y_{nk} & \text{if} \quad y_{nk} \leqslant 1 \\ 0 & \text{if} \quad y_{nk} \geqslant 2 \end{cases}, \quad 1 \leqslant k \leqslant n,$$

and

$$y'_n = y'_{n1} + \ldots + y'_{nn}.$$

It follows from the hypotheses made that

$$\mathsf{P}(\lim_{n\to\infty} y'_n = y) = 1.$$

If we set $p_{nk} = \mathsf{P}(y'_{nk} = 1)$, then $p_{nk} = \mathsf{P}(y_{nk} = 1)$, therefore $p_{nk} \to 0$ $(n \to \infty)$ uniformly with respect to k. Notice that $y'_{n1}, \ldots, y'_{nn}$ are independent since $y_{n1}, \ldots, y_{nn}$ are. Proceeding further, we distinguish three cases:

a) $\sum_{k=1}^{n} p_{nk} \to \lambda(s, t) < \infty \quad (n \to \infty)$.

In this case we have

$$\mathsf{E} \exp i\theta\, y = \lim_{n\to\infty} \mathsf{E} \exp i\theta\, y'_n = \lim_{n\to\infty} \prod_{k=1}^{n} \mathsf{E} \exp i\theta\, y'_{nk} =$$

$$= \lim_{n\to\infty} \prod_{k=1}^{n} (1 - p_{nk} + e^{i\theta} p_{nk}) = \lim_{n\to\infty} \prod_{k=1}^{n} (1 + p_{nk}(e^{i\theta} - 1)).$$

The relations

$$\lim_{n\to\infty} \sum_{k=1}^{n} p_{nk} = \lambda(s, t),$$

$$\sum_{k=1}^{n} p_{nk}^2 \leqslant \max_{1\leqslant k\leqslant n} p_{nk} \sum_{k=1}^{n} p_{nk} \to 0 \ (n \to \infty)$$

imply

$$\mathsf{E} \exp i\theta\, y = \exp\{\lambda\,(s, t)\,(e^{i\theta} - 1)\}.$$

It is easy to deduce that $\lambda\,(s, t) = \lambda\,(0, t) - \lambda\,(0, s)$. Thus ξ is the process $P\,(\lambda\,(0, .))$.

b) The sequence $\left(\sum_{k=1}^{n} p_{nk}\right)_{n\in N^*}$ has a finite limit point.
This case does not differ from the previous one.

c) $\sum_{k=1}^{n} p_{nk} \to \infty \ (n \to \infty)$.
It follows from the uniform convergence with respect to k of p_{nk} to zero as $n \to \infty$ that for every $\lambda > 0$ there exists a $K = K\,(n, \lambda)$ such that $\sum_{k=1}^{K} p_{nk} \to \lambda\,(n \to \infty)$. We conclude as in the case a) that

$$|\mathsf{E} \exp i\theta\, y| \leqslant \lim_{n\to\infty} \left| \prod_{k=1}^{K} \mathsf{E} \exp i\theta\, y'_{nk} \right| = |\exp\{\lambda(e^{i\theta} - 1)\}| = \exp\{\lambda(\cos\theta - 1)\}.$$

Since λ may be chosen arbitrarily large, we get $\mathsf{E} \exp i\theta y = 0$ for $0 < \theta < 2\pi$. Whence, letting $\theta \to 0$, we deduce $1 = 0$, therefore case c) is not possible. ◇

2.2.2.1.4. The reader will be able to verify that the last two theorems lead us to the following characterization of the process $P\,(\lambda)$.

Proposition 2.2.9. *The separable homogeneous Poisson process $P\,(\lambda)$ is the only stochastically continuous N-valued process with independent increments satisfying*

$$\mathsf{P}\,(\xi\,(t) - \xi\,(s) = 0) = 1 - \lambda\,(t - s) + o\,(t - s)$$

$$\mathsf{P}\,(\xi\,(t) - \xi\,(s) = 1) = \lambda\,(t - s) + o\,(t - s)$$

$$\mathsf{P}\,(\xi\,(t) - \xi\,(s) = k) = o\,(t - s),\ k \geqslant 2,\ t - s \to 0.$$

For another characterization of $P(\lambda)$ see DOOB (1953, pp. 401—402).

2.2.2.1.5. Let us consider the time ζ_n, $n \in N^*$, at which the nth jump occurs, that is,

$$\zeta_n = \inf(t: \xi(t) = n).$$

It follows that ζ_1 is the time spent in state 0, $\zeta_2 - \zeta_1$ the time spent in state 1, etc.

Theorem 2.2.10. *For the homogeneous Poisson process* $P(\lambda)$

i) *The conditional law of jumps in* $[s, t)$, *given that* n *occurred, is that of* n *independent random variables uniformly distributed in* $[s, t)$.

ii) *The sequence* $(\zeta_n)_{n\in N}$, $\zeta_0=0$, *is a chain with independent and identically distributed increments*:

$$\mathsf{P}(\zeta_n - \zeta_{n-1} > t) = e^{-\lambda t}, \; n \in N^*, \; t \geqslant 0.$$

Proof. The numbers of jumps $y = \xi(u) - \xi(s)$ and $z = \xi(t) - \xi(u)$ respectively in $[s, u)$ and in $[u, t)$ $(s < u < t)$ are independent Poisson random variables with parameters $\lambda(u-s)$ and $\lambda(t-u)$ and $y+z$ is also a Poisson random variable with parameter $\lambda(t-s)$. By elementary computations we obtain

$$\mathsf{P}(y = m \mid y + z = n) = \frac{\mathsf{P}(y = m, z = n - m)}{\mathsf{P}(y + z = n)} =$$

$$= \binom{n}{m}\left(\frac{u-s}{t-s}\right)^m \left(1 - \frac{u-s}{t-s}\right)^{n-m}, \quad m \leqslant n,$$

which proves i) by taking $m = n$. Furthermore, for $t > t_1 + t_2$, $t_1 = u_0 < u_1 < \ldots < u_m = t - t_2$, we have

$$\mathsf{P}(\zeta_1 > t_1, \zeta_2 - \zeta_1 > t_2 \mid \xi(t) = n) =$$

$$= \lim_{\substack{\max(u_{i+1}-u_i)\to 0\\ 0\leqslant i<m}} \sum_{i=0}^{m-1} \mathsf{P}\Big(\xi(u_i) = 0, \xi(u_{i+1}) - \xi(u_i) = 1,$$

$$\xi(u_{i+1} + t_2) - \xi(u_{i+1}) = 0, \; \xi(t) - \xi(u_{i+1} + t_2) = n - 1\Big) \Big/ \mathsf{P}(\xi(t)=n) =$$

$$= \lim_{\substack{\max(u_{i+1}-u_i)\to 0\\ 0\leqslant i<m}} \frac{n}{t} \sum_{i=0}^{m-1} \left(1 - \frac{t_2 + u_{i+1}}{t}\right)^{n-1} (u_{i+1} - u_i) =$$

$$= \frac{n}{t} \int_{t_1}^{t-t_2} \left(1 - \frac{u + t_2}{t}\right)^{n-1} du = \left(1 - \frac{t_1 + t_2}{t}\right)^n = \alpha^n,$$

and

$$\mathsf{P}(\zeta_1 > t_1, \zeta_2 - \zeta_1 > t_2) = e^{-\lambda t}\left(1 + \lambda t\alpha + \frac{(\lambda t\,\alpha)^2}{2!} + \cdots\right) = e^{-\lambda t_1} e^{-\lambda t_2},$$

which proves ii) for $\zeta_1 - \zeta_0$ and $\zeta_2 - \zeta_1$; similar computations yield ii) for any number of $\zeta_n - \zeta_{n-1}$. ◇

Corollary. *For any numbers* t_i, $1 \leqslant i \leqslant n$, *satisfying* $0 \leqslant t_1 \leqslant \ldots \leqslant t_n < t$,

$$\mathsf{P}(\zeta_i \leqslant t, 1 \leqslant i \leqslant n \mid \xi(t) = n) =$$

$$= \frac{n!}{t^n} \int_0^{t_1} \int_{x_1}^{t_2} \cdots \int_{x_{n-1}}^{t_n} dx_n \ldots dx_1,$$

which is the distribution of the order statistics from a sample of n observations taken from the uniform distribution on $[0, t]$[10].

Proof. On account of ii) Theorem 2.2.10 we have

$$\mathsf{P}(\zeta_i \leqslant t_i,\ 1 \leqslant i \leqslant n,\ \xi(t) = n) =$$

$$\mathsf{P}(\zeta_i \leqslant t_i, 1 \leqslant i \leqslant n,\ \zeta_{n+1} > t) =$$

$$\int_0^{t_1} \int_0^{t_2 - u_1} \cdots \int_0^{t_n - (u_1 + \cdots + u_{n-1})} \int_{t - (u_1 + \cdots + u_n)}^{\infty} \lambda^{n+1} e^{-\lambda(u_1 + \cdots + u_{n+1})} du_{n+1} \cdots$$

$$\cdots du_1 = \lambda^n e^{-\lambda t} \int_0^{t_1} \int_0^{t_2 - u_1} \cdots \int_0^{t_n - (u_1 + \cdots + u_{n-1})} du_n \ldots du_1.$$

By introducing the new variables $x_i = u_1 + \ldots + u_i$, $1 \leqslant i \leqslant n$, we can write the last expression as

$$\lambda^n e^{-\lambda t} \int_0^{t_1} \int_{x_1}^{t_2} \cdots \int_{x_{n-1}}^{t_n} dx_n \cdots dx_1.$$

Therefore

$$\mathsf{P}(\zeta_i \leqslant t_i,\ 1 \leqslant i \leqslant n \mid \xi(t) = n) =$$

$$= \frac{\mathsf{P}(\zeta_i \leqslant t_i,\ 1 \leqslant i \leqslant n,\ \xi(t) = n)}{\mathsf{P}(\xi(t) = n)} = \frac{n!}{t^n} \int_0^{t_1} \int_{x_1}^{t_2} \cdots \int_{x_{n-1}}^{t_n} dx_n \ldots dx_1. \ ◇$$

---

[10] If we consider n independent observations of a random variable uniformly distributed on $[0, t]$, and $X_1 \leqslant \cdots \leqslant X_n$ denote them arranged in increasing order, then the joint distribution of $X_1, \cdots, X_n$ is precisely that given in the corollary.

The above corollary is the starting point for applications of Poisson processes to order statistics and related topics. For details see KARLIN (1966, Ch. 9) and RÉNYI (1962, Ch. 8, §§ 9—10).

2.2.2.1.6. It follows from ii) Theorem 2.2.10 that the homogeneous Poisson process $P(\lambda)$ can be constructed as follows: let $(\tau_n)_{n \in N^*}$ be a sequence of independent and identically distributed random variables with $\mathsf{P}(\tau_1 \leqslant t) = 1 - e^{-\lambda t}$, $t > 0$; then

$$\xi(t) = \inf(n - 1 : \tau_1 + \ldots + \tau_n > t)$$

is the process $P(\lambda)$.

By generalizing this construction we shall define the *compound* Poisson process $P(\lambda, F)$, where F is a distribution function on R. Let $(u_n)_{n \in N^*}$ be a sequence of independent and identically distributed random variables, with distribution function F. Suppose that the sequences $(\tau_n)_{n \in N^*}$ and $(u_n)_{n \in N^*}$ are also independent. The process ξ_F defined by

$$\xi_F(t) = \begin{cases} 0 & \text{if } \tau_1 > t \\ \sum_{i=1}^{n} u_i & \text{if } \sum_{j=1}^{n} \tau_j \leqslant t < \sum_{j=1}^{n+1} \tau_j \end{cases} = \sum_{j=1}^{\xi(t)} u_i$$

is called a compound Poisson process and is denoted by $P(\lambda, F)$[11]. It is still a homogeneous process with independent increments. If F is continuous at 0, then almost all the trajectories of the process $P(\lambda, F)$ are step-functions whose successive jumps of magnitudes $u_1, u_2, \ldots$ are separated by time-intervals of lengths $\tau_1, \tau_2, \ldots$.

It is easily seen that

$$\mathsf{P}(\xi_F(t) - \xi_F(s) < x) = \sum_{n \in N} e^{-\lambda(t-s)} \frac{\lambda^n (t-s)^n}{n!} F_n(x)$$

where F_n, $n \in N^*$, is the n-fold convolution of F with itself and $F_0(x)$ is 0 or 1 according as $x < 0$ or $\geqslant 0$.

2.2.2.1.7. Now, following K. ITÔ, we shall indicate how a family of Poisson processes can be associated with every separable and stochastically continuous process with independent increments.

Let ξ be a separable and stochastically continuous process with independent increments. According to Proposition 2.2.5 ξ has no discontinuities of the second kind. Let $\mathcal{B}_\varepsilon$ be the σ-algebra of Borel sets in

[11] The Poisson process $P(\lambda)$ is obtained by setting

$$F(x) = \begin{cases} 0 \text{ if } x < 1 \\ 1 \text{ if } x \geqslant 1. \end{cases}$$

$R - [-\varepsilon, \varepsilon]$ for a fixed but arbitrarily given $\varepsilon > 0$. The absence of discontinuities of the second kind implies that for every $A \in \mathcal{B}_\varepsilon$, the number $\nu(t, A)$ of values $s \in (0, t)$, $t < \infty$, such that $\xi(s) - \xi(s-) \in A$, is finite with probability 1. The process $(\nu(t, A))_{t \in T}$ is a process with independent increments; independence of the increments of $\nu(\cdot, A)$ on disjoint time-intervals follows from the independence of the corresponding increments of ξ. It is also easily seen that stochastic continuity of ξ implies that of $\nu(\cdot, A)$ for all $A \in \mathcal{B}_\varepsilon$. Obviously, $\nu(t, \cdot)$ is a completely additive measure on $\mathcal{B}_\varepsilon$ for all $t \in T$.

Theorem 2.2.11 (K. Itô). i) *For all $t \in T$, $A \in \mathcal{B}_\varepsilon$ and $r > 0$,*

$$\mathsf{E}\,[\nu(t, A)]^r < \infty.$$

ii) *If A_i, $1 \leqslant i \leqslant n$, are pairwise disjoint sets from $\mathcal{B}_\varepsilon$, then the processes $(\nu(t, A_i))_{t \in T}$, $1 \leqslant i \leqslant n$, are independent.*

iii) *For every $A \in \mathcal{B}_\varepsilon$, $(\nu(t, A))_{t \in T}$ is a Poisson process*[12] *$P(\Pi(\cdot\,; A))$ where*

$$\Pi(t, A) = \mathsf{E}\,\nu(t, A).$$

The function $\Pi(\cdot, A)$ is continuous and increasing for all $A \in \bigcup_{\varepsilon > 0} \mathcal{B}_\varepsilon$ and

$$\lim_{\varepsilon \to 0} \int_{\varepsilon < |x| < 1} x^2\, \Pi(t, \mathrm{d}x) < \infty, \quad t \in T.$$

Proof. See, e.g. Gihman and Skorohod (1965, Ch. 6, § 1). ◇

2.2.2.1.8. In this subparagraph we discuss increments of Poisson processes for arbitrary Borel sets. Let ξ be a separable stochastically continuous Poisson process $P(\lambda(\cdot))$. For every Borel set $A \subset T$ let $\sigma_\xi(A)$ denote the number of jumps occurring at points of A. If A is bounded, then $\sigma_\xi(A)$ is a non-negative integer with probability 1. In particular, for every interval $[s, t]$ we have $\sigma_\xi([s, t]) = \xi(t) - \xi(s)$ with probability 1.

It is obvious that $\sigma_\xi(\cdot)$ is a completely additive measure on the Borel sets in T.

One can show [see Cramér (1951), Marczewski (1953), Ryll-Nardzewski (1953)], that the random variable $\sigma_\xi(A)$ has, for each bounded Borel set $A \subset T$, a Poisson distribution with parameter $\mu_\lambda(A)$ where μ_λ is the measure induced on the Borel sets in T by $\lambda(\cdot)$ [see Loève (1963, p. 96)]. Moreover, if A_i, $1 \leqslant i \leqslant n$, are pairwise disjoint bounded Borel sets in T, then the random variables $\sigma_\xi(A_i)$, $1 \leqslant i \leqslant n$, are independent.

---

[12] Here the degenerate case $P(0)$ is not excluded.

This might suggest the definition of a stochastically continuous Poisson process as a random completely additive set-function on bounded Borel sets in T having the above properties [13].

However, as shown by RÉNYI (1967 a), this definition is redundant, the assumption of complete additivity and independence being unnecessary. More precisely, let $\sigma(\cdot)$ be a random additive set-function on the family $\mathfrak{J}$ of all subsets of T which can be obtained as the union of a finite number of disjoint finite intervals closed to the right and open to the left. If for each $A \in \mathfrak{J}$, $\sigma(A)$ has a Poisson distribution with parameter $\mu(A)$, where μ is a non-atomic measure on the Borel sets in T which is finite for all sets in $\mathfrak{J}$, then $\xi(t) = \sigma([0, t))$ is a stochastically continuous Poisson process. It is not possible to replace $\mathfrak{J}$ by the set of all intervals closed to the right and open to the left. This was shown by a counterexample of L. SHEPP [see GOLDMAN (1967 a)].

2.2.2.2. The Wiener process

2.2.2.2.1. We begin with a formal definition

Definition 2.2.12. A process with independent increments is said to be a *Wiener process* $W(m(\cdot), v(\cdot))$ if the increments $\xi(t) - \xi(s)$ are normally distributed with mean $m(t) - m(s)$ and variance $v(t) - v(s)$, where m and v are real-valued functions defined on T, the last being increasing. In case $m(t) = mt$ and $v(t) = vt$, $(v > 0)$, the process ξ is called a *homogeneous* Wiener process $W(m, v)$[14].

Thus, for the process $W(m(\cdot), v(\cdot))$ we have

$$\mathsf{P}(\xi(t) - \xi(s) < x) =$$

$$= \frac{1}{\sqrt{2\pi(v(t) - v(s))}} \int_{-\infty}^{\infty} \exp\left\{-\frac{(u - m(t) + m(s))^2}{2(v(t) - v(s))}\right\} du, \quad x \in R,$$

$$\mathsf{E}(\xi(t) - \xi(s)) = m(t) - m(s)$$

$$\mathsf{D}(\xi(t) - \xi(s)) = v(t) - v(s)$$

$$\varphi_{st}(\theta) = \exp\left\{i\,\theta[m(t) - m(s)] - \frac{\theta^2}{2}[v(t) - v(s)]\right\}.$$

On account of Proposition 2.2.4, the process $W(m(\cdot), v(\cdot))$ is stochastically continuous iff $m(\cdot)$ and $v(\cdot)$ are continuous functions on T.

2.2.2.2.2. Like Poisson processes, stochastically continuous Wiener processes may be characterized by a property of their trajectories.

---

[13] An extensive study from this point of view of processes with independent increments whose trajectories are non-decreasing was carried out by KINGMAN (1967a).

[14] It is still an open problem whether or not the normality of the increments $\xi(t) - \xi(s)$ implies their mutual independence. See LÉVY (1969), RÉNYI (1967 a).

Theorem 2.2.13. *If a process $W(m(\cdot), v(\cdot))$ is separable and stochastically continuous, then almost all its trajectories are continuous functions (equivalently: the process is continuous).*

Proof. Without any loss of generality we may suppose that $m \equiv 0$ (otherwise, we consider the process $\xi - m(\cdot)$). According to Proposition 2.2.5, the considered process ξ has no discontinuities of the second kind. Denote by $\nu_\varepsilon (0 \leqslant \nu_\varepsilon \leqslant \infty)$ the number of values $t \in [0, \alpha]$, α finite $< a$, such that $|\xi(t+) - \xi(t-)| > 2\varepsilon$. Let

$$0 = t_0^{(n)} < t_1^{(n)} < \ldots < t_{m_n}^{(n)} = \alpha$$

be a sequence of partitions of the interval $[0, \alpha]$ with norms tending to 0 as $n \to \infty$. Denote by $\nu_\varepsilon^{(n)}$ the number of integers k, $1 \leqslant k \leqslant m_n$, for which $|\xi(t_k^{(n)}) - \xi(t_{k-1}^{(n)})| > \varepsilon$. Obviously

$$\nu_\varepsilon \leqslant \underline{\lim_{n \to \infty}} \, \nu_\varepsilon^{(n)}.$$

On the other hand

$$\mathsf{E}\, \nu_\varepsilon^{(n)} = \sum_{k=1}^{m_n} \mathsf{P}(|\xi(t_k^{(n)}) - \xi(t_{k-1}^{(n)})| > \varepsilon) \leqslant$$

$$\leqslant \frac{1}{\varepsilon^4} \sum_{k=1}^{m_n} \mathsf{E}\, |\xi(t_k^{(n)}) - \xi(t_{k-1}^{(n)})|^4 \leqslant \frac{3}{\varepsilon^4} \sum_{k=1}^{m_n} [v(t_k^{(n)}) - v(t_{k-1}^{(n)})]^2 \leqslant$$

$$\leqslant \frac{3(v(\alpha) - v(0))}{\varepsilon^4} \max_{1 \leqslant k \leqslant m_n} [v(t_k^{(n)}) - v(t_{k-1}^{(n)})] \underset{(n \to \infty)}{\to} 0.$$

By Fatou's lemma

$$\mathsf{E}\, \nu_\varepsilon \leqslant \mathsf{E} \underline{\lim_{n \to \infty}} \, \nu_\varepsilon^{(n)} \leqslant \underline{\lim_{n \to \infty}} \, \mathsf{E}\, \nu_\varepsilon^{(n)} = 0,$$

thus $\nu_\varepsilon = 0$ almost surely for all $\varepsilon > 0$. It follows that, with probability 1, $\xi(t-) = \xi(t+)$ for $t \in [0, \alpha]$. By separability, we have $\xi(t-) = \xi(t+) = \xi(t)$, that is ξ is continuous on $[0, \alpha]$. The continuity on $[0, a)$ follows by considering a sequence $\alpha_n \to a$ as $n \to \infty$. ◇

2.2.2.2.3. A converse theorem also holds.

Theorem 2.2.14. *Let ξ be a continuous process with independent increments. Then ξ is a separable Wiener process.*

Proof. [Itô (1960, p. 63)]. Separability follows from the continuity of the trajectories on account of Proposition 2.1.6. Thus it remains to show that the distribution of the random variable $\xi(t) - \xi(s)$, $s < t$, is normal. The trajectories being continuous, for all $\varepsilon > 0$ there is a $\delta = \delta(\varepsilon)$ such that

$$\mathsf{P}(|\xi(u') - \xi(u'')| < \varepsilon \quad \text{for} \quad |u' - u''| < \delta, u', u'' \in [s, t]) > 1 - \varepsilon.$$

Consider a sequence $\varepsilon_n \downarrow 0$ $(n \to \infty)$ and choose $p(n) \in N^*$ such that

$$\frac{t - s}{p(n)} < \delta(\varepsilon_n).$$

Divide the segment $[s, t]$ into $p(n)$ equal parts and denote by $y_{n1}, \ldots, y_{n,\, p(n)}$ the increments of ξ on the partial segments. We have $y = \xi(t) - \xi(s) = y_{n1} + \ldots + y_{n,\, p(n)}$. Set

$$y'_{nk} = \begin{cases} y_{nk} & \text{if} \quad |y_{nk}| < \varepsilon_n \\ 0 & \text{if} \quad |y_{nk}| \geqslant \varepsilon_n \end{cases} \quad ,1 \leqslant k \leqslant p(n),$$

and

$$y'_n = y'_{n1} + \ldots + y'_{n,\, p(n)}.$$

The random variables y'_{nk}, $1 \leqslant k \leqslant p(n)$, are also independent and

$$\mathsf{P}(y \neq y'_n) < \varepsilon_n, \qquad n \in N^*.$$

Put

$$m_{nk} = \mathsf{E}(y'_{nk}),\ v_{nk} = \mathsf{D}(y'_{nk}), \quad 1 \leqslant k \leqslant p(n),$$

$$m_n = \sum_{k=1}^{p(n)} m_{nk},\ v_n = \sum_{k=1}^{p(n)} v_{nk}.$$

We now distinguish four cases.

a) $m_n \to m(s, t) < \infty$ and $v_n = v(s, t) < \infty$ $(n \to \infty)$.
In this case we have

$$\mathsf{E} \exp \mathrm{i}\,\theta y = \lim_{n\to\infty} \mathsf{E} \exp \mathrm{i}\,\theta\, y'_n = \lim_{n\to\infty} \prod_{k=1}^{p(n)} \mathsf{E} \exp \mathrm{i}\,\theta\, y'_{nk} =$$

$$= \lim_{n\to\infty} \exp(\mathrm{i}\theta\, m_n) \prod_{k=1}^{p(n)} \mathsf{E}[\exp \mathrm{i}\theta\,(y'_{nk} - m_{nk})] =$$

$$= \exp(\mathrm{i}\theta\, m(s, t)) \lim_{n\to\infty} \prod_{k=1}^{p(n)} \left(1 - \frac{v_{nk}}{2}\,\theta^2 + v_{nk} O(\varepsilon_n)\right) =$$

$$= \exp(\mathrm{i}\theta\, m(s, t)) \lim_{n\to\infty} \prod_{k=1}^{p(n)} \exp\left(-\frac{v_{nk}}{2}\,\theta^2 + v_{nk} O(\varepsilon_n)\right) =$$

$$= \exp\left(\mathrm{i}\theta\, m(s, t) - \frac{\theta^2}{2}\, v(s, t)\right).$$

We conclude immediately that $m(s, t) = m(0, t) - m(0, s)$ and $v(s, t) = v(0, t) - v(0, s)$. Consequently, ξ is the Wiener process $W(m(0, \cdot), v(0, \cdot))$.

b) $m_{n_j} \to m(s, t) < \infty$ and $v_{n_j} \to v(s, t) < \infty$ $(n_j \to \infty)$.

This case does not differ from the previous one.

c) $v_{n_j} \to v(s, t) < \infty$ $(n_j \to \infty)$.

In this case, reasoning as in case a), we conclude that the distribution of the random variable $y'_{n_j} - m_{n_j}$ approaches the normal distribution with parameters 0 and $v(s, t)$ as $j \to \infty$. But the distribution of y'_n converges to that of y and this implies the convergence of the sequence (m_{n_j}) to a finite value, say $m(s, t)$. We thus arrive at case b).

d) $v_n \to \infty$ $(n \to \infty)$.

Since $|v_{nk}| \leqslant 4\varepsilon_n^2$, $1 \leqslant k \leqslant p(n)$, for all $v > 0$ there is a $K = K(n, v)$ such that

$$\sum_{k=1}^{K} v_{nk} \to v (n \to \infty).$$

We conclude as in case a)

$$\mathsf{E}|\exp i\theta y| \leqslant \lim_{n\to\infty} \prod_{k=1}^{K} |\mathsf{E} \exp i\theta y'_{nk}| = \exp\left(-\frac{v}{2}\theta^2\right).$$

Letting $v \to \infty$, we obtain $\mathsf{E} \exp i\theta\, y = 0$, therefore case d) is not possible. ◇

2.2.2.3. Brownian motion

2.2.2.3.1. The separable process $W(0, 1)$, usually denoted by w, is known as the *Brownian motion* process (other nomenclatures less frequently used are: the Wiener-Bachelier, the Wiener-Lévy, or the Wiener-Einstein process). The process is named after the English botanist ROBERT BROWN who was the first to notice in 1827 the erratic movement of microscopic particles suspended in a fluid under the impact of the fluid molecules. A mathematical description of this phenomenon was first obtained from statistical mechanical considerations by A. EINSTEIN in 1905 and was then considerably generalized and extended by the Polish physicist MARJAN SMOLUCHOWSKI.

For values of t large compared with the intervals between successive collisions, the Brownian motion process w has been found to be a very good description of the motion of the projection on a straight line of a free particle (that is, one on which no forces other than those due to the molecules of the surrounding medium are acting). Initially, the physical theory advanced faster[15] than the mathematical one because

[15] For historical references see KAC (1947) and NELSON (1967).

the precise mathematical definition of the Brownian motion involves a measure on the trajectory space. The first rigorous formulation of the theory is due to WIENER (1923), after the ideas of BOREL, LEBESGUE and DANIELL had appeared. WIENER's work was preceded by a heuristic treatment by BACHELIER (1900).

2.2.2.3.2. In recent years the multi-dimensional Brownian motion has acquired a special importance. An r-dimensional Brownian motion process is an R^r-valued process $\mathbf{w}$ with independent increments for which the probability density of the increments $\mathbf{w}(t) - \mathbf{w}(s)$ is

$$[2\pi(t-s)]^{-\frac{r}{2}} \exp\left\{-\frac{1}{2(t-s)} \sum_{i=1}^{r} x_i^2\right\}.$$

In other words, a multi-dimensional Brownian motion is a vector-valued process whose components are independent one-dimensional Brownian motions.

S. KAKUTANI observed in 1944 that the basic quantities of logarithmic potential theory could be probabilistically interpreted in terms of two-dimensional Brownian motion. After the extension of KAKUTANI's work by J. L. DOOB and G. A. HUNT, it gradually became clear that in a certain sense Brownian motion and classical Newtonian potential theory were really the same subject in dimension $r \geqslant 3$[16]. The details of this connection can be found in KNAPP (1965).

We note also that some problems in mathematical analysis (e.g. Dirichet, Neumann, or Lauricella problem) can be solved by exploiting the connections between Brownian motion and classical potential theory. See e.g. DYNKIN (1965, Ch. 13), KAC (1951, 1952, 1959), HELMS (1967).

2.2.2.3.3. In a sense to be made more precise, Brownian motion is the limit of a simple random walk. This fact was in fact used by SMOLUCHOWSKI in his development.

For the sake of simplicity, we shall consider only the one-dimensional case in the following.

For every $k \in N^*$ let $(\eta_n^{(k)})_{n \in N^*}$ be a sequence of independent and identically distributed random variables such that

$$\mathsf{P}(\eta_n^{(k)} = 2^{-k}) = \mathsf{P}(\eta_n^{(k)} = -2^{-k}) = \frac{1}{2}.$$

---

[16] HUNT (1957–58) showed that one could develop a potential theory for a general Markov process and, conversely, for a potential theory one can construct a Markov process associated with this theory. A proof of the equivalence of a general class of Markov processes with axiomatic potential theory was given by MEYER (1963).

Then

$$\xi_n^{(k)} = \sum_{j=1}^{n} \eta_j^{(k)}, \quad n \in N^*,$$

is a simple random walk with a change in scale.

Proposition 2.2.15. *If t is a diadic rational and if $\alpha, \beta \in R$, then*

$$\lim_{k\to\infty} \mathsf{P}(\xi_{4^k t}^{(k)} \in (\alpha, \beta)) = \mathsf{P}(w(t) \in (\alpha, \beta)).$$

Proof. This result is an easy consequence of the central limit theorem. Let $m = 4^k t \in N^*$. The random variable $\xi_m^{(k)}$ is the sum of m independent random variables each with mean 0 and variance 4^{-k}. Hence, the central limit theorem implies

$$\lim_{k\to\infty} \mathsf{P}(\xi_{4^k t}^{(k)} \in (\alpha, \beta)) = \lim_{k\to\infty} \mathsf{P}\left(\frac{\alpha}{\sqrt{t}} < \frac{\xi_m^{(k)}}{\sqrt{t}} < \frac{\beta}{\sqrt{t}}\right) =$$

$$= \frac{1}{\sqrt{2\pi t}} \int_\alpha^\beta \exp\left(-\frac{u^2}{2}\right) du = \mathsf{P}(w(t) \in (\alpha, \beta)). \quad \diamond$$

The demonstrated connection between Brownian motion and the simple random walk allows us to expect that the former will have some special properties, reflecting those of the latter. We shall see that it is in fact so.

2.2.2.3.4. A most useful tool in studying Brownian motion is the so-called *reflection principle* (Compare with Subparagraph 1.2.1.2.2). For $a > 0$ let τ_a be the first (random) value of t for which $w(t) = a$ (the existence of τ_a follows from continuity of the trajectories). For $t > \tau_a$ we reflect the trajectory in the line $w = a$ to obtain

$$w_a(t) = \begin{cases} w(t) & \text{for} \quad t \leqslant \tau_a \\ 2a - w(t) & \text{for} \quad t > \tau_a. \end{cases}$$

We may reason heuristically [17] as follows: since the changes in trajectories of w after τ_a are independent of the changes before τ_a, and are equally likely to be positive or negative, the probabilities will not change under the reflection considered. Therefore (and this is just the reflection principle) $w(\cdot, \omega) \to w_a(\cdot, \omega)$ is a one-to-one measure preserving transformation of trajectory space into itself. We note that the reflection principle is in fact of wider applicability since only the symmetric distribution of the increments, and not their specific form, is invoked. An immediate consequence of the reflection principle is the formula

$$\mathsf{P}(\max_{0\leqslant s\leqslant t} w(s) \geqslant a) = 2\mathsf{P}(w(t) > a) \tag{2.2.3}$$

[17] A rigorous treatment would require the strong Markov property applied to the Markov time τ_a. See Paragraph 2.3.1.3.

where $a \geqslant 0$, $t > 0$, a result first obtained by L. BACHELIER. Indeed, the set $(\max_{0 \leqslant s \leqslant t} w(s) > a)$ is the union of three disjoint sets

$$(\max_{0 \leqslant s \leqslant t} w(s) \geqslant a,\ w(t) > a),$$

$$(\max_{0 \leqslant s \leqslant t} w(s) \geqslant a,\ w(t) = a),$$

$$(\max_{0 \leqslant s \leqslant t} w(s) \geqslant a,\ w(t) < a).$$

The second of these sets has probability 0 while the other two are mapped onto one another by reflection in the line $w = a$ after the time τ_a. Hence the probability in question is twice the probability of the first set, which is identical to $(w(t) > a)$.

With the help of (2.2.3) we can determine the distribution of τ_a. Clearly,

$$\mathsf{F}(\tau_a < t) = \mathsf{P}(\max_{0 \leqslant s \leqslant t} w(s) \geqslant a) =$$

$$= \frac{2}{\sqrt{2\pi t}} \int_a^\infty \mathrm{e}^{-\frac{x^2}{2t}}\, \mathrm{d}x = \frac{a}{\sqrt{2\pi}} \int_0^t x^{-\frac{3}{2}} \mathrm{e}^{-\frac{a^2}{2x}}\, \mathrm{d}x, \quad t > 0.$$

It is not difficult to see that $(\tau_a)_{a \geqslant 0}$ is a stable process with independent increments.

A skilful use of the reflection principle [see e.g. GIHMAN and SKOROHOD (1965), p. 372] yields the following result due to P. LÉVY: If $a_1 < 0 < a_2$ and $[b_1, b_2] \subset [a_1, a_2]$, then

$$\mathsf{P}(\min_{0 \leqslant u \leqslant t} w(u) > a_1,\ \max_{0 \leqslant u \leqslant t} w(u) < a_2,\ w(t) \in [b_1, b_2]) =$$

$$= \frac{1}{\sqrt{2\pi t}} \sum_{k \in Z} \int_{b_1}^{b_2} \left[\exp\left\{-\frac{1}{2t}(x + 2k(a_2 - a_1))^2\right\} - \right.$$

$$\left. - \exp\left\{-\frac{1}{2t}(x - 2a_2 + 2k(a_2 - a_1))^2\right\}\right] \mathrm{d}x.$$

For further results concerning distributions of various quantities associated with Brownian motion we refer the reader to LÉVY (1965, Ch. 6) and SKOROHOD (1964, Ch. 6).

2.2.2.3.5. In this subparagraph we state a theorem showing that Brownian motion is the limit of a sequence of suitable (random) polygonal functions.

Let ξ_{ni}, $1 \leqslant i \leqslant k_n$, for all $n \in N^*$, be independent random variables with distribution functions F_{ni} such that $\mathsf{E}\xi_{ni} = 0$, and $\sum^{k_n} \mathsf{D}\,\xi_{ni} = 1$.

Let ξ_n be the polygonal function obtained by linearly interpolating $S_{nl} = \sum_{i=1}^{l} \xi_{ni}$ at $t_{nl} = \sum_{i=1}^{l} \mathsf{D}\xi_{ni}$, $0 \leqslant l \leqslant k_n$, $S_{n0} = t_{n0} = 0$, that is

$$\xi_n(t) = S_{nl} + \frac{t - t_{nl}}{t_{n,l+1} - t_{nl}} (S_{n,l+1} - S_{nl})$$

for $t \in [t_{nl}, t_{n,l+1}]$, $0 \leqslant l < k_n$.

Theorem 2.2.16 (M. DONSKER, JU. V. PROHOROV). *If*

$$\lim_{n\to\infty} \sum_{i=1}^{k_n} \int_{|u|>\varepsilon} u^2 \,\mathrm{d}F_{ni}(u) = 0$$

for all $\varepsilon > 0$ *(Lindeberg's condition), then the sequence of stochastic processes* $((\xi_n(t))_{t\in[0,1]})_{n\in N^*}$ *converges weakly to the Brownian motion* w*; for every functional* f *on* $C[0,1]$ *almost surely continuous with respect to the measure induced in* $C[0, 1]$ *by* w*, the random variable* $f((\xi_n(t))_{t\in[0,1]})$ *converges in law to* $f((w(t))_{t\in[0,1]})$ *as* $n \to \infty$.

Proof. See GIHMAN and SKOROHOD (1965, Ch. 9, § 3). ◇

The above theorem allows us to obtain limiting distributions of functionals of sequences of independent random variables. To consider an example, let $(\eta_n)_{n\in N^*}$ be a sequence of independent identically distributed random variables such that $\mathsf{E}\eta_n = 0$, $\mathsf{D}\eta_n = 1$, $n \in N^*$. Put $\xi_{ni} = n^{-1/2}\eta_i$, $1 \leqslant i \leqslant n$, and notice that $\mathsf{P}(\xi_{ni} < u) = \mathsf{P}(\eta_i < \sqrt{n}\,u)$ $(= F(\sqrt{n}\,u))$. We have

$$\sum_{i=1}^{n} \int_{|u|>\varepsilon} u^2 \,\mathrm{d}F_{ni}(u) = \sum_{i=1}^{n} \int_{|u|>\varepsilon} u^2 \,\mathrm{d}F(\sqrt{n}\,u) =$$

$$= \int_{|u|>\varepsilon} nu^2 \,\mathrm{d}F(\sqrt{n}\,u) = \int_{|u|>\varepsilon\sqrt{n}} u^2 \,\mathrm{d}F(u) \to 0 \ (n \to \infty).$$

Set $M_n = \max_{0\leqslant k\leqslant n} S_k$ with $S_0 = 0$, $S_n = \sum_{k=1}^{n} \xi_k$, $n \in N^*$, and note that $n^{-1/2} M_n = \sup_{0\leqslant t\leqslant 1} \xi_n(t)$. Since $\sup_{0\leqslant t\leqslant 1}$ is a continuous functional on $C[0, 1]$, we may apply Theorem 2.2.16. On account of (2.2.3) we get

$$\lim_{n\to\infty} (M_n < a\sqrt{n}) = \frac{2}{\sqrt{2\pi}} \int_0^a \mathrm{e}^{-\frac{u^2}{2}} \,\mathrm{d}u,$$

a result first obtained by P. ERDÖS and M. KAC.

For further details, examples and developments, the reader is referred to KOLMOGOROV and PROHOROV (1956), PROHOROV (1956), and SKOROHOD (1964, Ch. 8).

2.2.2.3.6. The question of the growth of an r-dimensional Brownian motion when $t \to \infty$ or $t \to 0$ is answered by Hinčin's law of the iterated logarithm[18]:

$$\mathsf{P}\left(\overline{\lim_{t\to\infty}} \frac{|\mathbf{w}(t)|}{\sqrt{2t \ln \ln |t|}} = 1\right) = 1,$$

$$\mathsf{P}\left(\overline{\lim_{t\to 0}} \frac{|\mathbf{w}(t)|}{\sqrt{2t \ln \ln |t|}} = 1\right) = 1.$$

Here $|\mathbf{w}(t)|$ is the Euclidean length of the vector $\mathbf{w}(t)$. It is not difficult to deduce from the latter equation that almost all trajectories of a one-dimensional Brownian motion have infinite upper derivatives. This may also be deduced from the results of DVORETZKY, ERDÖS and KAKUTANI (1961) who proved that for all $a \in R$, with probability 1, the function $w(t) + at$ is nowhere increasing (decreasing)[19]. It follows (for $a = 0$) that with probability 1 the one-dimensional Brownian trajectory is nowhere increasing (decreasing), that is

$\mathsf{P}(w(t)$ has at least one point of increase$) = 0$,

$\mathsf{P}(w(t)$ has at least one point of decrease$) = 0$.

Thus, with probability 1, there are no points t for which $w(t+h, \omega) - w(t, \omega) + ah$ has the same sign as h for all $|h| < h(t, \omega)$ with $h(t, \omega) > 0$. Hence

$$\underline{\lim_{h\to 0}} \frac{w(t+h) - w(t)}{h} \leqslant -a, \quad a \in R.$$

We deduce analogously that

$$\overline{\lim_{h\to 0}} \frac{w(t+h) - w(t)}{h} \geqslant a, \quad a \in R.$$

It follows that, with probability 1, the lower derivative of the one-dimensional Brownian trajectory is everywhere $-\infty$ and the upper derivative $+\infty$, that is

$$\mathsf{P}\left(\underline{\lim_{h\to 0}} \frac{w(t+h) - w(t)}{h} = -\infty, \ \overline{\lim_{h\to 0}} \frac{w(t+h) - w(t)}{h} = \infty, \ t \in T\right) = 1.$$

---

[18] HINČIN considered only the one-dimensional case. A treatment of the general case may be found in SKOROHOD (1964, § 31).

[19] A real-valued function f is said to increase (decrease) at t_0 if for some $\delta > 0$

$$\max_{t_0-\delta \leqslant t \leqslant t_0} (\min) f(t) = f(t_0) = \min_{t_0 \leqslant t \leqslant t_0+\delta} (\max) f(t).$$

2.2.2.3.7. It is possible to prove [see e.g. DOOB (1953, p. 395] that, with probability one, the one-dimensional Brownian trajectory is not a function of bounded variation on any finite interval. It follows that, with probability one, every arc of an r-dimensional Brownian motion has infinite length.

2.2.2.3.8. The question of multiple points [20] of r-dimensional Brownian trajectories is completely resolved by results of DVORETZKY, ERDÖS and KAKUTANI (1950, 1954, 1958) and DVORETZKY, ERDÖS, KAKUTANI and TAYLOR (1957). Thus the following holds with probability 1: For $r \geqslant 4$, the trajectory contains only simple points, for $r = 3$ it contains an infinite number of double points but no triple points, while for $r = 2$ (and also, of course, for $r = 1$) it contains an infinite number of n-multiple points for each $n \geqslant 2$ and, moreover, it contains points of multiplicity $\mathfrak{c}$ (the power of the continuum) whose set is everywhere dense in the entire plane.

2.2.2.3.9. In analogy to a property of the simple random walk, the r-dimensional ($r \geqslant 3$) Brownian trajectory, after leaving a ball of R^r, returns to it with probability less than one, whereas for $r = 1$ or 2 it returns infinitely often with probability one. For a proof, see ITÔ and MCKEAN (1965, p. 236).

On the other hand, with probability 1, the r-dimensional ($r \geqslant 2$) Brownian trajectory never hits a point singled out in advance at a positive time. [ITÔ and MCKEAN (1965, p. 62)].

2.2.2.3.10. It follows from the previous subparagraph that, with probability 1, the two-dimensional Brownian trajectory is everywhere dense in the plane. Nevertheless, P. LÉVY showed that, with probability 1, the Lebesgue measure of the two-dimensional Brownian trajectory equals 0. For a proof see ITÔ and MCKEAN (1965, p. 235).

In an attempt to explain the extreme irregularity of two-dimensional Brownian trajectories, LÉVY (1969) formulated the following conjecture: Let $\mathbf{x} = f(t)$ be a plane curve, f continuous ($t \in R$, $\mathbf{x} \in R^2$). Denote by $m(t, u)$ the two-dimensional Lebesgue measure of the closure of the image of the interval $(t, t + u)$. Assume that to every interval $\mathfrak{i} \subset R$ there corresponds a c and an l such that, for every $t \in \mathfrak{i}$ and $0 < u < l$, $m(t, u) \geqslant cu$. Assume further that the two-dimensional Lebesgue measure of the set $(f(t): t \in R)$ is 0. The conjecture is that under these conditions $\mathbf{x} = f(t)$ has points of multiplicity $\mathfrak{c}$.

2.2.3. General properties

In this subsection we give without proof two ways of representing general processes with independent increments. Details and proofs

[20] A point $\mathbf{x} \in R^r$ is said to be k-multiple if the equation $\mathbf{w}(t) = \mathbf{x}$ has at least k distinct solutions.

may be found in Doob [(1953), Ch. 8, § 6], Gihman and Skorohod [(1965), Ch. 6, § 3], Itô [(1960), § 12], Loève [(1963), § 37], and Skorohod (1964).

2.2.3.1. Integral decomposition

2.2.3.1.1. We use the notation introduced in Subparagraph 2.2.2.1.7. Set

$$\tilde{\nu}(t, A) = \nu(t, A) - \Pi(t, A).$$

Theorem 2.2.17 (K. Itô). *For every separable and stochastically continuous process ξ with independent increments there exists a Wiener process ξ_0 independent of the processes $\nu(., A), A \in \bigcup_{\varepsilon>0} \mathcal{B}_\varepsilon$ such that*

$$\xi(t) = \xi_0(t) + \int_{|x|>1} x\nu(t, dx) + \int_{|x|\leqslant 1} x\tilde{\nu}(t, dx)^{21)} \tag{2.2.4}$$

Proof. See e.g. Skorohod [(1964), Ch. 3. § 6]. ◇

2.2.3.1.2. The above representation leads to a new (equivalent) form for the characteristic functions φ_{st} [see (2.2.2)], namely,

$$\log \varphi_{0t}(\theta) = \mathrm{i}\, m(t)\, \theta - \frac{\theta^2}{2} v(t) + \int_{|x|>1} (e^{\mathrm{i}\theta x} - 1)\, \Pi(t, dx) +$$

$$+ \int_{0<|x|\leqslant 1} (e^{\mathrm{i}\theta x} - 1 - \mathrm{i}\theta x)\, \Pi(t, dx),$$

where $m(\cdot)$ and $v(\cdot)$ are continuous functions, the latter being increasing.

The representation (2.2.4) is useful in the study of trajectories. The logarithm of the characteristic function φ_{0t} corresponding to a process ξ with independent increments whose trajectories are step-functions has the form

$$\int_R (e^{\mathrm{i}\theta x} - 1)\, \Pi(t, dx).$$

To the case of increasing trajectories there corresponds the characteristic function φ_{0t}, whose logarithm equals

$$\mathrm{i}m(t)\, \theta + \int_0^\infty (e^{\mathrm{i}\theta x} - 1)\, \Pi(t, dx),$$

---

[21] The integrals are to be understood as the limits in probability of the corresponding Riemannian sums.

with

$$\int_0^1 x\Pi(t, dx) < \infty$$

and increasing $m(.)$.

2.2.3.2 The three parts decomposition

2.2.3.2.1. It is obvious that if ξ is a process with independent increments, then $\xi - f(.)$ is also for an arbitrary real-valued function f defined on T. P. LÉVY showed that f may be chosen such that the process $\xi - f(.)$ has certain continuity properties.

A function f is said to be a *centering* function if (i) to each $t \in T$ there correspond two random variables ζ_t^- and ζ_t^+ [22] such that, with probability 1,

$$\lim_{s \to t-} \zeta(s) = \zeta_t^-, \quad \lim_{s \to t+} \zeta(s) = \zeta_t^+,$$

where $\zeta(s) = \xi(s) - f(s)$; (ii) whenever the differences $\zeta_t^+ - \zeta(t)$ or $\zeta(t) - \zeta_t^-$ are constant then they are null; (iii) except possibly for the points of at most a denumerable set of values $t \in T$, one has with probability 1

$$\zeta_t^- = \zeta(t) = \zeta_t^+.$$

A sufficient condition for f to be a centering function is

$$\mathsf{E}\,[\arctan(\xi(t) - f(t))] = 0, \quad t \in T.$$

2.2.3.2.2. The process $\xi - f(.)$ is said to be a *centered* process with independent increments if f is a centering function for ξ.

Theorem 2.2.18 (P. LÉVY). *Every process ξ with independent increments can be represented as*

$$\xi(t) = f(t) + \alpha(t) + \beta(t), \quad t \in T,$$

where f is a centering function, α is a centered process with independent increments with almost all trajectories continuous except at the points of a fixed and at most denumerable set, β is a centered process with independent increments without discontinuities of the second kind. Moreover, the processes α and β are independent and

$$\mathsf{E}\,(\arctan \alpha(t)) = \mathsf{E}\,(\arctan \beta(t)) = 0, \ t \in T.$$

Proof. See DOOB [(1953), pp. 407—417] or ITÔ [(1960), pp. 49—54]. ◇

[22] Except for $t = 0$ when we require only the existence of ζ_0^+.

2.3. Markov processes

Continuous parameter Markov processes were rigorously introduced by A. N. KOLMOGOROV in 1931 who also raised and partly solved the problem of determining their transition functions from suitable infinitesimal characteristics. Important contributions to this problem are due to W. FELLER during the period 1936—1959. The study of the trajectories of Markov processes was initiated by W. DOEBLIN and developed by J. L. DOOB and P. LÉVY. In recent years, E. B. DYNKIN has developed a method of investigating Markov processes by connecting probabilistic ideas with purely analytic considerations.

2.3.1. Preliminaries

2.3.1.1. Transition functions

2.3.1.1.1. Let $(X, \mathcal{X})$ be a measurable space. Paralleling the discrete parameter case, we base the construction of continuous parameter Markov processes on the transition functions. To be precise, let $P(.,.;.,.)$ be a real-valued function defined on $X \times \{(s, t): s, t \in T, s < t\} \times \mathcal{X}$ and satisfying:

(i) $P(s, x; t, .)$ is a probability on $\mathcal{X}$ for fixed $s, t \in T, x \in X$;

(ii) $P(s, .; t, A)$ is an $\mathcal{X}$-measurable function for fixed $s, t \in T, A \in \mathcal{X}$;

(iii) for arbitrary $s < u < t$, $x \in X$, $A \in \mathcal{X}$ the Chapman-Kolmogorov equation

$$P(s, x; t, A) = \int_X P(s, x; u, \mathrm{d}y) P(u, y; t, A)$$

holds.

A function P satisfying (i)—(iii) is said to be a *stochastic* [23] *transition function* on $(X, \mathcal{X})$.

It is convenient to define

$$P(s, x; s, A) = \delta_x(A) = \begin{cases} 1 & \text{if } x \in A \\ 0 & \text{if } x \notin A \end{cases}$$

for all $s \in T$, $x \in X$, $A \in \mathcal{X}$ so that the Chapman-Kolmogorov equation will also hold for $u = s$ or $u = t$.

2.3.1.1.2. Sometimes instead of (i) one considers the weaker condition

(i′) $P(s, x; t, .)$ is a non-negative completely additive set function for fixed $s, t \in T$, $x \in X$ such that $P(s, x; t, X) \leqslant 1$.

[23] The nomenclatures *Markovian, proper,* and *honest,* are also used.

A function P satisfying (i'), (ii), and (iii) is said to be a *substochastic (submarkovian, improper,* or *dishonest) transition function* on $(X, \mathscr{X})$.

With any substochastic transition function P one can always associate a stochastic transition function $\overline{P}$ as follows: let $\mathfrak{d}$ be an arbitrary object, $\mathfrak{d} \notin X$, and consider the measurable space $(\overline{X}, \overline{\mathscr{X}})$ where $\overline{X} = X \cup \{\mathfrak{d}\}$ and $\overline{A} \in \overline{\mathscr{X}}$ iff $\overline{A} \cap X \in \mathscr{X}$. Then set

$$\overline{P}(s, x; t, A) = P(s, x; t, A) \quad , \quad x \in X, A \in \mathscr{X},$$

$$\overline{P}(s, x; t, \{\mathfrak{d}\}) = 1 - P(s, x; t, X), \quad x \in X,$$

$$\overline{P}(s, \mathfrak{d}; t, \{\mathfrak{d}\}) = 1$$

It is easy to see that $\overline{P}$ satisfies (i) — (iii), and therefore it is a stochastic transition function on $(\overline{X}, \overline{\mathscr{X}})$.

In the following, unless otherwise stated, transition functions will be stochastic.

2.3.1.2. Definition and existence theorem

2.3.1.2.1. From the many possibilities of defining continuous parameter Markov processes we choose one which imitates the Markov property for chains. For other possibilities see Cuculescu (1968, p. 61), Dynkin (1961, 2.1), Hunt (1966, pp. 75—80) and Loève (1963, p. 562—566).

Definition 2.3.1. A process $(\xi(t))_{t \in T}$ on the probability space $(\Omega, \mathcal{K}, \mathsf{P})$ is said to be a *Markov process* with state space $(X, \mathscr{X})$ and transition function P[24] if for all $t_i \in T$, $0 \leqslant i \leqslant n$, such that $t_0 < \ldots < t_n$, $n \in N^*$, and $A \in \mathscr{X}$ we have

$$\mathsf{P}(\xi(t_n) \in A \mid \xi(t_0), \ldots, \xi(t_{n-1})) = \mathsf{P}(\xi(t_n) \in A \mid \xi(t_{n-1})) =$$
$$= P(t_{n-1}\, \xi(t_{n-1}); t_n, A)$$

P — a.s. for all $s, t \in T$, $s < t$, $A \in \mathscr{X}$.

Clearly, Definition 2.3.1 amounts to the fact that for every increasing sequence $(t_n)_{n \in N}$, $t_n \in T$, the sequence of random variables $(\xi(t_n))_{n \in N}$ is a Markov chain.

---

[24] A more general concept of a Markov process which does not involve transition functions can be obtained by suppressing the latter term in the equations below. For details see Loève (1963, pp. 567—568).

Definition 2.3.2. A Markov process is said to be *homogeneous* if its transition function P is time homogeneous, i.e. the dependence of P on the time arguments s, t reduces to dependence upon their differences $t-s$ only. That is, $P(s, x; t, A) = P(t-s; x, A)$.

We leave to the reader the task of working out the continuous parameter analogues of the variants of the Markov property given for chains in Paragraph 1.1.1.1.

2.3.1.2.2. Under rather weak conditions one can prove an existence theorem for a Markov process with given transition function.

Theorem 2.3.3. *Let X be a σ-compact topological space and $\mathcal{X}$ the σ-algebra of Borel sets in X. Given a transition function P on $(X, \mathcal{X})$, there exist a probability space $(\Omega, \mathcal{K}, \mathsf{P})$ and a Markov process on it with state space $(X, \mathcal{X})$ and transition function P.*

Proof. We proceed as in the discrete parameter case from the very definition of a Markov process. We shall take $\Omega = X^T$ (the set of X-valued functions defined on T), $\mathcal{K} = \mathcal{X}^T$. Let p be an arbitrary probability on $\mathcal{X}$ which will serve as the initial distribution of the process. We define the finite dimensional distributions by

$$F_{0t_1\ldots t_n}(A_0, A_1, \ldots, A_n) =$$

$$\int_{A_0} p(\mathrm{d}x_0) \int_{A_1} P(0, x_0; t_1, \mathrm{d}x_1) \ldots \int_{A_n} P(t_{n-1}, x_{n-1}; t_n, dx_n)$$

for $0 < t_1 < \ldots < t_n$, $t_i \in T$, $1 \leqslant i \leqslant n$, $A_j \in \mathcal{X}$, $0 \leqslant j \leqslant n$, $n \in N^*$. Set

$$\xi(t, \omega) = \mathrm{pr}_t\, \omega = x_t$$

if $\omega = (x_t)_{t \in T}$. The existence of a probability P on $\mathcal{K}$ yielding the above finite dimensional distributions is assured by Theorem 2.1.4 The fact that $(\xi(t))_{t \in T}$ is a Markov process is obvious by the very definition of its finite dimensional distributions. ◇

When we want to emphasize the initial distribution p we shall write P_p, in particular P_x if $p = \delta_x$.

We note that the existence of a first element in T is essential for Theorem 2.3.3 to hold. For details see Cuculescu (1968, pp. 15—16).

2.3.1.2.3. Let X be a locally compact but not compact topological space and $\mathcal{X}$ the σ-algebra of Borel sets in X.

With a substochastic transition function P on $(X, \mathcal{X})$ one can associate a Markov process as follows: Consider the measurable space $(\overline{X}, \overline{\mathcal{X}})$ and the transition probability $\overline{P}$ on $(\overline{X}, \overline{\mathcal{X}})$ introduced in Subparagraph 2.3.1.1.2. We agree that $\mathfrak{d}$ is adjoined to X as the point at infinity if X is non-compact and as an isolated point otherwise. Then Theorem 2.3.3 applies to $(\overline{X}, \overline{\mathcal{X}})$ and $\overline{P}$ yielding a Markov process $(\overline{\xi}(t))_{t \in T}$ with state space $(\overline{X}, \overline{\mathcal{X}})$ and transition function $\overline{P}$. It is clear that if $\xi(s, \omega) = \mathfrak{d}$, then $\xi(t, \omega) = \mathfrak{d}$ for all $t \geqslant s$. Put $\zeta(\omega) = \inf(t: \xi(t, \omega) = \mathfrak{d})$, with $\zeta(\omega) = a$ if the set within parentheses is empty. Thus ζ is the *lifetime* of $(\xi(t))_{t \in T}$, i.e. the duration of its stay in X.[25]

The above construction provides a probabilistic interpretation of substochastic transition functions.

Notice that these considerations also provide a variant of the existence theorem for a locally compact but not compact state space and a (stochastic) transition function.

2.3.1.3. The strong Markov property

2.3.1.3.1. The investigation of the behaviour of trajectories of a Markov process $(\xi(t))_{t \in T}$ requires a strengthening of the Markov property appearing in Definition 2.3.1. As soon as trajectory questions arise, we have to consider random times, e.g., the time of the first discontinuity of a trajectory. We might expect the family $(\xi(t + \tau))_{t \in T}$, where τ is a random time, to still be a Markov process[26]. The consideration of random times raises, however, immediate difficulties: such a time may assume infinite values or may not be measurable. Moreover, if τ is finite and measurable, $\xi(\tau)$ may not be measurable.

2.3.1.3.2. To eliminate these difficulties, we adopt the following precise definitions.

Let us consider a Markov process $(\xi(t))_{t \in T}$ with state space $(X, \mathscr{X})$ (an arbitrary measurable space) and transition function P. A *random time* τ is a $\mathcal{K}$-measurable mapping of Ω into $[0, a]$ such that $\mathsf{P}(\tau < a) > 0$. A *Markov time* is a random time τ such that for all $t \in T$, the set $(\tau \leqslant t)$ belongs to $\mathcal{K}_t$ — the σ-algebra generated by $\xi(s)$, $0 \leqslant s \leqslant t$. Let $\Omega^\tau = (\tau < a)$, $\mathcal{K}^\tau = (\Omega^\tau \cap K: K \in \mathcal{K})$, $\mathsf{P}^\tau = \mathsf{P}/\mathsf{P}(\Omega^\tau)$. Denote by $\mathcal{K}_\tau$ the σ-algebra of sets $K \in \mathcal{K}^\tau$ such that $K \cap (\tau < t) \in \mathcal{K}_t$ for all $t \in T$.

Definition 2.3.4. The process $(\xi(t))_{t \in T}$ is said to have the *strong Markov property* (or to be a *strong* Markov process) if it is measurable and

(j) $P(\cdot, \cdot; t, A)$ is a $\mathscr{B}_t \times \mathscr{X}$ — measurable function for fixed $t \in T$, $A \in \mathscr{X}$. Here $\mathscr{B}_t$ is the σ-algebra of Borel sets in $[0, t]$.

(jj) For an arbitrary Markov time τ of $(\xi(t))_{t \in T}$

$$\mathsf{P}^\tau(\xi(t + \tau) \in A \mid \mathcal{K}_\tau) = P(\tau, \xi(\tau); \tau + t, A)$$

P^τ — a.s. for all $t \in T$, $A \in \mathscr{X}$.

---

[25] We note that a theory of Markov processes for which the trajectories have both random terminal and starting times was first considered by HUNT (1960).

[26] For example, if we wish to compute the distribution of $\xi(t + \tau)$ given $\xi(\tau) = x$, it seems reasonable that we should be able to invoke the Markov property at the random time τ.

The systematic study of the strong Markov property began in 1955—1956 with the work of E.B. DYNKIN, A. A. JUŠKEVIČ, and D. RAY. Before their work there is only a pioneering paper by J. L. DOOB. As we already noted in Paragraph 1.1.1.4, not all continuous parameter Markov processes are strong Markov ones. See a counterexample in LOÈVE (1963, p. 578). The basic distinction concerning the strong Markov property between discrete and continuous parameter processes is that Markov times for discrete parameter processes can take only countably many values.

2.3.1.3.3. We now indicate a criterion for the strong Markov property.

Proposition 2.3.5. (E. B. DYNKIN, A. A. JUŠKEVIČ). *Let X be a topological space and $\mathcal{X}$ the σ-algebra of Borel sets in X. Suppose that for every open set $U \subset X$ there exists a real-valued continuous function f on X such that $f(x) \neq 0$ iff $x \in U$. A sufficient condition for a right continuous Markov process with state space $(X, \mathcal{X})$ and transition function P satisfying* (j) *of Definition 2.3.4 to be strong Markov is the following: for an arbitrary real-valued, bounded and continuous function h on X, the function*

$$F(s) = \int_X P(s, \xi(s, \omega); t, \mathrm{d}z)\, h(z)$$

is right-continuous for all fixed $\omega \in \Omega$, $t > s$.

Proof. See DYNKIN (1961, 5.18). ◇

In the homogeneous case, the continuity condition in the above proposition can be modified as follows: for an arbitrary bounded real-valued function h on X the function

$$G(s) = \int_X P(t; \xi(s, \omega), \mathrm{d}y)\, h(y)$$

is right-continuous for all fixed $\omega \in \Omega$, $t \geqslant 0$.

For details concerning the strong Markov property we refer the reader to BLUMENTHAL and GETOOR (1968, pp. 37—44), CUCULESCU (1968, Ch. 14), DYNKIN [(1961, Ch. 5), (1965, Ch. 3)], HUNT (1966, § 11), ITÔ (1963), LOÈVE (1963, § 38), and MEYER (1967, Ch. 13).

2.3.1.4. The semi-group approach to homogeneous Markov processes

2.3.1.4.1. Let $(X, \mathcal{X})$ be an arbitrary measurable space and $P(.\,;.,.)$ a time-homogeneous substochastic transition function. Thus $P(t; x, A)$, $t \in T = [0, \infty)$, $x \in X$, $A \in \mathcal{X}$ is $\mathcal{X}$-measurable in x, completely additive in A, with $P(t; x, X) \leqslant 1$, $P(0; x, A) = \delta_x(A)$ and it satisfies the homogeneous Chapman-Kolmogorov equation

$$P(s+t; x, A) = \int_X P(s; x, \mathrm{d}y)\, P(t; y, A) \tag{2.3.1}$$

for $s, t \geqslant 0$.

With $(X, \mathscr{X})$ we associate two Banach spaces $B(X, \mathscr{X})$ and $V(X, \mathscr{X})$. The first one consists of all real-valued, bounded, and $\mathscr{X}$-measurable functions f defined on X with the norm

$$\|f\| = \sup_{x \in X} |f(x)|;$$

the second one consists of all real-valued completely additive signed measures μ defined on $\mathscr{X}$ with the norm

$$\|\mu\| = \operatorname{var} \mu.$$

The equation

$$(T_t f)(x) = \int_X P(t; x, \mathrm{d}y) f(y), \quad t \geqslant 0,$$

defines a family of linear operators on $B(X, \mathscr{X})$. Similarly, the equation

$$(U_t \mu)(A) = \int_X P(t; x, A)\, \mu(\mathrm{d}x), \quad t \geqslant 0,$$

defines a family of linear operators on $V(X, \mathscr{X})$. The Chapman-Kolmogorov equation and the inequality $P(t; x, X) \leqslant 1$ lead us to the conclusion that the families $(T_t)_{t \geqslant 0}$ and $(U_t)_{t \geqslant 0}$ are contraction semigroups, i.e.,

$$T_{s+t} = T_s T_t, \quad U_{s+t} = U_s U_t, \quad s, t \geqslant 0,$$

$$\|T_t f\| \leqslant \|f\|, \ \|U_t \mu\| \leqslant \|\mu\|, \ t \geqslant 0.$$

The simple remark above is at the root of the use of Hille-Yosida semigroup theory in investigating homogeneous Markov processes. For details see DYNKIN (1965, Chs. 1 and 2).

2.3.1.4.2. Suppose now that X is a topological space and that $\mathscr{X}$ is the σ-algebra of Borel sets in X. W. FELLER who developed the semigroup approach to homogeneous Markov processes was concerned with the space $C(X) \subset B(X, \mathscr{X})$ of real-valued bounded continuous functions defined on X. This led him to consider only transition functions such that $T_t f \in C(X)$ for all $t \geqslant 0$, $f \in C(X)$. These are now called *Feller* transition functions and their semigroup theory is relatively simple. For a concise account we refer the reader to ITÔ (1963).

2.3.2. Markov jump processes. I. General theory

In this and the next two subsections we shall be concerned with a class of Markov processes which are characterized by the fact that in a small time interval such a process remains with large probability in the state originally occupied and jumps to another state with only small probability. These are the so-called classical Markov jump processes. Generally speaking, a process $(\xi(t))_{t\in T}$ is said to be a *jump process* if for almost all $\omega \in \Omega$ and all $t \in T$ there exists an $h > 0$ such that $\xi(t, \omega) = \xi(t + u, \omega)$ for $0 \leqslant u < h$, equivalently, if almost all the trajectories are right-continuous in the discrete topology (in the state space). The Markov processes we now study are characterized by the existence of transition intensity functions, which allows them to be represented as jump processes. For a class of Markov processes which are an extension of those considered here we refer the reader to MOYAL (1957).

Throughout this subsection we consider only state spaces $(X, \mathcal{X})$ such that all the singletons $\{x\}$, $x \in X$, belong to $\mathcal{X}$.

2.3.2.1. Transition intensity functions

2.3.2.1.1. We are led naturally to the notion of a transition intensity function by the study of the differentiability of a substochastic transition function P with respect to the time arguments.

Let us introduce the function

$$F(s_1, s_2; x, A) = \frac{P(s_1, x; s_2, A) - \delta_x(A)}{s_2 - s_1}$$

which for all given $s_1 < s_2$ is

completely additive with respect to $A \in \mathcal{X}$;

and

$\mathcal{X}$-measurable with respect to $x \in X$.

Notice that

$$F(s_1, s_2; x, A) \begin{cases} \leqslant 0 & \text{if } x \in A \\ \geqslant 0 & \text{if } x \notin A \\ = 0 & \text{if } A = \emptyset. \end{cases}$$

By taking $u = s_2$ in the Chapman-Kolmogorov equation we get

$$P(s_2, x; t, A) - P(s_1, x; t, A) =$$

$$= -\int_X [P(s_1, x; s_2, dy) - P(s_1, x; s_1, dy)]\, P(s_2, y; t, A),$$

which yields

$$\frac{P(s_2, x; t, A) - P(s_1, x; t, A)}{s_2 - s_1} = -\int_X F(s_1, s_2; x, \mathrm{d}y)\, P(s_2, y; t, A). \tag{2.3.2}$$

Analogously, taking $u = t_1$ and replacing t by $t_2 > t_1$ in the Chapman-Kolmogorov equations, we find

$$\frac{P(s, x; t_2, A) - P(s, x; t_1, A)}{t_2 - t_1} = \int_X P(s, x; t_1, \mathrm{d}y)\, F(t_1, t_2; y, A). \tag{2.3.3}$$

It follows that the existence of the derivative $\partial P/\partial s$, respectively, $\partial P/\partial t$ is equivalent to the existence of the limit

$$\lim_{\substack{h+k\to 0+\\(h\equiv 0 \text{ or } k\equiv 0)}} \int_X F(s-h, s+k; x, \mathrm{d}y)\, P(s+k, y; t, A),$$

respectively,

$$\lim_{\substack{h+k\to 0+\\(h\equiv 0 \text{ or } k\equiv 0)}} \int_X P(s, x; t-h, \mathrm{d}y)\, F(t-h, t+k; y, A).$$

It is to be expected that the existence of these limits is connected with the existence of the limit of $F(s-h, s+k;\ x, A)$ as $h + k \to 0+$ with either $h \equiv 0$ or $k \equiv 0$. Although this assumption is not sufficient to imply completely satisfactory results for our problem, it delineates an interesting class of transition functions.

2.3.2.1.2. Suppose [27] that there exists

$$\lim_{\substack{h+k\to 0+\\(h\equiv 0 \text{ or } k\equiv 0)}} F(s-h, s+k; x, A) = Q(s, x, A) \tag{2.3.4}$$

for all $s \in T$, $x \in X$, $A \in \mathscr{X}$; the function Q is called the *transition intensity function* (shortly, the *intensity*) corresponding to P and it is easily seen that

i) Q is completely additive with respect to $A \in \mathscr{X}$;
ii) Q is $\mathscr{X}$-measurable with respect to $x \in X$;

---

[27] In special cases (see Subparagraphs 2.3.4.2.3 and 2.3.4.3.6) this assumption represents no restriction, being automatically satisfied.

and

iii)

$$Q(s, x, A)\begin{cases} \leqslant 0 & \text{if } x \in A \\ \geqslant 0 & \text{if } x \notin A \\ = 0 & \text{if } A = \emptyset. \end{cases}$$

It is convenient to represent Q in terms of two other functions which have an interesting probabilistic interpretation. Set

$$q(s, x) = -Q(s, x), \{x\}), \text{ and}$$

$$\sigma(s, x, A) = \begin{cases} \delta_x(A) & \text{if } q(s, x) = 0 \\ \dfrac{Q(s, x, A - \{x\})}{q(s, x)} & \text{if } q(s, x) \neq 0. \end{cases}$$

The properties of Q imply that

q is non-negative and $\mathscr{X}$-measurable with respect to $x \in X$;

σ is non-negative, $\mathscr{X}$-measurable with respect to $x \in \mathscr{X}$ and completely additive with respect to $A \in \mathscr{X}$ with $\sigma(s, x, X) \leqslant 1$.

We obviously have

$$Q(s, x, A) = q(s, x)[-\delta_x(A) + \sigma(s, x, A)].$$

Concerning the probabilistic interpretation of q and σ we have first

$$P(s, x; t, \{x\}) = 1 - (q(s, x) + o(1))(t - s);$$

consequently $q(s, x)(t - s) + o(t - s)$ is the probability of no longer being in x at time t given that the process was in x at time s. Secondly, if $q(s, x) \neq 0$ we may write

$$\sigma(s, x, A) = \lim_{t \to s+} \frac{P(s, x; t, A - \{x\})}{1 - P(s, x; t, \{x\})}.$$

Thus $\sigma(s, x, A)$ might be thought of as the probability of a jump from x at time s into the set $A - \{x\}$ given that a jump has in fact occurred. This interpretation will be justified by the corollary to Theorem 2.3.16.

2.3.2.2. The Kolmogorov-Feller equations

2.3.2.2.1. We can now prove

Proposition 2.3.6. *Under condition* (2.3.4), *the partial left-derivative* $\partial^- P(s, x; t, A)/\partial s$ *exists for all* $s, t \in T$, $x \in X$, $A \in \mathcal{X}$ *and is given by*

$$\frac{\partial^- P(s, x; t, A)}{\partial s} = -\int_X Q(s, x, \mathrm{d}y)\, P(s, y; t, A).$$

Proof. According to (2.3.2) we may write for $h > 0$

$$\frac{P(s-h, x; t, A) - P(s, x; t, A)}{-h} = -\int_X F(s-h, s; x, \mathrm{d}y)\, P(s, y; t, A)$$

or, equivalently,

$$\frac{P(s-h, x; t, A) - P(s, x; t, A)}{-h} = -P(s, x; t, A)\, F(s-h, s; x, \{x\})$$

$$-\int_{X-\{x\}} F(s-h, s; x, \mathrm{d}y)\, P(s, y; t, A). \tag{2.3.5}$$

Now by (2.3.4)

$$F(s-h, s; x, \{x\}) \to -q(s, x)$$

and

$$F(s-h, s; x, B) \to q(s, x)\, \sigma(s, x, B)$$

as $h \to 0+$ for all sets $B \in \mathcal{X}$ not containing x. Then for fixed s and x the integral in (2.3.5) will approach [28)]

$$-\int_{X-\{x\}} -q(s, x)\, \sigma(s, x, \mathrm{d}y)\, P(s, y; t, A)$$

as $h \to 0+$. Therefore (2.3.5) leads to the stated result. ◇

[28)] Here we use the following proposition : *If finite measures* μ_n, $n \in N^*$, *on* $\mathcal{X}$ *converge to a finite measure* μ *on* $\mathcal{X}$ (i. e. $\lim_{n\to\infty} \mu_n(A) = \mu(A)$ *for all* $A \in \mathcal{X}$) *and* f *is a real-valued, bounded, and* $\mathcal{X}$*-measurable function defined on* X, *then* $\lim_{n\to\infty}\int_X f(x)\mu_n(\mathrm{d}x) = \int_X f(x)\mu(\mathrm{d}x)$ [see LOÈVE (1963, p.593)]

Corollary. *Under condition* (2.3.4) *the partial derivative* $\partial P(s,x;t,A)/\partial s$ *exists for almost all* $s \in T$ *(the set of excepted values of* s *may depend on* $x \in X$, $t \in T$, $A \in \mathcal{X}$*) and satisfies the equation*

$$\frac{\partial P(s, x; t, A)}{\partial s} = -\int_X Q(s, x, dy)\, P(s, y; t, A). \qquad (2.3.6)$$

Proof. Use the well-known Denjoy theorem [see Saks (1937), p. 269]. ◇

The equation (2.3.6) is called the *backward* Kolmogorov-Feller equation because it involves differentiation with respect to the earlier time s.

2.3.2.2.2. An analogous result concerning the partial right-derivative $\partial^+ P(s, x; t, A) / \partial t$ may be obtained if we impose further restrictions.

Proposition 2.3.7. *Under condition* (2.3.4) *and the boundedness of* $q(s, .)$ *for every* $s \in T$ *the partial right-derivative* $\partial^+ P(s, x; t, A) / \partial t$ *exists for all* $s, t \in T$, $x \in X$, $A \in \mathcal{X}$ *and is given by*

$$\frac{\partial^+ P(s, x; t, A)}{\partial t} = \int_X P(s, x; t, dy)\, Q(t, y, A).$$

Proof. According to (2.3.3) we may write for $h > 0$

$$\frac{P(s, x; t + k, A) - P(s, x; t, A)}{k} =$$

$$\int_X P(s, x; t, dy)\, F(t, t + k; y, A). \qquad (2.3.7)$$

Taking into account the boundedness of $q(t, .)$ the integral can be shown to approach

$$\int_X P(s, x; t, dy)\, Q(t, y, A)$$

as $k \to 0+$ and therefore (2.3.7) leads to the stated result. ◇

Corollary. *Under the assumptions of Proposition 2.3.7, the partial derivative* $\partial P(s, x; t, A) / \partial t$ *exists for almost all* $t \in T$ *(the set of excepted values of* t*, may depend on* $s \in T$, $x \in X$, $A \in \mathcal{X}$*) and satisfies the equation*

$$\frac{\partial P(s, x; t, A)}{\partial t} = \int_X P(s, x; t, dy)\, Q(t, y, A). \qquad (2.3.8)$$

Proof. The same argument as in the proof of the corollary to Proposition 2.3.6. ◇

The equation (2.3.8) is called the *forward* Kolmogorov-Feller equation because it involves differentiation with respect to the later time t.

2.3.2.2.3. To obtain equations (2.3.6) and (2.3.8) without excluding some values of s and t, stronger restrictions must in general be imposed on P and the limit in (2.3.4). Thus, to deduce the backward equation it suffices to assume that $P(., x; t, A)$ is right-continuous uniformly with respect to $x \in X$ and that the limit in (2.3.4) is uniform with respect to $A \in \mathscr{X}$. Correspondingly, to obtain the forward equation it suffices to assume that $P(s, x; ., A)$ is left-continuous uniformly with respect to $A \in \mathscr{X}$ and that the limit in (2.3.4) is uniform with respect to $x \in X$. In some special cases (see Subparagraphs 2.3.3.1.4, 2.3.4.3.3 and 2.3.4.3.4), the Kolmogorov-Feller equations hold under weaker restrictions.

2.3.2.2.4. Let us note the simplifications arising in the homogeneous case for which $P(s, x; t, A) = P(t - s; x, A)$. In this case, the transition intensity function does not depend on $s \in T$, it being defined by

$$\lim_{t \to 0+} \frac{P(t; x, A) - \delta_x(A)}{t} = Q(x, A). \tag{2.3.4'}$$

The same will be true for the functions q and σ.

Note that for (2.3.4') to hold it is necessary that

$$\lim_{t \to 0+} P(t; x, A) = \delta_x(A)$$

for all $x \in X$, $A \in \mathscr{X}$. Consequently, throughout this and the next two subsections we will always suppose that in the homogeneous case the above continuity condition holds.

Finally, note that the Kolmogorov-Feller equations become

$$\frac{\partial P(t; x, A)}{\partial t} = \int_X Q(x, \mathrm{d}y)\, P(t; y, A) \tag{2.3.6'}$$

$$\frac{\partial P(t; x, A)}{\partial t} = \int_X P(t; x, \mathrm{d}y)\, Q(y, A). \tag{2.3.8'}$$

2.3.2.3. Determining a transition function from its intensity

2.3.2.3.1. We turn now to the problem of finding P given Q. More precisely, we are confronted with the problem of determining (E) whether or not to any function Q satisfying i) — iii) in Subparagraph 2.3.2.1.2

there corresponds a transition probability function P whose intensity is Q, and if one does correspond, (U) whether or not this P is uniquely determined. We shall see that it is possible to solve the existence part of our problem completely while the uniqueness part will receive only a partial answer. Our treatment follows with some improvements FELLER (1940).

In the following we shall adopt a slighty less restrictive point of view concerning the limit in (2.3.4); namely, we shall require that (2.3.4) hold for almost all $s \in T$ (the set of excepted values of s may depend on $x \in X$ and $A \in \mathcal{X}$).

Let us list a number of conditions to be imposed on the functions q and σ in the following.

C_1. q is $\mathcal{B}_T \times \mathcal{X}$-measurable ($\mathcal{B}_T$ = the σ-algebra of Borel sets in T), and $q(., x)$ is Lebesgue-integrable on every finite interval contained in T for all fixed $x \in X$.

C_1'. q is $\mathcal{B}_T \times \mathcal{X}$-measurable, and $q(., x)$ is continuous on T for all fixed $x \in X$.

C_2. $\sigma(., ., A)$ is $\mathcal{B}_T \times \mathcal{X}$-measurable for all fixed $A \in X$.

C_2'. $\sigma(., ., A)$ is $\mathcal{B}_T \times \mathcal{X}$-measurable for all fixed $A \in X$ and $\sigma(., x, A)$ is continuous uniformly with respect to $A \in \mathcal{X}$ on every finite interval contained in T.

2.3.2.3.2. The problem (E) is solved by

Theorem 2.3.8. *Under C_1 and C_2 there exists a (substochastic) transition function $P_{\min}$[29] whose intensity is Q and which satisfies the integral equations*

$$P_{\min}(s, x; t, A) = \delta_x(A) + \int_s^t \mathrm{d}u \int_X P_{\min}(u, y; t, A)\, Q(u, x, \mathrm{d}y) \tag{2.3.9}$$

and

$$P_{\min}(s, x; t, A) = \delta_x(A) + \int_s^t \mathrm{d}u \int_X Q(u, y, A)\, P_{\min}(s, x; u, \mathrm{d}y). \tag{2.3.10}$$

Proof. Set

$$\psi(s, t, x) = \exp\left(-\int_s^t q(u, x)\, \mathrm{d}u\right),$$

$$\tilde{q}(s, x, A) = q(s, x)\, \sigma(s, x, A).$$

[29] The subscript min (= minimum) will be explained later.

Consider the functions

$$P_0(s, x; t, A) = \delta_x(A)\, \psi(s, t, x),$$

$$P_{n+1}(s, x; t, A) =$$

$$\int_s^t du \int_X P_n(u, y; t, A)\; \psi(s, u, x)\, \tilde{q}(u, x, dy),$$

$n \in N$, $s \leqslant t \in T$, $A \in \mathcal{X}$. It will be seen later that $P_n(s, x; t, A)$ may be thought of as the probability of n jumps in the interval $[s, t]$ in the course of going from x at time s into the set A at time t. It is obvious that the functions P_n, $n \in N$, are nonnegative, completely additive with respect to $A \in \mathcal{X}$, and $\mathcal{X}$-measurable with respect to $x \in X$. If we set

$$P^n_{\min} = \sum_{k=0}^{n} P_k, \quad n \in N,$$

it is easy to see that $P^{n+1}_{\min} \geqslant P^n_{\min}$, $n \in N$, and that

$$P^{n+1}_{\min}(s, x; t, A) = \delta_x(A)\, \psi(s, t, x) +$$

$$+ \int_s^t du \int_X P^n_{\min}(u, y; t, A)\, \psi(s, u, x)\, \tilde{q}(u, x, dy). \tag{2.3.11}$$

By supposing that $P^n_{\min} \leqslant 1$ (which holds for $n = 0$), we deduce

$$P^{n+1}_{\min}(s, x; t, A) \leqslant \psi(s, t, x) + \int_s^t q(u, x)\, \psi(s, u, x)\, du = 1.$$

It follows by induction that for every $n \in N$

$$0 \leqslant P^n_{\min} \leqslant 1,$$

and thus $P_{\min} = \lim_{n \to \infty} P^n_{\min} = \sum_{n \in N} P_n$ exists with

$$0 \leqslant P_{\min} \leqslant 1,$$

$$P_{\min}(s, x; s, A) = \delta_x(A).$$

The function $P_{\min}$ is completely additive with respect to $A \in \mathcal{X}$ and $\mathcal{X}$-measurable with respect to $x \in X$ because the functions $P^n_{\min}$, $n \in N$, have these properties; therefore, to show that $P_{\min}$ is a substochastic transition function, it remains to verify that it satisfies the Chapman-Kolmogorov equation. We note that

$$\int_X P_{\min}(s, x; u, \mathrm{d}y)\, P_{\min}(u, y; t, A) =$$

$$= \sum_{l,k \in N} \int_X P_l(s, x; u, \mathrm{d}y)\;\; P_k(u, y; t, A) =$$

$$= \sum_{n \in N} \sum_{l=0}^{n} \int_X P_l(s, x; u, \mathrm{d}y)\, P_{n-l}(u, y; t, A),\;\; s \leqslant u \leqslant t,$$

and thus if we can show that

$$\sum_{l=0}^{n} \int_X P_l(s, x; u, \mathrm{d}y)\, P_{n-l}(u, y; t, A) = P_n(s, x; t, A), \tag{2.3.12}$$

the verification will be concluded. For $n = 0$, relation (2.3.12) is an identity. Supposing that (2.3.12) is true for an $n \in N$, we have

$$\sum_{l=0}^{n+1} \int_X P_l(s, x; u, \mathrm{d}y)\, P_{n+1-l}(u, y; t, A) =$$

$$= \psi(s, u, x)\, P_{n+1}(u, x; t, A) +$$

$$+ \sum_{l=1}^{n+1} \int_X \int_s^t \mathrm{d}v \int_X P_{l-1}(v, z; u, \mathrm{d}y)\, \psi(s, v, x)\, \tilde{q}(v, x, \mathrm{d}z)\, P_{n+1-l}(u, y; t, A) =$$

$$= \psi(s, u, x) \int_u^t \mathrm{d}v \int_X P_n(v, z; t, A)\;\; \psi(u, v, x)\, \tilde{q}(v, x, \mathrm{d}z) +$$

$$+ \int_s^u \mathrm{d}v \int_X P_n(v, z; t, A)\;\; \psi(s, v, x)\, \tilde{q}\,(v, x, \mathrm{d}z) =$$

$$= \int_u^t \mathrm{d}v \int_X P_n(v, z; t, A)\;\; \psi(s, v, x)\, \tilde{q}\,(v, z, \mathrm{d}z) +$$

$$+ \int_s^u \mathrm{d}v \int_X P_n(v, z; t, A) \psi(s, v, x)\, \tilde{q}\,(v, x, \mathrm{d}z) = P_{n+1}(s, x; t, A).$$

To prove the fact that the intensity of $P_{\min}$ is Q let us assume that one of equations (2.3.9) and (2.3.10), say (2.3.9), holds. By iterating it we get

$$P_{\min}(s, x; t, A) = \delta_x(A) + \int_s^t Q(u, x, A)\, du +$$

$$+ \int_s^t du \int_u^t dv \int_{X^2} Q(u, x, dy)\, Q(v, y, dz)\, P_{\min}(v, z; t, A), \qquad (2.3.13)$$

and well-known properties of Lebesgue integral imply

$$\lim_{\substack{h+k \to 0+ \\ (h \equiv 0 \text{ or } k \equiv 0)}} \frac{P_{\min}(s-h, x; s+k, A) - \delta_x(A)}{h+k} = Q(s, x, A)$$

for almost all $s \in T$. Notice that if $Q(., x, A)$ is continuous, then the above equality holds for *all* $s \in T$.

Now let us prove that $P_{\min}$ satisfies equation (2.3.9). Letting $n \to \infty$ in (2.3.11), the monotone convergence of $P^n_{\min}$ leads to the integral equation

$$P_{\min}(s, x; t, A) = \delta_x(A)\, \psi(s, t, x) +$$

$$+ \int_s^t du \int_X P_{\min}(u, y; t, A)\, \psi(s, u, x)\, \tilde{q}(u, x, dy). \qquad (2.3.14)$$

We shall show that this equation is equivalent to (2.3.9). Taking into account that

$$\psi(s, t, x) + \int_s^t q(u, x)\, \psi(u, t, x)\, du = 1,$$

(2.3.14) can be written as

$$P_{\min}(s, x; t, A) = \delta_x(A) + \int_s^t du \Big[- \delta_x(A)\, q(u, x)\, \psi(u, t, x) +$$

$$\int_X P_{\min}(u, y; t, A)\, \psi(s, u, x)\, \tilde{q}(u, x, dy) \Big]. \qquad (2.3.15)$$

By (2.3.14)

$$\delta_x(A) \psi(u, t, x) = P_{\min}(u, x; t, A) -$$

$$\int_u^t dv \int_X P_{\min}(v, y; t, A)\, \psi(u, v, x)\, \tilde{q}(v, x, dy)$$

and (for fixed t and A) with

$$f(u) = \int_X P_{\min}(u, y; t, A)\, \tilde{q}(u, x, dy),$$

(2.3.15) becomes

$$P_{\min}(s, x; t, A) = \delta_x(A) - \int_s^t q(u, x)\, P_{\min}(u, x; t, A)\, du +$$

$$+ \int_s^t q(u, x)\, du \int_u^t \psi(u, v, x) f(v)\, dv + \int_s^t \psi(s, u, x) f(u)\, du.$$

But

$$\int_s^t q(u, x)\, du \int_u^t \psi(u, v, x) f(v)\, dv =$$

$$= \int_s^t f(v)\, dv \int_s^v q(u, x)\, \psi(u, v, x)\, du =$$

$$= \int_s^t f(v)\, dv - \int_s^t \psi(s, v, x) f(v)\, dv$$

(the justification of interchange of the order of integration is obvious). Therefore we have

$$P_{\min}(s, x; t, A) = \delta_x(A) - \int_s^t q(u, x)\, P_{\min}(u, x; t, A)\, du +$$

$$+ \int_s^t du \int_X P_{\min}(u, y; t, A)\, \tilde{q}(u, x, dy) = \delta_x(A) +$$

$$+ \int_s^t du \int_X P_{\min}(u, y; t, A)\, Q(u, x, dy).$$

The reader will be able to deduce (2.3.14) from (2.3.9) for himself.

The demonstration of (2.3.10) requires some preparation. Let us set

$$\Psi(s, x; t, A) = \int_A \psi(s, t, y)\, \sigma(s, x, dy)$$

and consider the functions

$$P_0'(s, x; t, A) = \delta_x(A)\, \psi(s, t, x)$$

$$P_{n+1}'(s, x; t, A) = \int_s^t du \int_X q(u, y) \Psi(u, y; t, A)\, P_n'(s, x; u, dy),$$

for $n \in N$, $s \leqslant t \in T$, $A \in \mathcal{X}$. We shall prove that

$$P'_n \equiv P_n \,, n \in N.$$

For $n = 0$ the assertion is obviously true and for $n = 1$ easy to verify. Suppose that $P'_k \equiv P_k$ for $0 \leqslant k \leqslant n$ with $n \geqslant 1$. Then, we have

$$P'_{n+1}(s, x; t, A) = \int_s^t du \int_X q(u, y) \Psi(u, y; t, A) \, P_n(s, x; u, dy) =$$

$$= \int_s^t du \int_X q(u, y) \Psi(u, y; t, A) \times$$

$$\times \int_s^u dv \int_X P_{n-1}(v, z; u, dy) \psi(s, v, x) \, \tilde{q}(v, x, dz) =$$

$$= \int_s^t \psi(s, v, x) \, dv \int_X \tilde{q}(v, x, dz) \times$$

$$\times \int_v^t du \int_X q(u, y) \Psi(u, y; t, A) \, P'_{n-1}(v, z; u, dy) =$$

$$= \int_s^t \psi(s, v, x) \, dv \int_X \tilde{q}(v, x, dz) \, P'_n(v, z; u, A) =$$

$$= \int_s^t dv \int_X P_n(v, z; u, A) \, \psi(s, v, x) \tilde{q}(v, x, dz) = P_{n+1}(s, x; t, A)$$

(the justification of interchange of the order of integration is obvious).

It follows by induction, that $P'_n \equiv P_n$ for every $n \in N$. This implies that

$$P^{n+1}_{\min}(s, x; t, A) = \delta_x(A) \, \psi(s, t, x) +$$

$$\int_s^t du \int_X q(u, y) \Psi(u, y; t, A) \, P^n_{\min}(s, x; u, dy)$$

for every $n \in N$. Letting $n \to \infty$, the monotone convergence of $P^n_{\min}$ leads to the integral equation

$$P_{\min}(s, x; t, A) = \delta_x(A) \psi(s, t, x) +$$

$$+ \int_s^t du \int_X q(u, y) \, \psi(u, y; t, A) P_{\min}(s, x; u, dy). \qquad (2.3.16)$$

We shall show that this equation is equivalent to equation (2.3.10) Taking into account the fact that

$$\psi(s, t, x) + \int_s^t q(u, x)\, \psi(s, u, x)\, du = 1,$$

(2.3.16) may be written as

$$P_{\min}(s, x; t, A) = \delta_x(A)\left[1 - \int_s^t q(u, x)\, \psi(s, u, x)\, du\right] +$$

$$+ \int_s^t du \int_X q(u, y)\left[\int_A \left(1 - \int_u^t q(v, z)\, \psi(u, v, z)\, dv\right)\sigma(u, y, dz)\right] \times$$

$$\times P_{\min}(s, x; u, dy) = \delta_x(A) + \int_s^t du \int_X \tilde{q}(u, y, A)\, P_{\min}(s, x; u, dy) -$$

$$- \delta_x(A) \int_s^t q(v, x)\, \psi(s, v, x)\, dv - \int_s^t du \int_X q(u, y)\, P_{\min}\ (s, x; u, dy) \times$$

$$\times \int_A \sigma(u, y, dz) \int_u^t q(v, z)\, \psi(u, v, z)\, dv.$$

But the last integral equals

$$-\int_s^t dv \int_A q(v, z) \int_s^v du \int_X q(u, y)\, \psi\, (u, v, z)\, \sigma\, (u, y, dz)\, P_{\min}(s, x; u, dy)$$

(the justification of interchange of the order of integration is again obvious) and then, it is easily seen that by virtue of (2.3.16), the sum of the last two integrals equals

$$-\int_s^t dv \int_A q(v, z)\, P_{\min}(s, x; v, dz).$$

Therefore we have found that

$$P_{\min}(s, x; t, A) = \delta_x(A) + \int_s^t du \int_X \tilde{q}(u, y, A)\, P_{\min}(s, x; u, dy)$$

$$-\int_s^t du \int_A q(u, z)\, P_{\min}(s, x; u, dz) =$$

$$= \delta_x(A) + \int_s^t du \int_X Q(u, y, A)\, P_{\min}(s, x; u, dy).$$

We leave the derivation of (2.3.16) from (2.3.10) to the reader. ◇

Corollary. *Under assumptions* C_1 *and* C_2 *the partial derivatives* $\partial P_{\min}(s, x; t, A)/\partial s$ *and* $\partial P_{\min}(s, x; t, A)/\partial t$, *respectively, exist for almost all* s *(the set of values* s *excepted may depend on* x, t, *and* A*) and for almost all* t *(the set of values* t *excepted may depend on* s, x *and* A*). They satisfy respectively the backward and the forward equations* [30].

Proof. This corollary is implied by (2.3.9) and (2.3.10) and an elementary property of the Lebesgue integral. ◇

2.3.2.3.3. We shall now justify the use of the subscript "min".

Proposition 2.3.9. *Let* C_1 *and* C_2 *hold. Every substochastic transition function* P *satisfying either* (2.3.9) *or* (2.3.10) *has the property* $P \geqslant P_{\min}$.

Proof. Let us assume that P satisfies the equation

$$P(s, x; t, A) = \delta_x(A) + \int_s^t du \int_X Q(u, x, dy)\, P(u, y; t, A).$$

From the proof of Theorem 2.3.8 we know that this equation is equivalent to

$$P(s, x; t, A) = \delta_x(A)\, \psi(s, t, x) + \int_s^t du \int_X P(u, y; t, A)\, \psi(s, u, x)\, \tilde{q}(u, x, dy). \tag{2.3.17}$$

Comparing this equation with (2.3.11) we find by an easy induction that

$$P(s, x; t, A) \geqslant P^n_{\min}(s, x; t, A)$$

for all $n \in N$. Letting $n \to \infty$, we get $P \geqslant P_{\min}$ as asserted. ◇

Notice that substochastic transition functions P with given intensity Q and such that $P(., x; t, A)$ is absolutely continuous on $[0, t]$ for all fixed $x \in X$, $t \in T$, $A \in \mathcal{X}$ must satisfy (2.3.17). Indeed the backward equation (2.3.6) implies that

$$\int_X \tilde{q}(u, x, dy)\, P(u, y; t, A) = q(u, x)\, P(u, x; t, A) - \frac{\partial P(u, x; t, A)}{\partial u}$$

for fixed $x \in X$, $t \in T$, $A \in \mathcal{X}$ and almost all $u \leqslant t$; multiplying by $\psi(s, u, x)$ yields

$$\psi(s, u, x) \int_X \tilde{q}(u, x, dy)\, P(u, y; t, A) = \frac{\partial}{\partial u}\left(-\psi(s, u, x)\, P(u, x; t, A)\right),$$

and hence integrating on $[s, t]$ with respect to u gives (2.3.17).

---

[30] It is to be noted that $P_{\min}$ does not satisfy all the hypotheses leading to the backward and the forward equations in Subparagraphs 2.3.2.2.1 and 2.3.2.2.2.

The class of substochastic transition functions considered in Proposition 2.3.9 is therefore not as narrow as it might seem.

2.3.2.3.4. Let us turn now to the uniqueness part of our problem. First we have

Proposition 2.3.10. *Let C_1 and C_2 hold. If $P_{\min}$ is stochastic (i.e., $P_{\min}(s, x; t, X) = 1$ for all $s, t \in T$, $x \in X$), then $P_{\min}$ is the unique transition function satisfying both* (2.3.9) *and* (2.3.10).

Proof. Suppose to the contrary that there is a transition function $P \neq P_{\min}$ which satisfies equation (2.3.9) (or (2.3.10)). Then there exist s, $t \in T$, $x \in X$ and $A \in \mathcal{X}$ such that $P(s, x; t, A) \neq P_{\min}(s, x; t, A)$. By Proposition 2.3.9 we must have

$$P(s, x; t, A) > P_{\min}(s, x; t, A)$$

$$P(s, x; t, X - A) \geqslant P_{\min}(s, x; t, X - A),$$

from which we find that $P(s, x; t, X) > 1$. Thus we are led to a contradiction. ◇

It is possible to provide a necessary and sufficient condition that $P_{\min}$ be stochastic. Namely, we have

Proposition 2.3.11. *Let C_1 and C_2 hold and suppose that $\sigma(u, y, X) = 1$ for all $u \in [s, t]$, $y \in X$. Then $P_{\min}(s, x; t, X) = 1$ iff*

$$\lim_{n \to \infty} \int_s^t du \int_X q(u, y)\, P_n(s, x; u, dy) = 0.$$

Proof. We set

$$L_n(s, t, x) = \int_s^t du \int_X q(u, y)\, P_n(s, x; u, dy),\ n \in N.$$

Taking into account the relation

$$\psi(u, t, x) + \int_u^t q(v, x)\, \psi(s, v, x)\, dv = 1, \tag{2.3.18}$$

we deduce

$$P_0(s, x; t, X) + L_0(s, t, x) = 1.$$

We shall prove that

$$P_{n+1}(s, x; t, X) + L_{n+1}(s, t, x) = L_n(s, t, x)$$

for $n \in N$. Indeed, using (2.3.18) again we obtain

$$P_{n+1}(s, x; t, X) + L_{n+1}(s, t, x) =$$

$$= \int_s^t du \int_X q(u, y) \Psi(u, y; t, X) P_n(s, x; u, dy) +$$

$$+ \int_s^t du \int_X q(u, y) \int_s^u dv \int_X q(v, z) \Psi(v, z; u, dy) P_n(s, x; v, dz) =$$

$$= \int_s^t du \int_X q(u, y) P_n(s, x; u, dy) -$$

$$- \int_s^t du \int_X q(u, y) P_n(s, x; u, dy) \int_X \sigma(u, y, dz) \int_u^t q(v, z) \psi(u, v, z) dv +$$

$$+ \int_s^t dv \int_v^t du \int_X q(v, z) P_n(s, x; v, dz) \int_X q(u, y) \psi(v, u, y) \sigma(v, z, dy) =$$

$$= L_n(s, t, x).$$

It follows that

$$P^n_{\min}(s, x; t, X) = \sum_{k=0}^{n} P_k(s, x; t, X) = 1 - L_n(s, t, x).$$

Since $P_n \geqslant 0$ we have $L_{n+1} \leqslant L_n$, and therefore the limit

$$L(s, t, x) = \lim_{n \to \infty} L_n(s, t, x)$$

exists. Hence

$$P_{\min}(s, x; t, X) = 1 - L(s, t, x),$$

which ends the proof. ◇

Corollary. *Let* C_1 *and* C_2 *hold. Suppose that* $\sigma(., ., X) = 1$ *and that there exists a non-negative function k defined on T which is (Lebesgue) integrable on every finite interval contained in T and* $q(s, x) \leqslant k(s)$ *for all* $s \in T$, $x \in X$. *Under these conditions* $P_{\min}$ *is a stochastic transition function.*

Proof. It is easily seen that

$$L_n(s, t, x) \leqslant \int_s^t k(t_1)\,\mathrm{d}t_1 \int_s^{t_1} k(t_2)\,\mathrm{d}t_2 \ldots \int_s^{t_n} k(t_{n+1})\,\mathrm{d}t_{n+1} =$$

$$= \frac{1}{(n+1)!}\left[\int_s^t k(u)\,du\right]^{n+1}$$

for all $n \in N$. It remains to note that

$$\lim_{n\to\infty} \frac{1}{(n+1)!}\left[\int_s^t k(u)\,\mathrm{d}u\right]^{n+1} = 0$$

for all $s, t \in T$. ◇

Under the assumptions of the above corollary we can say more about $P_{\min}$. We have

Proposition 2.3.12. *If the assumption of the corollary to Proposition 2.3.11 hold, then the transition probability function* $P_{\min}$ *is given by*

$$P_{\min}(s, x\ ;\ t, A) = \sum_{n\in N} \Pi_n(s, t\ ;\ x, A), \tag{2.3.19}$$

where

$$\Pi_0(s, t; x, A) = \delta_x(A)$$

$$\Pi_1(s, t\ ;\ x, A) = \int_s^t Q(u, x, A)\,\mathrm{d}u,$$

and for $n \geqslant 2$

$$\Pi_n(s, t; x, A) = \int_s^t \mathrm{d}s_1 \int_{s_1}^t \mathrm{d}s_2 \ldots$$

$$\ldots \int_{s_{n-1}}^t \mathrm{d}s_n \int_{X^{n-1}} Q(s_1, x, \mathrm{d}y_1)\, Q(s_2, y_1, \mathrm{d}y_2) \ldots Q(s_n, y_{n-1}, A)$$

$$\Bigg(= \int_s^t \mathrm{d}t_1 \int_s^{t_1} \mathrm{d}t_2 \ldots$$

$$\ldots \int_s^{t_{n-1}} \mathrm{d}t_n \int_{X^{n-1}} Q(t_n, x, \mathrm{d}y_{n-1})\, Q(t_{n-1}, y_{n-1}, \mathrm{d}y_{n-2}) \ldots Q(t_1, y_1, A)\Bigg).$$

The series is absolutely and uniformly convergent with respect to $s, t \in T_1, x \in X, A \in \mathscr{X}$, *for every finite interval* $T_1 \subset T$.

Proof: Proposition 2.3.10 and the corollary to Proposition 2.3.11 imply that $P_{\min}$ is the unique stochastic transition function which satisfies (2.3.9) and (2.3.10). In what follows we actually reprove the uniqueness of the solution excepting the fact that it is a transition function. By iterating (2.3.9) or (2.3.10) we get

$$P_{\min}(s, x\,;\, t, A) = \sum_{l=0}^{n} \Pi_l(s, t; x, A) + R_n(s, t; x, A)$$

where

$$R_n(s, t\,;\, x, A) = \int_s^t ds_1 \int_{s_1}^t ds_2 \ldots$$

$$\ldots \int_{s_{n-1}}^t ds_n \int_{X^n} Q(s_1, x, dy_1) \ldots Q(s_n, y_{n-1}, dy_n)\, P_{\min}(s_n, y_n\,;\, t, A)$$

$$\Big(= \int_s^t dt_1 \int_s^{t_1} dt_2 \ldots$$

$$\ldots \int_s^{t_{n-1}} dt_n \int_{X^n} P_{\min}(s, x\,;\, t_n, dy_n)\, Q(t_n, y_n\, dy_{n-1}) \ldots Q(t_1, y_1, A) \Big).$$

It is easy to see that

$$|\,\Pi_n(s, t : x, A)\,| \;,\; |R_n(s, t\,;\, x, A)\,| \leqslant$$

$$2^n \int_s^t k(s_1)\, ds_1 \int_{s_1}^t k(s_2)\, ds_2 \ldots \int_{s_{n-1}}^t k(s_n)\, ds_n =$$

$$2^n \int_s^t k(t_1)\, dt_1 \int_s^{t_1} k(t_2)\, dt_2 \ldots \int_s^{t_{n-1}} k(t_n)\, dt_n \leqslant$$

$$\leqslant \frac{2^n}{n!} \left[\int_s^t k(u)\, du \right]^n$$

for all $s, t \in T$, $x \in X$, $A \in \mathscr{X}$. Therefore the series (2.3.19) is dominated by the series

$$\sum_{n \in N} \frac{2^n}{n!} \left[\int_s^t k(u)\, du \right]^n = \exp\left\{ 2 \int_s^t k(u)\, du \right\},$$

which concludes the proof. ◇

2.3.2.3.5. In the homogeneous case, a direct test, involving only the functions q and σ, of whether or not $P_{\min}$ is stochastic is available. Namely, we have

Theorem 2.3.13 (G. E. H. Reuter). *Let C_1 and C_2 hold and suppose that $\sigma(x, X) = 1$ for all $x \in X$. A necessary and sufficient condition ensuring that $P_{\min}$ be a stochastic transition function is the following: For some $\lambda > 0$ the equation*

$$(\lambda + q(x)) f(x) = q(x) \int_X f(y)\, \sigma(x, \mathrm{d}y) \quad , \quad x \in X, \qquad (2.3.20)$$

has no bounded solution other than $f = 0$.

Proof. Set

$$P_n(t; x, A) = P_n(0, x; t, A) \quad , \quad n \in N,$$

$$L(t, x) = L(0, t, x) = 1 - P_{\min}(t; x, A)$$

By (2.3.9) we have $L(0, x) = 0$, $x \in X$, and

$$-\frac{\partial L(t, x)}{\partial t} = q(x) L(t, x) - \int_X L(t, y)\, q(x)\, \sigma(x, \mathrm{d}y)$$

for almost all $t \geqslant 0$. If we put

$$z(\lambda, x) = \lambda \int_0^\infty e^{-\lambda t} L(t, x)\, \mathrm{d}t \quad , \quad \lambda > 0,$$

the above equation leads to

$$(\lambda + q(x))\, z(\lambda, x) = \int_X f(y)\, q(x)\, \sigma(x, \mathrm{d}y).$$

Now let us prove that every solution of (2.3.20) such that $\sup_{x \in X} |f(x)| \leqslant 1$ satisfies

$$-z(\lambda, x) \leqslant f(x) \leqslant z(\lambda, x)$$

for arbitrary $\lambda > 0$. Notice that

$$z(\lambda, x) = \lim_{n \to \infty} z_n(\lambda, x),$$

where

$$z_n(\lambda, x) = \lambda \int_0^\infty e^{-\lambda t} \left(1 - \sum_{i=0}^{n-1} P_i(t; x, X)\right) dt.$$

Set

$$p_n(\lambda, x) = \lambda \int_0^\infty e^{-\lambda t} P_n(t; x, X)\, dt \quad , \quad n \in N^*.$$

We have

$$p_0(\lambda, x) = \frac{\lambda}{\lambda + q(x),},$$

$$p_n(\lambda, x) = q(x) \int_X p_{n-1}(\lambda, y) \frac{\sigma(x, dy)}{\lambda + q(x),}, \quad n \in N^*,$$

$$z_n(\lambda, x) = 1 - \sum_{i=0}^{n} p_i(\lambda, x) =$$

$$= 1 - \frac{\lambda}{\lambda + q(x)} - q(x) \int_X (1 - z_{n-1}(\lambda, y)) \frac{\sigma(x, dy)}{\lambda + q(x)},$$

whence

$$(\lambda + q(x))\, z_n(\lambda, x) = q(x) \int_X z_{n-1}(\lambda, y)\, \sigma(x, dy).$$

Since $-z_0(\lambda, x) = -1 \leqslant f(x) \leqslant 1 = z_0(\lambda, x)$, by comparing the latter equation with (2.3.20) one can show by an easy induction argument that

$$-z_n(\lambda, x) \leqslant f(x) \leqslant z_n(\lambda, x)$$

for all $n \in N^*$, and hence we deduce the desired result.

Now, if $P(t; x, X) = 1$ for all $t \geqslant 0$, $x \in X$, then $z(\lambda, x) = 0$ for all $\lambda > 0$, $x \in X$, thus (2.3.20) has no bounded solution other than $f = 0$. Conversely, if for some $\lambda > 0$ (2.3.20) has no bounded solution other than $f = 0$, as $z(\lambda, x)$ is a bounded solution of (2.3.20), it must vanish identically, therefore $P(t; x, X) = 1$ for all $t \geqslant 0$, $x \in X$. ◇

2.3.2.4. The minimal process

2.3.2.4.1. In this paragraph we shall, unless otherwise stated, suppose that X is either a compact metric space or a separated locally compact (but not compact) complete metric space with metric ρ. We shall denote by $\mathscr{X}$ the σ-algebra of Borel sets in X. On account of Subparagraphs 2.3.1.2.2 and 2.3.1.2.3, there exists a Markov process $\xi_{\min} = (\xi_{\min}(t))_{t \in T}$ with transition function $P_{\min}$ which we shall call the *minimal* process associated with the given transition intensity function Q.

In what follows the main properties of the minimal process are studied. The reader will observe that these properties are actually valid for every Markov process whose transition function satisfies equation (2.3.9).

2.3.2.4.2. We start with

Proposition 2.3.14. *Let C_1 and C_2 hold. The Markov process $\xi_{\min}$ is stochastically continuous.*

Proof. For $s < t$ we have

$$\mathsf{P}(\rho(\xi_{\min}(t), \xi_{\min}(s)) \geqslant \varepsilon) =$$

$$\mathsf{E}\,\mathsf{P}(\rho(\xi_{\min}(t), \xi_{\min}(s)) \geqslant \varepsilon \mid \xi_{\min}(s)) \leqslant$$

$$\leqslant \mathsf{E}\,\mathsf{P}(\xi_{\min}(t) \neq \xi_{\min}(s) \mid \xi_{\min}(s)).$$

But

$$\mathsf{P}(\xi_{\min}(t) \neq \xi_{\min}(s) \mid \xi_{\min}(s) = x) = P_{\min}(s, x; t, X - \{x\}),$$

and (2.3.9) (or the fact that $P_{\min}$ has Q as transition intensity function) implies that $P_{\min}(s, x; t, X - \{x\}) \to 0$ as $t - s \to 0$ for every fixed $x \in X$. Then by the dominated convergence theorem

$$\mathsf{E}\,\mathsf{P}(\xi_{\min}(t) \neq \xi_{\min}(s) \mid \xi_{\min}(s)) \to 0$$

as $t - s \to 0$, which concludes the proof. ◇

2.3.2.4.3. Without any loss of generality we shall assume from now on that $\xi_{\min}$ is separable.

Theorem 2.3.15. *Let C_1' and C_2 hold. The probability that starting in x at time s, $\xi_{\min}$ stays in x during time $t - s$ is given by* [31]

$$\mathsf{P}(\xi_{\min}(u) = x,\ s < u \leqslant t \mid \xi_{\min}(s) = x) = \exp\left(-\int_s^t q(u, x)\, du\right).$$

---

[31] It should be noted that the duration of stay in a given state x for an arbitrary homogeneous separable Markov process always has an exponential (perhaps degenerate) distribution. For, if we set $p(t) = \mathsf{P}(\xi(u),\ 0 \leqslant u \leqslant t \mid \xi(0) = x)$, then the Markov property implies easily that $p(s + t) = p(s)\,p(t)$ hence $p(t) = \exp(-\lambda t)$ for some $0 \leqslant \lambda \leqslant \infty$.

Proof. Since $\xi_{\min}$ is separable and stochastically continuous, on account of Theorem 2.1.9 we can take as separability set any countable set S dense in $[s, t]$, say

$$S = \left\{ t_{nk} = s + \frac{k}{2^n}(t-s) \ : \ 0 \leqslant k \leqslant 2^n, \quad n \in N \right\}.$$

Thus

$$\mathsf{P}(\xi_{\min}(u) = x, \ s < u \leqslant t \mid \xi_{\min}(s) = x) =$$

$$\mathsf{P}(\xi_{\min}(u) = x, u \in S \mid \xi_{\min}(s) = x) = p.$$

Note that

$$p = \lim_{n\to\infty} \mathsf{P}(\xi_{\min}(t_{nk}) = x, \ 0 \leqslant k \leqslant 2^n \mid \xi_{\min}(s) = x)$$

since the sequence of events with general term $E_n = (\xi_{\min}(t_{nk}) = x,$ $0 \leqslant k \leqslant 2^n)$ is not decreasing and $\bigcap_{n\in N} E_n = (\xi_{\min}(u) = x, u \in S)$. Furthermore,

$$p_n = \mathsf{P}(\xi_{\min}(t_{nk}) = x, \ 0 \leqslant k \leqslant 2^n \mid \xi_{\min}(s) = x) =$$

$$= \prod_{k=1}^{2^n} P_{\min}(t_{n,k-1}, \ x \ ; \ t_{nk}, \ \{x\}) =$$

$$= \exp \sum_{k=1}^{2^n} \log P_{\min}(t_{n,k-1}, \ x; t_{nk}, \ \{x\}).$$

Now consider the step function f_n defined on $[s, t)$ by

$$f_n(u) = \frac{\log P_{\min}(t_{n,k-1}, \ x; t_{nk}, \ \{x\})}{t_{nk} - t_{n,k-1}}$$

for $t_{n,k-1} \leqslant u < t_{nk}$.

Noting that C_1' and (2.3.13) imply

$$\lim_{\substack{s_2 - s_1 \to 0 \\ s_1 \leqslant s \leqslant s_2}} \frac{1 - P_{\min}(s_1, \ x \ ; \ s_2, \ \{x\})}{s_2 - s_1} = q(s, x)$$

uniformly with respect to s on every finite closed interval contained in T, we can write

$$f_n(u) = \frac{\log P_{\min}(t_{n,k-1}, x; t_{nk}, \{x\})}{1 - P_{\min}(t_{n,k-1}, x; t_{nk}, \{x\})} \cdot$$

$$\cdot \frac{1 - P_{\min}(t_{n,k-1}, x; t_{nk}, \{x\})}{t_{nk} - t_{n,k-1}} \to q(u, x)$$

uniformly with respect to $u \in [s, t)$ as $n \to \infty$.

Thus

$$\log p_n = \int_t^t f_n(u)\, du \to -\int_s^t q(u, x)\, du$$

as $n \to \infty$. The proof is complete. ◇

Corollary. *If $\xi_{\min}$ assumes the value x at some fixed time s, then almost all trajectories remain constant for some positive time after s.*

Proof. Let $E_n = \left(\xi_{\min}(u) = x, s \leqslant u \leqslant s + \frac{1}{n}\right)$, $n \in N^*$. By separability, the E_n are events and so is their limit $E = \bigcup_{n \in N^*} E_n$ — the set to which correspond those trajectories which remain constant for some positive time after s. Since the sequence of the E_n is not decreasing, we have

$$\mathsf{P}(E^c \mid \xi(s) = x) = \lim_{n \to \infty} \mathsf{P}(E_n^c \mid \xi(s) = x) =$$

$$= 1 - \lim_{n \to \infty} \exp\left(-\int_s^{s + \frac{1}{n}} q(u, x)\, du\right) = 0.$$

Therefore $\mathsf{P}(E \mid \xi(s) = 1$. ◇

At first sight, we should expect in the light of the above corollary that almost all trajectories remain constant for some positive time, then jump and again remain constant for some positive time, and so on. (Here the value $+\infty$ for occupation times is not excluded). Therefore, we might believe that almost all trajectories would consist of a finite or infinite sequence of isolated jumps. However, under the conditions assumed, we shall see that more complicated discontinuities may occur. To avoid this possibility, further restrictions must be imposed.

2.3.2.4.4. We shall now study those trajectories $\xi_{\min}(., \omega)$ whose first discontinuity $\tau^{(s)}(\omega)$ after a fixed but arbitrarily given $s \in T$ is such that the trajectory remains constant between s and $\tau^{(s)}(\omega)$ and then jumps to a different value at time $\tau^{(s)}(\omega)$, remaining there for a length of time $h(\omega)$. We thus wish to investigate the first isolated jump after time s.

Theorem 2.3.16. *Suppose that C_1', and C_2' hold and let $\xi_{\min}$ be in state x at time s. The probability that the first discontinuity in the interval $(s, s+t)$ consist of an isolated jump into $A \not\ni x$ is given by*

$$\int_0^t \psi(s, s+u, x)\, \tilde{q}(s+u, x, A)\, du.$$

Proof. Put

$$D_h = (\xi_{\min}(u) = x,\ s \leqslant u < s + \tau^{(s)} < s + t) \cap$$

$$\cap\, (\xi_{\min}(v) = y \in A,\ s + \tau^{(s)} < v < s + \tau^{(s)} + h),$$

$$D = \bigcup_{h>0} D_h.$$

Notice that $D_h \uparrow D$ as $h \downarrow 0$. We have to compute $\mathsf{P}(D \mid \xi_{\min}(s) = x)$. Let

$$D_{n,h} = \bigcup_{k=1}^{n-1} \left(\xi_{\min}(u) = x,\ s \leqslant u \leqslant s + \frac{kt}{n} \right) \cap$$

$$\cap \left(\xi_{\min}(v) = y \in A,\ s + \frac{(k+1)t}{n} \leqslant v \leqslant s + \frac{(k+1)t}{n} + h \right),$$

$$\underline{D}_h = \varliminf_{n\to\infty} D_{n,h}\ ,\quad \overline{D}_h = \varlimsup_{n\to\infty} D_{n,h}$$

and note that

$$\underline{D}_h \uparrow \underline{D} = \bigcup_{n\in N^*} \underline{D}_{1/n}\ ,\quad \overline{D}_h \uparrow \overline{D} = \bigcup_{n\in N^*} \overline{D}_{1/n}$$

as $h \downarrow 0$. It is easy to prove that $\overline{D}_h \subset D_h$ and $D_h \subset D_{n,h/2}$ for n sufficiently large so that $\overline{D} \subset D$ and $D \subset \underline{D}$, and therefore

$$\underline{D} = D = \overline{D}. \tag{2.3.21}$$

According to Theorem 2.3.15

$$\mathsf{P}(D_{n,h} \mid \xi_{\min}(s) = x) =$$

$$= \sum_{k=1}^{n-1} \psi\left(s, s + \frac{kt}{n}, x\right) \int_A P_{\min}\left(s + \frac{kt}{n}, x; s + \frac{(k+1)t}{n}, \mathrm{d}y\right) \times$$

$$\times \psi\left(s + \frac{(k+1)t}{n}, s + \frac{(k+1)t}{n} + h, y\right).$$

Note that by C_1', C_2' and (2.3.13)

$$P_{\min}\left(s + \frac{kt}{n}, x; s + \frac{(k+1)t}{n}, A\right) = \frac{t}{n}\left(\tilde{q}\left(s + \frac{(k+1)t}{n}, x, A\right) + \varepsilon_n\right)$$

where $\varepsilon_n \to 0$ as $n \to \infty$ uniformly with respect to $A \in \mathcal{X}$ and $1 \leqslant k \leqslant n - 1$. Note also that $\psi(s, s + u, x)$ is uniformly continuous with respect to u on $[0, t]$. On account of these facts it is easily seen that

$$\lim_{n\to\infty} \mathsf{P}(D_{n,h} \mid \xi_{\min}(s) = x) =$$

$$= \int_0^t \psi(s, s + u, x)\, \mathrm{d}u \int_A \tilde{q}(s + u, x, \mathrm{d}y)\ \psi(s + u, s + u + h, y)$$

and then

$$\lim_{h\to 0} \lim_{n\to\infty} \mathsf{P}(D_{n,h} \mid \xi_{\min}(s) = x) = \int_0^t \psi(s, s + u, x)\ \tilde{q}(s + u, x, A)\, \mathrm{d}u.$$

Fatou 's lemma implies

$$\mathsf{P}(\underline{D}_h \mid \xi_{\min}(s) = x) \leqslant \lim_{n\to\infty} \mathsf{P}(D_{n,h} \mid \xi_{\min}(s) = x) \leqslant \mathsf{P}(\overline{D}_h \mid \xi_{\min}(s) = x),$$

and therefore (2.3.21) leads us to

$$\mathsf{P}(D \mid \xi_{\min}(s) = x) = \int_0^t \psi(s, s + u, x)\, \tilde{q}(s + u, x, A)\, \mathrm{d}u. \diamond \qquad (2.3.22)$$

Corollary. *Suppose that C_1' and C_2' hold and let $\xi_{\min}$ be in state x at time s. If $\sigma(.,.,X) = 1$, then the conditional probability given $\tau^{(s)}$*

that the first discontinuity following s *be an isolated jump into* $A \not\ni x$ *is* $\sigma(s + \tau^{(s)}, x, A)$.

Proof. It follows from (2.3.22) that

$$\mathsf{P}(\tau^{(s)} < t \mid \xi_{\min}(s) = x) =$$

$$= \int_0^t \psi(s, s + u, x)\, q(s + u, x)\, du = 1 - \psi(s, s + t, x),$$

and we may write successively

$$\mathsf{P}(D \mid \xi_{\min}(s) = x) =$$

$$= \int_0^t \sigma(s + u, x, A) \frac{\mathrm{d}}{\mathrm{d}u} \mathsf{P}(\tau^{(s)} < u \mid \xi_{\min}(s) = x)\, \mathrm{d}u,$$

$$\mathsf{P}(\xi(\tau^{(s)}+) \in A \mid \xi_{\min}(s) = x) = \sigma(s + \tau^{(s)}, x, A). \quad \diamond$$

2.3.2.4.5. We come now to the probabilistic interpretation of the functions $P_n(s, x; t, A)$, $n \in N$, in the proof of Theorem 2.3.8. In what follows we suppose that $\sigma(., ., X) = 1$.

Consider the event

$J_n(s, t, A) = (\xi_{\min}$ has exactly n jumps in $[s, t]$ and $\xi_{\min}(t) \in A)$.

We have

$$\mathsf{P}(J_0(s, t, A) \mid \xi_{\min}(s) = x) = \delta_x(A) = P_0(s, x; t, A).$$

Suppose then that

$$\mathsf{P}(J_n(s, t, A) \mid \xi_{\min}(s) = x) = P_n(s, x; t, A)$$

for an $n \in N$. On account of Theorem 2.3.16 we may formally write

$$\mathsf{P}(J_{n+1}(s, t, A) \mid \xi_{\min}(s) = x) =$$

$$= \mathsf{P}(\xi_{\min}(u) = x, \quad s \leqslant u < s + \tau^{(s)} < t, \quad \xi_{\min}(s + \tau^{(s)}) \neq x,$$

$$J_n(s + \tau^{(s)}, t, A) \mid \xi_{\min}(s) = x) =$$

$$= \mathsf{E}(P_n(s + \tau^{(s)}, \quad \xi_{\min}(s + \tau^{(s)}); \quad t, A) \mid \xi_{\min}(s) = x) =$$

$$= \int_s^t \mathrm{d}u \int_X P_n(u, y; \; t, A)\, \psi(s, u, x)\, \tilde{q}(u, x, \mathrm{d}y) = P_{n+1}(s, x; t, A).$$

It would thus follow that

$$P_n(s, x; t, A) = \mathsf{P}(J_n(s, t, A) | \xi_{\min}(s) = x) \tag{2.3.23}$$

for all $n \in N$.

The above computation would be rigorous if $\xi_{\min}$ were a strong Markov process with right-continuous trajectories. This happens under further conditions so that the given interpretation of $P_n(s, x; t, A)$ can be considered valid with this reservation. It is, however, possible to construct under just conditions C_1 and C_2 a strong Markov process $\xi'_{\min}$ with an arbitrary state space $(X, \mathscr{X})$ and transition function $P_{\min}$ for which (2.3.23) holds (with $\xi'_{\min}$ instead of $\xi_{\min}$). This possibility will be considered after we examine a special case, to which the following subparagraph is devoted.

2.3.2.4.6. We begin by showing that under a previously used condition (see Corollary to Proposition 2.3.11), the trajectories of $\xi_{\min}$ can be considered right-continuous.

Proposition 2.3.17. *Let C_1 and C_2 hold. Suppose that there is a non-negative function k defined on T which is integrable on every finite interval contained in T and that $q(s, x) \leqslant k(s)$ for all $s \in T$, $x \in X$. Under these conditions $\xi_{\min}$ has no discontinuities of the second kind.*

Proof. With no loss of generality it is sufficient to prove that $\xi_{\min}$ has no discontinuities of the second kind on every finite interval $[0, t] \subset T$. By (2.3.14)

$$\alpha(\varepsilon, \delta) = \sup_{\substack{s \leqslant u \leqslant s+\delta \\ s \leqslant t,\, x \in X}} P_{\min}(s, x; u, X - \{x\}) \leqslant$$

$$\leqslant \sup_{s \leqslant t,\, x \in X} \left(1 - \exp\left[-\int_s^{s+\delta} q(u, x)\,\mathrm{d}u\right]\right) \leqslant$$

$$\leqslant \sup_{s \leqslant t,\, x \in X} \left(1 - \exp\left[-\int_s^{s+\delta} k(u)\,\mathrm{d}u\right]\right) \to 0$$

as $\delta \to 0$ by absolute continuity of Lebesgue integral. Now we apply Theorem 2.1.16. ◇

Propositions 2.3.14, 2.3.17 and 2.1.17 allow us to assume that almost all trajectories of $\xi_{\min}$ are right-continuous. In particular, if X is at most a countable set we may endow it with the discrete topology which is metrizable by $\rho(x, y) = 1$ or 0 according as $x \neq y$ or $x = y$. The right-continuity of trajectories amounts then to the fact that $\xi_{\min}$ is a jump process.

Proposition 2.3.18. *Let C_1' and C_2' hold. Suppose that* $\sigma(.,.,X)=1$ *and there is a non-negative function k defined on T which is integrable on every finite interval contained in T and $q(s,x)\leqslant k(s)$ for all $s\in T$, $x\in X$. Under these conditions almost all trajectories of $\xi_{\min}$ are step functions.*

Proof. It is easy to see that the hypotheses of Proposition 2.3.5 are fulfilled so that $\xi_{\min}$ is a strong Markov process. This fact and Theorem 2.3.16 imply the existence of a sequence $(\tau_n)_{n\in N^*}$ of positive (possibly infinite) random times such that if $\tau_n(\omega)=\infty$ then $\tau_m(\omega)=\infty$ for $m>n$ and almost all trajectories $\xi_{\min}(.,\omega)$ are constant on $[0,\tau_1(\omega))$, $[\tau_1(\omega), \tau_1(\omega)+\tau_2(\omega))$, ... with different values in any two consecutive non-degenerate intervals. It remains to prove that $\sum_{n\in N^*}\tau_n=\infty$ a.s., that is, almost all trajectories have a finite number of jumps in every finite interval. This is an immediate consequence of the corollary to Proposition 2.3.11 combined with (2.3.23). ◇

Assume now that for all $x\in X$, $q(.,x)$ is not identically zero on any semi-infinite interval $[t,\infty)$, $t\geqslant 0$. Then it is easily seen that under the assumptions of Proposition 2.3.18 we have $\mathsf{P}(\tau_i=\infty)=0$, $i\in N^*$. Let us set $\tau_0=0$, $\xi_0=\xi(0)$, and

$$\xi_n(\omega)=\xi\left(\sum_{i=1}^{n}\tau_i(\omega),\ \omega\right)$$

if $\sum_{i=1}^{n}\tau_i(\omega)<\infty$ for $n\in N^*$. Thus the ξ_n are defined a.s. on Ω. The strong Markov property of $\xi_{\min}$ and Theorems 2.3.15 and 2.3.16 imply that

$$\mathsf{P}(\xi_n\in A\mid \xi_0,\ldots,\xi_{n-1},\tau_1,\ldots,\tau_n)=\sigma\left(\sum_{i=1}^{n}\tau_i,\ \xi_{n-1},\ A\right)$$

$$\mathsf{P}(\tau_n\geqslant t\mid \xi_0,\ldots,\xi_{n-1},\tau_0,\ldots,\tau_{n-1})=$$

$$=\psi\left(\sum_{i=0}^{n-1}\tau_i,\ t+\sum_{i=0}^{n-1}\tau_i,\ \xi_{n-1}\right)$$

for all $n\in N^*$, $A\subset X$, $t\geqslant 0$.

Therefore in the homogeneous case $(\xi_n)_{n\in N}$ is a homogeneous Markov chain with transition matrix $(\sigma(x,\{y\}))_{x,y\in X}$.

2.3.2.4.7. The above considerations suggest a probabilistic construction of the minimal process in case of an arbitrary state space $(X,\mathscr{X})$. Let the functions q and σ satisfy C_1 and C_2, and $\sigma(.,.,X)=1$.

In what follows the random variables τ and ξ with affixes are $[0, \infty]$- and X-valued, respectively, and defined on a suitable probability space $(\Omega', \mathcal{K}', \mathsf{P}')$. We take $\tau_0 \equiv 0$, ξ_0 arbitrary and if τ_i, ξ_i, $0 \leqslant i \leqslant n$, are already known, we choose τ_{n+1} and ξ_{n+1} so that

$$\mathsf{P}'(\tau_{n+1} < t \mid \tau_0, \xi_0, \ldots, \tau_n, \xi_n) = 1 - \psi\left(\sum_{k=0}^{n} \tau_k,\ t + \sum_{k=0}^{n} \tau_k,\ \xi_n\right),$$

$$\mathsf{P}'(\xi_{n+1} \in A \mid \tau_0, \xi_0, \ldots, \tau_n, \xi_n, \tau_{n+1}) = \sigma\left(\sum_{k=0}^{n} \tau_k, \xi_n, A\right).$$

If $\sum_{k=0}^{n} \tau_k(\omega') = \infty$, we put $\tau_m(\omega') = \infty$ and $\xi_m(\omega') = \xi_n(\omega')$ for $m > n$. We set

$$\xi'_{\min}(t, \omega') = \xi_n(\omega')$$

for $t \in \left[\sum_{k=0}^{n} \tau_k(\omega'), \sum_{k=0}^{n+1} \tau_k(\omega')\right)$. Thus $\xi'_{\min}(t, \omega')$ is defined for all $t \in \left[0, \sum_{k=0}^{\infty} \tau_k(\omega')\right)$. If $\mathsf{P}'\left(\sum_{k=0}^{\infty} \tau_k < \infty\right) < 1$, then to define $\xi'_{\min}(t, \omega')$ for all $t \in [0, \infty)$ we add the fictitious point $\mathfrak{d}$ to the state space X and set $\xi'_{\min}(t, \omega') = \mathfrak{d}$ for $t \geqslant \sum_{k=0}^{\infty} \tau_k(\omega')$.

Theorem 2.3.19. *The $X \cup \{\mathfrak{d}\}$-valued process $\xi'_{\min}$ is Markov and its trajectories are right-continuous (in the discrete topology in $X \cup \{\mathfrak{d}\}$). Given $\xi'_{\min}(s) = x$, the probability that $\xi'_{\min}$ has exactly n jumps in $[s, t]$ and $\xi'_{\min}(t) \in A \in \mathcal{X}$ is $P_n(s, x; t, A)$. The transition function of $\xi'_{\min}$ is*

$$\mathsf{P}'(\xi'_{\min}(t) \in A \mid \xi'_{\min}(s) = x) = \begin{cases} P_{\min}(s, x; t, A) & \text{if } A \in \mathcal{X}, \\ 1 - P_{\min}(s, x; t, X) & \text{if } A = \{\mathfrak{d}\} \end{cases}$$

for $x \neq \mathfrak{d}$ and

$$\mathsf{P}'(\xi'_{\min}(t) \in A \mid \xi'_{\min}(s) = \mathfrak{d}) = \delta_{\mathfrak{d}}(A).$$

Proof. The trajectories of $\xi'_{\min}$ are right-continuous by definition. Let us prove that $\xi'_{\min}$ is a Markov process. We first note that

$$\mathsf{P}'^{\mathcal{X}_n}\left(\tau_{n+1} \geqslant t + u - \sum_{k=0}^{n} \tau_k \,\middle|\, \tau_{n+1} \geqslant t - \sum_{k=0}^{n} \tau_k\right) =$$

$$= \psi(t, t + u, \xi_n),$$

where $\mathcal{K}_n$ is the σ-algebra generated by the random variables τ_i, ξ_i, $0 \leqslant i \leqslant n$. Indeed, we have

$$\mathsf{P}'^{\mathcal{K}_n}\left(\tau_{n+1} \geqslant t + u - \sum_{k=0}^{n} \tau_k \,\middle|\, \tau_{n+1} \geqslant t - \sum_{k=0}^{n} \tau_k\right) =$$

$$= 1 - \frac{\mathsf{P}'^{\mathcal{K}_n}\left(t - \sum_{k=0}^{n} \tau_k \leqslant \tau_{n+1} < t - \sum_{k=0}^{n} \tau_k + u\right)}{\mathsf{P}'^{\mathcal{K}_n}\left(\tau_{n+1} \geqslant t - \sum_{k=0}^{n} \tau_k\right)} =$$

$$= 1 - \frac{1 - \psi\left(\sum_{k=0}^{n} \tau_k,\ t + u,\ \xi_n\right) - 1 + \psi\left(\sum_{k=0}^{n} \tau_k,\ t,\ \xi_n\right)}{\psi\left(\sum_{k=0}^{n} \tau_k,\ t,\ \xi_n\right)} =$$

$$= \psi(t,\ t + u,\ \xi_n).$$

From this fact it is easy to deduce that

$$\mathsf{P}'^{\mathcal{K}_n}\left(\sum_{k=0}^{n+1} \tau_k - t < t_{n+1},\ \tau_i < t_i,\ n + 2 \leqslant i \leqslant n + r,\right.$$

$$\left.\xi_j \in A_j,\ n + 1 \leqslant j \leqslant n + r \,\middle|\, \sum_{k=0}^{n} \tau_k \leqslant t < \sum_{k=0}^{n+1} \tau_k\right)$$

does not depend on $\tau_0, \xi_0, \ldots, \tau_{n-1}; \xi_{n-1}, \tau_n$ but only on t and $\xi_n = \xi'_{\min}(t)$. This immediately implies that $\xi'_{\min}$ is a Markov process.

Further, consider the event $J'_n(s, t, A)$ defined previously with $\xi'_{\min}$ in place of $\xi_{\min}$. With this change the computation in Subparagraph 2.3.2.4.5 is valid and we may write

$$\mathsf{P}'(J'_n(s, t, A) \mid \xi'_{\min}(s) = x) = P_n(s, x; t, A).$$

This holds because $s + \tau^{(s)}$ equals $\sum_{k=0}^{n} \tau_k$ for some $n \in N$ and the joint distribution of $\tau_{n+1}, \ldots, \tau_{n+r}, \xi_{n+1}, \ldots, \xi_{n+r}$ given τ_i, ξ_i, $0 \leqslant i \leqslant n$, depends only upon ξ_n and $\sum_{k=0}^{n} \tau_k$.

The assertion concerning the transition function for $x \neq \mathfrak{d}$ follows from the above result and the construction of $\xi'_{\min}$. The case $x = \mathfrak{d}$ does not require any explanation. ◇

Corollary. *The process* $\xi'_{\min}$ *has with probability one a finite number of jumps on* $[s, t]$ *iff* $P_{\min}(s, x; t, X) = 1$.

2.3.3. Markov jump processes. II. Discrete state space

This subsection and the next one are concerned with the case in which the state space is finite or denumerable which arises in many applications. As we shall see, several of the results obtained in the previous subsection can be made more precise.

2.3.3.1. The case of a finite state space

2.3.3.1.1. Let X be a finite set, say $\{1, \ldots, m\}$. The transition function P will be determined by the *transition matrix function*

$$\mathbf{P}(.,.) = (p_{ij}(.,.))_{1 \leqslant i, j \leqslant m}$$

where

$$p_{ij}(.,.) = P(., i; ., \{j\}).$$

Clearly, we have

$$p_{ij}(.,.) \geqslant 0, \qquad \sum_{j=1}^{m} p_{ij}(.,.) = 1.$$

The Chapman-Kolmogorov equation amounts to

$$p_{ij}(s, t) = \sum_{h=1}^{m} p_{ih}(s, u) p_{hj}(u, t)$$

for $s \leqslant u \leqslant t$ with

$$p_{ij}(s, s) = \delta_{ij}, \qquad s \in T, \qquad i, j \in X\ ^{32)}.$$

32) Treating the conditions imposed on the p_{ij} as functional equations, FRÉCHET (1938, pp. 201–248) showed that any continuous p_{ij} has the form $p_{ij}(s, t) = \sum_{h=1}^{m} \overset{i}{\mu}_h(s) \overset{h}{\lambda}_j(t)$, where the $\overset{i}{\mu}_h$ are continuous functions subject to the requirement that $\det |\overset{i}{\mu}_h(s)| \neq 0$ for all $s \in T$; the $\overset{h}{\lambda}_j$ are then completely determined by the equations $\sum_{h=1}^{m} \overset{i}{\mu}_h(s) \overset{h}{\lambda}_j(s) = \delta_{ij}$. For a more recent treatment see ACZÉL (1966, pp. 359–364). On this basis VRANCEANU (1964) interprets an m-state Markov process as a system of m independent one-dimensional vector fields (i.e. parametric curves), the $\overset{i}{\mu}_h$ being contravariant, and the $\overset{h}{\lambda}_j$ covariant components of these fields.

If we set

$$F_{ij}(s_1, s_2) = F(s_1, s_2; i, \{j\}),$$

then (2.3.2) leads to

$$\frac{p_{ij}(s_2, t) - p_{ij}(s_1, t)}{s_2 - s_1} = -\sum_{h=1}^{m} F_{ih}(s_1, s_2) p_{hj}(s_2, t). \qquad (2.3.24)$$

Suppose $p_{ij}(\cdot, \cdot)$, $1 \leqslant i, j \leqslant m$, to be continuous functions. Then for t sufficiently near to s_2, the determinant $\det |p_{ij}(s_2, t)| \neq 0$ since det $|p_{ij}(t, t)| = 1$. Therefore the system (2.3.24) may be solved with respect to $F_{ij}(s_1, s_2)$, $1 \leqslant i, j \leqslant m$. It follows that the limits

$$\lim_{\substack{h+k\to 0+\\ (h\equiv 0 \text{ or } k\equiv 0)}} F_{ij}(s - h, s + k) = q_{ij}(s)$$

and the derivatives $\partial p_{ij}/\partial s$, $1 \leqslant i, j \leqslant m$, exist together. Moreover, using (2.3.3) instead of (2.3.2) we deduce that if $p_{ij}(\cdot, \cdot)$, $1 \leqslant i, j \leqslant m$ are continuous, then the limits $q_{ij}(s)$ and the derivatives $\partial p_{ij}/\partial t$, $1 \leqslant i, j \leqslant m$, exist also together.

The matrix function $\mathbf{Q}(\cdot) = (q_{ij}(\cdot))_{1\leqslant i, j\leqslant m}$ is said to be the *transition intensity matrix function* corresponding to $\mathbf{P}(\cdot, \cdot)$. We have obviously

$$q_{ij}(\cdot) \begin{cases} \leqslant 0 & \text{if } i = j \\ \geqslant 0 & \text{if } i \neq j, \end{cases}$$

$$\sum_{j=1}^{m} q_{ij}(\cdot) = 0, \qquad 1 \leqslant i, j \leqslant m.$$

Thus assuming the continuity of the p_{ij} and the existence of $\mathbf{Q}(\cdot)$ we can write the backward and the forward equations as follows

$$\begin{cases} \dfrac{\partial p_{ij}(s, t)}{\partial s} = -\displaystyle\sum_{h=1}^{m} q_{ih}(s) p_{hj}(s, t) \\ p_{ij}(t, t) = \delta_{ij}, \qquad 1 \leqslant i, \ j \leqslant m, \qquad s < t, \end{cases}$$

and

$$\begin{cases} \dfrac{\partial p_{ij}(s, t)}{\partial t} = \displaystyle\sum_{h=1}^{m} p_{ih}(s, t) q_{hj}(t) \\ p_{ij}(s, s) = \delta_{ij}, \qquad 1 \leqslant i, \ j \leqslant m, \qquad t > s. \end{cases}$$

2.3.3.1.2. According to the general theory in Paragraph 2.3.2.3, the above systems have a common solution. The specialisation to our present case is as follows.

Theorem 2.3.20. *Suppose* $\mathbf{Q}(\cdot)$, *to be measurable. If* $\max_{1 \leqslant i \leqslant m} |q_{ii}(\cdot)|$ *is integrable on every finite interval contained in* T, *then the minimal (stochastic) transition matrix function associated with* $\mathbf{Q}(\cdot)$ *is given by*

$$\mathbf{I} + \int_s^t \mathbf{Q}(u)\, du + \sum_{n \geqslant 2} \int_s^t dt_1 \int_s^{t_1} dt_2 \ldots \int_s^{t_{n-1}} \mathbf{Q}(t_n) \ldots \mathbf{Q}(t_1)\, dt_n,$$

where $\mathbf{I}$ *is the mth order identity matrix. The series is absolutely and uniformly convergent with respect to* $s, t \in T_1$ *for every finite interval* $T_1 \subset T$.

Proof. Proposition 2.3.12. ◇

For further considerations concerning the determination of finite state Markov processes by means of suitable infinitesimal characteristics we refer the reader to SENČENKO (1966). See also G. S. GOODMAN (1970).

2.3.3.1.3. Now we come to the structure of trajectories of finite state Markov jump processes.

It was proved by JUŠKEVIČ (1957) that the existence of a finite state Markov process with right-continuous trajectories (in the discrete topology) is equivalent to the relations

$$\lim_{t \to s+} p_{ii}(s, t) = 1, \qquad 1 \leqslant i \leqslant m, \qquad s \in T.$$

Thus, on account of Proposition 2.3.5, all finite state Markov jump processes are strong Markov processes. Consequently, if $\mathbf{Q}(\cdot)$ is continuous, the proof of Proposition 2.3.18 shows that almost all the trajectories are step functions iff the minimal transition matrix function is stochastic. In particular, this happens under the condition in Theorem 2.3.20; if this condition does not hold, then there exists a (not necessarily infinite) random time of accumulation of isolated jumps and almost all the trajectories are step functions up to this time.

2.3.3.1.4. It will be proved in Subparagraph 2.3.4.2.2 that in the homogeneous case, when $p_{ij}(s, t) = p_{ij}(t - s)$ the limits

$$q_{ij} = \lim_{t \to 0+} \frac{p_{ij}(t) - \delta_{ij}}{t}, \qquad 1 \leqslant i, j \leqslant m,$$

exist without any additional assumptions. Furthermore, the backward and forward equations always hold and their unique solution is the minimal transition matrix functions. Consequently, the problems (E) and (U) (see Subparagraph 2.3.2.3.1) are completely solved in this case.

2.3.3.2. The case of a denumerable state space

2.3.3.2.1. We now consider the case in which X is a countably infinite set. In the next two paragraphs some important examples will be discussed. As above, the transition function P will be determined by the (perhaps substochastic) transition matrix function

$$\mathbf{P}(\cdot,\cdot) = (p_{ij}(\cdot,\cdot))_{i,j\in X}$$

where

$$p_{ij}(\cdot,\cdot) = P(\cdot, i; \cdot, \{j\}).$$

Obviously we have

$$p_{ij}(\cdot,\cdot) \geqslant 0, \qquad \sum_{j\in X} p_{ij}(\cdot,\cdot) \leqslant 1.$$

Concerning the transition intensity function corresponding to P we shall denote

$$Q(\cdot,\ i,\ \{j\} = q_{ij}(\cdot),$$

$$q(\cdot,\ i) = -\ q_{ii}(.) = q_i(\cdot),$$

$$\sigma(\cdot,\ i,\ \{j\}) = \sigma_{ij}(\cdot), \qquad i,\ j \in X.$$

Thus

$$\sigma_{ij}(s) = \begin{cases} \delta_{ij} & \text{if} \quad q_i(s) = 0 \\ \dfrac{q_{ij}(s)}{q_i(s)}(1-\delta_{ij}) & \text{if} \quad q_i(s) \neq 0. \end{cases}$$

The matrix function $\mathbf{Q}(\cdot) = (q_{ij}(\cdot))_{i,j\in X}$ is said to be the *transition intensity matrix function* corresponding to $\mathbf{P}(\cdot,\cdot)$. We have obviously

$$q_{ij}(\cdot) \begin{cases} \leqslant 0 & \text{if} \quad i = j, \\ \geqslant 0 & \text{if} \quad i \neq j, \end{cases}$$

$$\sum_{j\in X} q_{ij}(\cdot) \leqslant 0, \qquad i,\ j \in X.$$

With this notation the backward and forward equations become

$$\begin{cases} \dfrac{\partial p_{ij}(s,\ t)}{\partial s} = q_i(s)[p_{ij}(s, t) - \sum\limits_{k \in X} \sigma_{ik}(s)\, p_{kj}(s, t)] \\ p_{ij}(s,\ s) = \delta_{ij}, \qquad s < t, \qquad i,\ j \in X, \end{cases}$$

and

$$\begin{cases} \dfrac{\partial p_{ij}(s,\ t)}{\partial t} = -\, q_j(t)\, p_{ij}(s,\ t) + \sum\limits_{k \in X} q_k(t)\, \sigma_{kj}(t) p_{ik}(s, t) \\ p_{ij}(t,\ t) = \delta_{ij}, \qquad t > s, \qquad i,\ j \in X. \end{cases}$$

These systems were first obtained in 1931 by A. N. Kolmogorov.

2.3.3.2.2. As in the case of a finite set of states, the theorem below emphasizes the minimal (common) solution to the Kolmogorov systems.

Theorem 2.3.21. *Suppose* $\mathbf{Q}(\cdot)$ *is measurable and* $\sum\limits_{j \in X} q_{ij}(\cdot) = 0$, $i \in X$. *If* $\sup\limits_{i \in X} |\, q_{ii}(\cdot)|$ *is integrable on every finite interval contained in* T, *then the minimal (stochastic) transition matrix function associated with* $\mathbf{Q}(\cdot)$ *is given by*

$$\mathbf{I} + \int_s^t \mathbf{Q}(u)\, \mathrm{d}u + \sum_{n \geqslant 2} \int_s^t \mathrm{d}t_1 \int_s^{t_1} \mathrm{d}t_2 \cdots \int_s^{t_{n-1}} \mathbf{Q}(t_n) \ldots \mathbf{Q}(t_1)\, \mathrm{d}t_n,$$

where $\mathbf{I}$ *is the identity matrix* $(\delta_{ij})_{i,\, j \in X}$. *The series is absolutely and uniformly convergent with respect to* $s, t \in T_1$ *on every finite interval* $T_1 \subset T$.

Proof. Proposition 2.3.12. ◇

2.3.3.2.3. We mention that it is usual in practice to specify a Markov jump process by writing down a transition intensity matrix function $\mathbf{Q}(\cdot)$ and then solving one of the Kolmogorov systems. The sole justification for this lies in the fact that in almost all practical applications the process is unique i.e. it coincides with the minimal process associated with $\mathbf{Q}(\cdot)$. If uniqueness does not hold, the Markov process is to be understood as the minimal one unless otherwise stated.

2.3.3.2.4. The structure of the trajectories of denumerable state Markov jump processes was touched on in Proposition 2.3.18. More information in the homogeneous case will be given later.

2.3.3.3. Poisson processes as Markov jump processes

2.3.3.3.1. According to Proposition 2.2.3 the Poisson process $P(\lambda(\cdot))$ is a Markov process with state space $X=N$ and transition matrix function $(p_{ij}(\cdot\,,\,\cdot))_{i,j\in N}$ given by

$$p_{ij}(s,t)=\begin{cases}\dfrac{[\lambda(t)-\lambda(s)]^{j-i}}{(j-i)\,!}\exp\,[\lambda(s)-\lambda(t)] & \text{if } j\geqslant i\\ 0 & \text{if } j<i.\end{cases} \tag{2.3.25}$$

It follows that if λ is continuous, then $P(\lambda(\cdot))$ is a classical Markov jump process with transition intensity function defined by

$$q_i(s)=\lambda'(s),\ i\in N, \tag{2.3.26}$$

$$\sigma_{i,i+1}(s)=1,\ \sigma_{ij}(s)=0,\quad i\in N,\ j\neq i+1.$$

(Notice that the derivative λ' exists almost everywhere on T since λ is a non-decreasing function).

2.3.3.3.2. On account of Theorem 2.3.21, the minimal transition matrix function associated with the intensity of $P(\lambda(\cdot))$ will be stochastic. It is interesting to note that this matrix may be different from (2.3.25)[33]. In fact, the forward equations are

$$\frac{\partial p_{i0}(s,t)}{\partial t}=-\lambda'(t)\,p_{i0}(s,t),\ i\in N,$$

$$\frac{\partial p_{ij}(s,t)}{\partial t}=-\lambda'(t)\,[p_{ij}(s,t)-p_{i,j-1}(s,t)],\ i\in N,\ j\in N^*.$$

Introducing the generating functions

$$G_i(z;s,t)=\sum_{j\in N}p_{ij}(s,t)\,z^j,\ i\in N,\ |z|<1,$$

leads to the differential equation

$$\frac{\partial G_i(z;s,t)}{\partial t}=-\lambda'(t)\,(1-z)\,G_i(z;s,t),$$

whence

$$\frac{\partial}{\partial t}\left\{G_i(z;s,t)\exp\left[(1-z)\int_s^t\lambda'(u)\,du\right]\right\}=0,$$

[33] The reader will be able to explain this apparent contradiction.

which ultimately yields

$$G_i(z; s, t) = z^i \exp\left\{-\left[(1-z)\int_s^t \lambda'(u)\,du\right]\right\}.$$

Therefore, the minimal transition matrix function associated with (2.3.26) will coincide with the transition matrix function (2.3.25) of $P(\lambda(\cdot))$ iff λ is absolutely continuous.

2.3.3.3.3. The reader will be able to show that the compound Poisson process $P(\lambda, F)$ (see Subparagraph 2.2.2.1.6) is a homogeneous Markov jump process with state space R and transition intensity function defined by

$$q(x) = \lambda, \quad x \in R,$$

$$\sigma(x, (-\infty, y)) = F(y - x), \quad x, y \in R.$$

2.3.3.4. Markov branching processes

2.3.3.4.1. Markov branching processes are the continuous parameter version of Galton-Watson chains. They can be defined in a manner paralleling the discrete parameter case (as was done by A. N. KOLMOGOROV and N. A. DMITRIEV) or as Markov jump processes of special type. The latter alternative was used by HARRIS (1963, p. 96) to whom we refer the reader for a detailed treatment with extensions and proofs.

Let $\mathfrak{b}$ and $\mathfrak{p}_i$, $i \in N$, be continuous, non-negative functions defined on T such that $\mathfrak{p}_1(\cdot) \equiv 0$ and $\sum_{i \in N} \mathfrak{p}_i(\cdot) \equiv 1$. A *(Markov) branching process* is the minimal process associated with the transition intensity function defined by

$$q_i(s) = i\mathfrak{b}(s), \; i \in N,$$

$$\sigma_{ij}(s) = \begin{cases} \mathfrak{p}_{j-i+1}(s) & \text{if } 0 \leqslant i \leqslant j+1, \\ \quad 0 & \text{if } i > j+1, \end{cases} \qquad i, j \in N.$$

According to the interpretation given in Subparagraph 2.3.2.1.2. $i\mathfrak{b}(s)\,h + o(h)$ is the probability of no longer being in state i at time $s+h$ given that the process was in i at time s. (It is convenient to think of a state $i \in N$ as the number of particles in a population). Further, $\mathfrak{p}_{j-i+1}(s)$ is the probability of entering the state $j \geqslant i-1$ given that a change of state occurred at time s when the state was i. (This explains why we postulated $\mathfrak{p}_1 \equiv 0$).

2.3.3.4.2. In our case the backward and forward equations become

$$\frac{\partial p_{ij}(s,t)}{\partial s} = i\mathfrak{b}(s)\left[p_{ij}(s,t) - \sum_{k=i-1}^{\infty} \mathfrak{p}_{k-i+1}(s)\, p_{kj}(s,t)\right], \quad i,j \in N, \tag{2.3.27}$$

$$\begin{cases} \dfrac{\partial p_{ij}(s,t)}{\partial t} = \mathfrak{b}(t)\left[-jp_{ij}(s,t) + \displaystyle\sum_{k=1}^{j+1} k\mathfrak{p}_{j-k+1}(t)\, p_{ik}(s,t)\right], & i \in N^*,\ j \in N, \\ \dfrac{\partial p_{0j}}{\partial t} = 0,\ j \in N. \end{cases} \tag{2.3.28}$$

Set

$$\mathfrak{f}(z,t) = \sum_{k \in N} \mathfrak{p}_k(t)\, z^k$$

and

$$F_i(z,s,t) = \sum_{k \in N} p_{ik}(s,t)\, z^k,\ |z| \leqslant 1,\ i \in N. \tag{2.3.29}$$

From (2.3.28) we deduce formally the equations

$$\frac{\partial F_i(z,s,t)}{\partial t} = \mathfrak{b}(t)\,[\mathfrak{f}(z,t) - z]\,\frac{\partial F_i(z,s,t)}{\partial z}, \quad |z| \leqslant 1,\ i \in N. \tag{2.3.30}$$

Since $p_{ij}(s,s) = \delta_{ij}$, we must have

$$F_i(z,s,s) = z^i,\ i \in N.$$

Theorem 2.3.22. *The minimal (substochastic) transition matrix function is the unique solution of* (2.3.28). *The generating functions* F_i *associated with it satisfy* (2.3.30) *and* $F_i = (F_1)^i$, $i \in N$.

Proof. See HARRIS (1963, pp. 99 and 119). ◇

Notice that the relation $F_i = (F_1)^i$ amounts to the fact that a corresponding population starting with i particles is distributed as the sum of i independent populations each starting with one particle. Thus our process is indeed a continuous parameter version of Galton-Watson chains.

Using the relation $F_i = (F_1)^i$ we deduce formally

$$\frac{\partial F_1(z,s,t)}{\partial s} = -\,\mathfrak{b}(s)[\mathfrak{f}(F_1(z,s,t),s) - F_1(z,s,t)],\ |z| \leqslant 1, \tag{2.3.31}$$

with $F_1(z,t,t) = z$.

Theorem 2.3.23. *The minimal (substochastic) transition matrix function is the unique solution of* (2.3.27) *whose generating functions* (2.3.29) *satisfy* $F_i = (F_1)^i$. *Moreover,* F_1 *satisfies* (2.2.31).

Proof. See HARRIS (1963, p. 99 and 119). ◇

2.3.3.4.3. A simple condition ensuring that the transition matrix function of the branching process is stochastic is that the series $\sum_{k \in N} k \mathfrak{p}_k(t)$ converge uniformly on every finite interval $\subset T$. Indeed, for a fixed t, set $g(s) = F_1(1, s, t)$, $s \leqslant t$. On account of (2.3.31) we have

$$g'(s) = -\mathfrak{b}(s)[f(g(s), s) - g(s)],$$

with $g(t) = 1$, $0 \leqslant g(s) \leqslant 1$, $s < t$. Under the condition assumed the right side of the above differential equation satisfies a Lipschitz condition in the region $|g| \leqslant 1$, $0 \leqslant s \leqslant t$. Therefore the solution is unique and since $g(s) \equiv 1$ is a solution we must have $F_1(1, s, t) \equiv 1$.

2.3.3.4.4. A well-known Markov branching process is the *non-homogeneous linear birth-and-death process* for which $\mathfrak{p}_i(.) \equiv 0$ for $i > 2$ and

$$\mathfrak{p}_0(t) = \frac{d(t)}{b(t) + d(t)}, \quad \mathfrak{p}_2(t) = \frac{b(t)}{b(t) + d(t)},$$

where b and d are continuous non-negative functions whose sum is strictly positive. In the birth-and-death process any particle existing at time t has probability $b(t)h + o(h)$ of remaining and giving birth to one new particle and a probability $d(t)\,h + o(h)$ of dying in the interval $(t, t + h)$. Since

$$\sum_{k \in N} k\mathfrak{p}_k(t) = \sum_{k=0}^{2} k\,\mathfrak{p}_k(t) = \frac{2b(t)}{b(t) + d(t)},$$

the matrix transition function of the birth-and-death process will be stochastic. Equation (2.3.31) becomes

$$\frac{\partial F_1}{\partial s} = -(1 - F_1)[d(s) - b(s)\,F_1],$$

that is an equation of the Riccati-type. KENDALL (1948) obtained its solution as

$$F_1(z, s, t) = \frac{u + (1 - u - v)z}{1 - vz},$$

where

$$u = 1 - \frac{\exp(-\rho)}{w}, \quad v = 1 - \frac{1}{w},$$

$$\rho(s, t) = \int_s^t [d(x) - b(x)] \, dx,$$

$$w(s, t) = \exp(-\rho(s, t)) \left[1 + \int_s^t \exp(\rho(s, x)) \, d(x) \, dx \right].$$

It follows that

$$p_{10} = u, \; p_{1n} = (1 - u)(1 - v) v^{n-1}, \; n \in N^*,$$

and hence the probability of extinction is 1 (i.e., $\lim_{t \to \infty} p_{10}(s, t) = 1$ for all $s \in T$) iff

$$\int_0^\infty d(t) \exp(\rho(0, t)) \, dt = \infty.$$

2.3.3.4.5. In the homogeneous case $\mathfrak{b}$ and $\mathfrak{p}_i$, $i \in N$, are independent of t, i.e., they are non-negative numbers such that $\mathfrak{p}_1 = 0$ and $\sum_{i \in N} \mathfrak{p}_i = 1$. Thus, a *homogeneous* Markov branching process is the minimal process associated with the transition intensity function defined by

$$q_i = \mathfrak{b} i, \quad i \in N,$$

$$\sigma_{ij} = \begin{cases} \mathfrak{p}_{j-i+1} & \text{if } \; 0 \leqslant i \leqslant j + 1 \\ 0 & \text{if } \; i > j + 1, \quad i, j \in N. \end{cases}$$

In this case $p_{ij}(s, t)$ are functions of $t - s$ only and the backward and forward equations become

$$p'_{ij}(t) = i\mathfrak{b} \left[-p_{ij}(t) + \sum_{k=i-1}^{\infty} \mathfrak{p}_{k-i+1} p_{kj}(t) \right], \quad i, j \in N, t \geqslant 0,$$

$$\begin{cases} p'_{ij}(t) = \mathfrak{b} \left[-j p_{ij}(t) + \sum_{k=1}^{j+1} k \mathfrak{p}_{j-k+1} p_{ik}(t) \right], & i \in N^*, \quad j \in N, \\ p'_{0j}(t) = 0, \; j \in N, \; t \geqslant 0. \end{cases}$$

Set

$$f(z) = \sum_{k \in N} \mathfrak{p}_k z^k$$

and

$$F_i(z, t) = \sum_{k \in N} p_{ik}(t) z^k, |z| \leqslant 1, \quad i \in N.$$

Obviously, Theorems 2.3.22 and 2.3.23 carry over immediately. Equations (2.3.30) and (2.3.31) become

$$\frac{\partial F_i(z,t)}{\partial t} = \mathfrak{b}[f(z) - z] \frac{\partial F_i(z, t)}{\partial z} \tag{2.3.32}$$

with the initial condition $F_i(z, 0) = z^i$ and

$$\frac{\partial F_1(z, t)}{\partial t} = \mathfrak{b}[f(F_1(z, t)) - F_1(z, t)] \tag{2.3.33}$$

with the initial condition $F_1(z, 0) = z$.

A necessary and sufficient condition that the transition function of the homogeneous Markov branching process be stochastic is given in

Theorem 2.3.24 (E.B. DYNKIN). *In order that $F_1(1, t) \equiv 1$ it is necessary and sufficient that for each $\varepsilon > 0$ the integral*

$$\int_{1-\varepsilon}^{1} \frac{\mathrm{d}u}{f(u) - u}$$

diverge.

Proof. See HARRIS (1963, p. 107). ◇

Corollary. *If $f'(1) < \infty$* [34]*, then $F_1(1, t) \equiv 1$.*
For, $f(x) - x = (f'(1) - 1)(x - 1) + o(x - 1)$ as $x \to 1-$.

2.3.3.4.6. The problem of extinction of a homogeneous branching process may be investigated in a way analogous to that for a Galton-Watson chain. We define the *extinction probability* q by

$$q = \lim_{t \to \infty} p_{10}(t).$$

This limit exists always since $p_{10}(t + h) \geqslant p_{10}(t) p_{00}(h) = p_{10}(t)$ that is p_{10} is a non-decreasing function. Let ξ denote our homogeneous branching process. It is easily seen that if $\xi(0) = 1$ and $f'(1) < \infty$,

[34] Notice that this condition corresponds to that appearing in the non-homogeneous case.

then the sequence $(\xi(nt_0))_{n\in N}$ is a Galton-Watson chain with generating function $F_1(z, t_0)$ for any $t_0 > 0$. According to Theorem 1.2.9, the extinction probability for this Galton-Watson chain is the smallest non-negative root of the equation $u = F_1(u, t_0)$. On the other hand, by definition this equals

$$\lim_{n\to\infty} \mathsf{P}(\xi(nt_0) = 0 \mid \xi(0) = 1) = \lim_{n\to\infty} p_{10}(nt_0) = q.$$

Therefore the extinction probability q for ξ is the smallest non-negative root of the equation $u = F_1(u, t)$, where t may have any positive value. The theorem below shows that q can be calculated from an equation that does not depend on time.

Theorem 2.3.25 (B. A. SEVAST'JANOV). *Let $f'(1) < \infty$. The probability of extinction q is the smallest non-negative root of the equation $u = f(u)$. Hence $q = 1$ iff $f'(1) \leqslant 1$.*

Proof. Since q satisfies $q = F_1(q, t_0)$ for any $t_0 > 0$, we have on account of (2.3.33) that $q = f(q)$. Since $f''(u) = \sum_{k\geqslant 2} k(k-1)\mathfrak{p}_k u^{k-2}$ (in the trivial case $\mathfrak{p}_0 = 1$ the theorem is obviously true), f is strictly convex in [0, 1], thus because $f(1) = 1$ the equation $u = f(u)$ can have at most one root in (0, 1). Notice that $\mathsf{E}\,\xi(t) > 1$ iff $f'(1) > 1$. Indeed, differentiating (2.3.32) with respect to z and interchanging the order of differentiation on the left side yields

$$\frac{\partial}{\partial t}\frac{\partial F_1(z,t)}{\partial z} = \mathfrak{b}\,[f(z) - z]\frac{\partial^2 F_1(z,t)}{\partial z^2} + \mathfrak{b}\,[f'(z) - 1]\frac{\partial F_1(z,t)}{\partial z}$$

whence for $z = 1$,

$$\frac{\mathrm{d}}{\mathrm{d}t}\mathsf{E}\,\xi(t) = \mathfrak{b}\,(f'(1) - 1)\,\mathsf{E}\,\xi(t),$$

therefore

$$\mathsf{E}\,\xi(t) = \exp\,[\mathfrak{b}\,(f'(1) - 1)\,t].$$

It follows that the Galton-Watson chain $(\xi(nt_0))_{n\in N}$ (and therefore ξ) becomes extinct for arbitrary $t_0 > 0$ with probability 1 iff $f'(1) \leqslant 1$. Therefore if $f'(1) > 1$, q must equal the unique root of $u = f(u)$ in [0, 1). Similarly we conclude that if $f'(1) \leqslant 1$, q must equal one. In either case q, is the smallest non-negative root of $u = f(u)$. ◇

2.3.3.4.7. In this subparagraph we briefly describe a generalization of homogeneous Markov branching processes, the so-called age-dependent branching processes first considered by R. BELLMAN and T. E. HARRIS. A particle born at $t = 0$ has a random life-time τ with distribution function $G(t) = \mathsf{P}(\tau \leqslant t)$, $t \geqslant 0$ [35]. At the end of its life it is replaced with probability $\mathfrak{p}_r$, $r \in N$, by r similar particles of age 0. The probabilities $\mathfrak{p}_r$ are assumed to be independent of time, of the age of a particle at the time it dies, and of the number of other particles present. The process continues as long as particles are present. Let $\xi(t)$ be the number of particles alive at time t. The process $\xi = (\xi(t))_{t \geqslant 0}$ is said to be an *age-dependent branching* process. The reason for the term "age-dependent" is that the probability $\mathrm{d}G(\tau)/1 - G(\tau)$ that a particle, living at age τ dies in the age interval $(\tau, \tau + \mathrm{d}\tau)$ is a non-constant function of the age τ (if $G(t) \neq 1 - \exp(-\mathfrak{b}t)$). An explicit description of the probability space on which ξ is defined will be deferred to II; 1.2.4.

In general ξ is not Markovian. It is, however, possible to prove that in the exponential case

$$G(t) = 1 - \exp(-\mathfrak{b}t)$$

if $\mathfrak{p}_1 = 0$, $\sum_{r \in N} r\mathfrak{p}_r < \infty$, then ξ is a homogeneous Markov branching process.

Let

$$F(z, t) = \sum_{r \in N} \mathsf{P}(\xi(t) = r)\, z^r, \quad t \geqslant 0,$$

$$f(z) = \sum_{r \in N} \mathfrak{p}_r z^r, \quad |z| \leqslant 1.$$

These generating functions satisfy the integral equation

$$F(z, t) = \int_{0-}^{t+} f[F(z, t-u)]\, \mathrm{d}G(u) + z(1 - G(t)). \tag{2.3.34}$$

If we assume $G(0+) = 0$ and $m = f'(1) = \sum_{r \in N} r\mathfrak{p}_r < \infty$, the differentiation of (2.3.34) leads to the following integral equation for the mean $M(t) = \mathsf{E}\xi(t)$:

$$M(t) = 1 - G(t) + m \int_{0-}^{t+} M(t-u)\, \mathrm{d}G(u). \tag{2.3.35}$$

---

[35] On account of Theorem 2.3.15 we have in the case of homogeneous Markov branching processes $G(t) = 1 - \exp(-\mathfrak{b}t)$

It is known (see Theorem 2.3.53) that $M(\cdot)$ is the only solution of (2.3.35) which is bounded on every interval.

Although age-dependent branching processes are in general not Markovian, they retain some important properties of homogeneous Markov branching processes. In particular, the probability of extinction (i.e. the event ($\xi(t) = 0$ for all sufficiently large t)) is the smallest non-negative root of the equation $u = f(u)$. Furthermore, if $G(0+) = 0$ and $m < \infty$ almost all trajectories of ξ are step functions.

For details and proofs the reader is referred to HARRIS (1963, Ch. 6). A survey of recent work in this area is to be found in SEVAST'JANOV (1968 a, b, 1971).

We conclude by noticing that the range of application of age-dependent branching processes is limited by the assumption that the death of a parent and the birth of an offspring must occur simultaneously. This assumption is not satisfied by many biological populations, e.g. human populations. A general age-dependent branching process which does not incorporate the above assumption was studied by CRUMP and MODE (1968—69).

2.3.4. Homogeneous Markov jump processes with discrete state space

In this subsection we consider the case in which the probabilities $p_{ij}(s, t)$ depend only on $t - s$, i.e. homogeneous Markov jump processes with countable state space. This case, which is the continuous parameter version of denumerable homogeneous Markov chains, enjoys many specific properties and has been thoroughly investigated.

2.3.4.1. Preliminaries

2.3.4.1.1. In the present case the homogeneous transition function P is determined by the *transition matrix function*

$$\mathbf{P}(\cdot) = (p_{ij}(\cdot))_{i,\, j \in X},$$

where

$$p_{ij}(\cdot) = P(\cdot\,; i, \{j\}).$$

Thus we have

$$p_{ij}(\cdot) \geqslant 0, \quad i,\ j \in X$$

$$\sum_{j \in X} p_{ij}(\cdot) \leqslant 1, \quad i \in X,$$

$$p_{ij}(s + t) = \sum_{k \in X} p_{ik}(s)\, p_{kj}(t), \quad i,\ j \in X,\ s, t \geqslant 0,$$

(equivalently

$$\mathbf{P}(t+s) = \mathbf{P}(t)\,\mathbf{P}(s)$$

for all $s, t \geqslant 0$).

As we mentioned in Subparagraph 2.3.2.2.4, we shall suppose that

$$\lim_{t \to 0+} p_{ij}(t) = p_{ij}(0) = \delta_{ij},\ i, j \in X\ [36]. \tag{2.3.36}$$

On account of Subparagraph 2.3.1.1.2, if P is substochastic we may adjoin to X an extra state denoted $\mathfrak{d}$ and define

$$p_{i\mathfrak{d}}(\cdot) = 1 - \sum_{j \in X} p_{ij}(\cdot),\ i \in X,$$

$$p_{\mathfrak{d}\mathfrak{d}}(\cdot) \equiv 1,\ p_{\mathfrak{d}j}(\cdot) \equiv 0,\ j \in X,$$

$$\bar{X} = X \cup \{\mathfrak{d}\}.$$

The transition matrix function $\bar{\mathbf{P}}(\cdot) = (p_{ij}(\cdot))_{i,j \in \bar{X}}$ is stochastic and satisfies the same relations as $\mathbf{P}(\cdot)$.

2.3.4.1.2. The requirements on the transition matrix function $\mathbf{P}(\cdot)$ may be described in terms of semi-group theory as follows. For all $t \geqslant 0$ define an operator $\mathfrak{P}_t$ on the Banach space l consisting of real sequences $\mathbf{x} = (x_i)_{i \in X}$ with $\|\mathbf{x}\| = \sum_{i \in X} |x_i| < \infty$ by setting

$$(\mathfrak{P}_t\, x)_i = \sum_{j \in X} x_j\, p_{ji}(t)\ [37].$$

Then the properties of $\mathbf{P}(\cdot)$ are equivalent to the following properties of the family $(\mathfrak{P}_t)_{t \geqslant 0}$:

$$\mathfrak{P}_t\, \mathbf{x} \geqslant 0 \quad \text{when } \mathbf{x} \geqslant \mathbf{0};$$

$$\|\mathfrak{P}_t\, \mathbf{x}\| \leqslant \|\mathbf{x}\| \quad \text{when } \mathbf{x} \geqslant \mathbf{0};\ [38]$$

$$\mathfrak{P}_{t+s} = \mathfrak{P}_t\, \mathfrak{P}_s,\ s, t \geqslant 0;$$

$$\|\mathfrak{P}_t\, \mathbf{x} - \mathbf{x}\| \to 0 \text{ as } t \to 0+ \text{ for all } \mathbf{x} \in l.$$

---

[36] A transition matrix function satisfying (2.3.36) is said to be *standard*. For a study of non-standard transition matrix functions see CHUNG (1967, pp. 119–128).

[37] Compare with Subparagraph 2.3.1.4.1.

[38] This relation can be replaced by $\|\mathfrak{P}_t\mathbf{x}\| = \|\mathbf{x}\|$ when $\mathbf{x} \geqslant 0$ iff P is stochastic.

It is known [see YOSIDA (1965, Ch. 9)] that the contraction semigroup $(\mathfrak{P}_t)_{t\geq 0}$ is uniquely determined by its infinitesimal generator $\mathfrak{A}$. This inear operator is defined by

$$\mathfrak{A}\mathbf{x} = \lim_{t\to 0+} \frac{\mathfrak{P}_t \mathbf{x} - \mathbf{x}}{t};$$

its domain $\mathcal{D}(\mathfrak{A})$ consists of those elements $\mathbf{x}$ for which the above (strong) limit exists. Thus $(\mathfrak{P}_t)_{t\geq 0}$, and therefore $\mathbf{P}(.)$ as well, can in principle be characterized by its infinitesimal behaviour.

2.3.4.2. Continuity and differentiability properties

2.3.4.2.1. The p_{ij} have many remarkable properties. We start with

Proposition 2.3.26. *For all $i, j \in \overline{X}$ the function p_{ij} is uniformly continuous on* $[0, \infty)$. *The modulus of continuity of p_{ij} does not exceed that of p_{ii} at zero.*

Proof. Consider an $h > 0$. We have

$$p_{ij}(t+h) - p_{ij}(t) = \sum_{k\in\overline{X}} p_{ik}(h)\, p_{kj}(t) - p_{ij}(t) =$$

$$= -(1 - p_{ii}(h))\, p_{ij}(t) + \sum_{k\neq i} p_{ik}(h)\, p_{kj}(t),$$

whence

$$-(1 - p_{ii}(h)) \leqslant p_{ij}(t+h) - p_{ij}(t) \leqslant \sum_{k\neq i} p_{ik}(h) = 1 - p_{ii}(h),$$

that is

$$|p_{ij}(t+h) - p_{ij}(t)| \leqslant 1 - p_{ii}(h).$$

For $h < 0$ set $h' = |h|$. If we replace t in the last inequality by $t - h' = t + h$ we get

$$|p_{ij}(t+h) - p_{ij}(t)| \leqslant 1 - p_{ii}(|h|).$$

Therefore for all $u \in R$ such that $t + u \geqslant 0$ we have

$$|p_{ij}(t+u) - p_{ij}(t)| \leqslant 1 - p_{ii}(|u|)$$

which proves the proposition. ◇

Notice that by the Chapman-Kolmogorov equation we can write

$$p_{ii}(t) \geqslant \left[p_{ii}\left(\frac{t}{n}\right)\right]^n$$

for all $n \in N^*$, $i \in \overline{X}$, $t \geqslant 0$. Then (2.3.36) implies that $p_{ii}(t) > 0$ for all $i \in \overline{X}$, $t \geqslant 0$. Further,

$$p_{ij}(t) \geqslant p_{ij}(t_0)\, p_{jj}(t - t_0)$$

for all $i, j \in \overline{X}$, $t \geqslant t_0$. It follows that if $p_{ij}(t_0) > 0$, then $p_{ij}(t) > 0$ for all $t \geqslant t_0$. The last assertion can be sharpened to a rather deep result. We have namely

Theorem 2.3.27 (P. Lévy, D. G. Austin, D. Ornstein). *For arbitrary $i, j \in \overline{X}$ the function p_{ij} is either identically zero or always positive on* $(0, \infty)$.

Proof. See Chung (1967, p. 126) or Freedman (1967). ◇

2.3.4.2.2. We shall now establish the existence of the righthand derivatives of the p_{ij} at zero which will be denoted simply by $p'_{ij}(0)$.

Theorem 2.3.28 (A. N. Kolmogorov). *For all $i \neq j \in \overline{X}$*

$$q_{ij} = p'_{ij}(0) = \lim_{t \to 0+} \frac{p_{ij}(t)}{t}$$

exists and is finite.

Proof. For a given $0 < \varepsilon < 1/3$ choose $t_\varepsilon > 0$ small enough so that

$$1 - p_{ii}(t) < \varepsilon,$$

$$1 - p_{jj}(t) < \varepsilon,$$

for $0 \leqslant t \leqslant t_\varepsilon$. Let $n \in N^*$ be such that $nt \leqslant t_\varepsilon$ and set [39]

$$a_k = \mathsf{P}(\xi(rt) \neq j,\ 1 \leqslant r < k,\ \xi(kt) = j \mid \xi(0) = i),$$

$$b_k = \mathsf{P}(\xi(rt) \neq j,\ 1 \leqslant r < k,\ \xi(kt) = i \mid \xi(0) = i)$$

---

[39] ξ is the process constructed in Theorem 2.3.3.

for $1 \leqslant k \leqslant n$. If $nt \leqslant u \leqslant t_\varepsilon$, then by Proposition 2.3.26 and the Chapman-Kolmogorov equation we have

$$\varepsilon \geqslant p_{ij}(u) \geqslant \sum_{k=1}^{n} a_k \, p_{jj}(u - kt) \geqslant (1 - \varepsilon) \sum_{k=1}^{n} a_k,$$

whence

$$\sum_{k=1}^{n} a_k \leqslant \frac{\varepsilon}{1 - \varepsilon}$$

and

$$b_k \geqslant p_{ii}(kt) - \sum_{r=1}^{k-1} a_r \geqslant 1 - \varepsilon - \frac{\varepsilon}{1 - \varepsilon} \geqslant \frac{1 - 3\varepsilon}{1 - \varepsilon}.$$

Therefore

$$p_{ij}(u) \geqslant \sum_{k=1}^{n} b_{k-1} \, p_{ij}(t) \, p_{jj}(u - kt) \geqslant$$

$$\geqslant n(1 - \varepsilon) \frac{1 - 3\varepsilon}{1 - \varepsilon} p_{ij}(t) = (1 - 3\varepsilon) \, n p_{ij}(t).$$

For given t and $u \leqslant t_\varepsilon$ choose $n = n(t, u)$ such that $nt \leqslant u < (n + 1) t$. The last inequality implies that

$$\frac{p_{ij}(u)}{u - t} \geqslant (1 - 3\varepsilon) \frac{p_{ij}(t)}{t},$$

whence

$$\overline{\lim_{t \to 0+}} \frac{p_{ij}(t)}{t} \leqslant \frac{p_{ij}(u)}{u} (1 - 3\varepsilon)^{-1} < \infty,$$

thus

$$\overline{\lim_{t \to 0+}} \frac{p_{ij}(t)}{t} \leqslant \underline{\lim_{t \to 0+}} \frac{p_{ij}(t)}{t} (1 - 3\varepsilon)^{-1}.$$

Since ε is arbitrary we deduce that

$$\lim_{t \to 0+} \frac{p_{ij}(t)}{t}$$

exists and is finite. ◇

Theorem 2.3.29. *For every* $i \in \overline{X}$

$$q_i = -p'_{ii}(0) = \lim_{t \to 0+} \frac{1 - p_{ii}(t)}{t}$$

exists and

$$\frac{1 - p_{ii}(t)}{t} \leqslant q_i \tag{2.3.37}$$

for all $t > 0$.

Proof. Set

$$\widetilde{q}_i = \overline{\lim_{t \to 0+}} \frac{1 - p_{ii}(t)}{t}.$$

If $\widetilde{q}_i = \infty$, then $q_i = \infty$. Suppose therefore that $\widetilde{q}_i < \infty$.

We shall first show that

$$\frac{1 - p_{ii}(t)}{t} \leqslant \widetilde{q}_i \tag{2.3.38}$$

for all $t > 0$. For given $\varepsilon > 0$ and $t > 0$, let $\eta = \eta(\varepsilon, t) > 0$ be such that

$$\frac{1 - p_{ii}(\eta)}{\eta} \leqslant \widetilde{q}_i + \frac{\varepsilon}{2}$$

and

$$1 - p_{ii}(s) < \frac{\varepsilon t}{2}$$

for $0 \leqslant s \leqslant \eta$. Choose $n = n(\eta, t) \in N$ such that $n\eta \leqslant t < (n+1)\eta$. Taking into account the fact that the Chapman-Kolmogorov equation implies $p_{ii}(u + v) \geqslant p_{ii}(u)\, p_{ii}(v)$ we can write

$$p_{ii}(t) \geqslant [p_{ii}(\eta)]^n\, p_{ii}(t - n\eta) \geqslant \left[1 - \eta\left(\widetilde{q}_i + \frac{\varepsilon}{2}\right)\right]^n \left(1 - \frac{\varepsilon t}{2}\right) \geqslant$$

$$\geqslant \left[1 - n\eta\left(\widetilde{q}_i + \frac{\varepsilon}{2}\right)\right]\left(1 - \frac{\varepsilon t}{2}\right) \geqslant 1 - \frac{\varepsilon t}{2} - t\left(\widetilde{q}_i + \frac{\varepsilon}{2}\right) =$$

$$= 1 - t(\widetilde{q}_i + \varepsilon).$$

Since, ε is arbitrary, we get (2.3.38). This inequality yields

$$\overline{\lim_{t\to 0+}} \frac{1 - p_{ii}(t)}{t} \leqslant \tilde{q}_i,$$

whence

$$\tilde{q}_i = \lim_{t\to 0+} \frac{1 - p_{ii}(t)}{t},$$

which concludes the proof. ◇

Theorems 2.3.28 and 2.3.29 were generalized to non-countable state spaces of a special type by KENDALL (1955) [cf. LOÈVE (1963, 39.1)].

2.3.4.2.3. It follows from the previous subparagraph that if $q_i < \infty$, $i \in X$, then $\mathbf{P}(\cdot)$ has a *transition intensity matrix* $\mathbf{Q} = (q_{ij})_{i,j\in X}$, where $q_{ii} = -q_i$ [40]. In general we have

$$q_{ij} \begin{cases} \geqslant 0 & \text{if } i \neq j \\ \leqslant 0 & \text{if } i = j, \end{cases}$$

$$\sum_{j\neq i} q_{ij} \leqslant -q_{ii} \leqslant \infty \text{ [41]}. \tag{2.3.40}$$

All four combinations of < and = can actually occur in the last line. [See KENDALL (1956), KENDALL and REUTER (1956), KOLMOGOROV (1951)].

It is to be noted that if X is a finite set, then we can have only

$$\sum_{j\neq i} q_{ij} = -q_{ii} < \infty, \quad i \in X.$$

The problem we considered in Subparagraph 2.3.2.3.1 will be reformulated now as follows: describe all (substochastic) transition matrix functions $\mathbf{P}(\cdot)$ such that $\mathbf{P}'(0) = \mathbf{Q}$, where the entries of $\mathbf{Q}$ verify (2.3.40).

---

40) The reader should observe that the existence of $\mathbf{Q}$ is not equivalent to the existence of a transition intensity function as the latter was defined in Subparagraph 2.3.2.1.2. This equivalence holds only for transition matrix functions for which

$$\lim_{t\to 0+} \frac{\sum_{j\in A} [p_{ij}(t) - \delta_{ij}]}{t} = \sum_{j\in A} q_{ij} \tag{2.3.39}$$

for any subset A of X.

41) We even have $\sum_{j\in \overline{X}-\{i\}} q_{ij} \leqslant -q_{ii}$, $i \in X$.

This is the so-called *complete construction problem* initiated by W. FELLER. It has been completely solved only for certain classes of matrices Q. For details see CHUNG (1963, 1966), DYNKIN (1967), FELLER (1954 b), and WILLIAMS (1964 a, b, 1967 a, b).

2.3.4.2.4. The next theorem establishes the continuous differentiability of the p_{ij} in $(0, \infty)$.

Theorem 2.3.30 (D. G. AUSTIN, D. ORNSTEIN). *For every $i, j \in \overline{X} = X \cup \{\mathfrak{d}\}$ the function p_{ij} has a continuous derivative in $(0, \infty)$ satisfying*

$$p'_{ij}(s+t) = \sum_{k \in \overline{X}} p'_{ik}(s)\, p_{kj}(t), \qquad s > 0,\, t \geqslant 0,$$

$$p'_{ij}(s+t) = \sum_{k \in \overline{X}} p_{ik}(s)\, p'_{kj}(t), \qquad s \geqslant 0,\, t > 0,$$

where the series are absolutely convergent. Furthermore, for each $i \in \overline{X}$

$$\sum_{j \in \overline{X}} |p'_{ij}(t)| < \infty, \qquad \sum_{j \in \overline{X}} p'_{ij}(t) = 0, \qquad t > 0.$$

If $i \neq j$, then p'_{ij} is also continuous at zero. If $q_i < \infty$, then p'_{ii} is also continuous at zero.

Proof. See CHUNG (1967, pp. 135—138, 208—211). ◇

Further differentiability properties of the p_{ij} are generally unknown. JUŠKEVIČ (1959) gave an example in which a finite second derivative does not exist [cf. KENDALL (1967 b)]. For other analytic problems concerning the p_{ij} see KENDALL (1967 a, b) and SPEAKMAN (1967).

2.3.4.3. The Kolmogorov differential equations

2.3.4.3.1. By letting $h \to 0+$ in the identities

$$\frac{p_{ij}(t+h) - p_{ij}(t)}{h} = \frac{p_{ii}(h) - 1}{h} p_{ij}(t) + \sum_{k \neq i} \frac{p_{ik}(h)}{h} p_{kj}(t),$$

$$\frac{p_{ij}(t+h) - p_{ij}(t)}{h} = p_{ij}(t) \frac{p_{jj}(h) - 1}{h} + \sum_{k \neq j} p_{ik}(t) \frac{p_{kj}(h)}{h},$$

we get from Theorems 2.3.28 and 2.3.29 together with Fatou's lemma the inequalities

$$p'_{ij}(t) \geqslant \sum_{k \in X} q_{ik} p_{kj}(t), \qquad t \geqslant 0, \tag{2.3.41}$$

$$p'_{ij}(t) \geqslant \sum_{k \in X} p_{ik}(t)\, q_{kj}, \qquad t \geqslant 0. \tag{2.3.42}$$

In general, these can be strict inequalities (see Subparagraph 2.3.4.7.4). In what follows we shall give conditions under which the backward and forward equations

$$p'_{ij}(t) = \sum_{k\in X} q_{ik} p_{kj}(t), \qquad t \geqslant 0, \qquad (\mathrm{B}_{ij})$$

$$p'_{ij}(t) = \sum_{k\in X} p_{ik}(t)\, q_{kj}, \qquad t \geqslant 0, \qquad (\mathrm{F}_{ij})$$

hold.

It is obvious that (B_{ij}) and (F_{ij}) hold for all $i, j \in X$ whenever X is a finite set so that from now on we shall assume that X is countably infinite.

2.3.4.3.2. It should be noted that when (B_{ij}) [or (F_{ij})] is known to hold only for almost all $t \geqslant 0$, then this implies that it holds for all $t \geqslant 0$. Indeed, let (B_{ij}) hold for almost all $t \geqslant 0$. Since the series $\sum_{k\in X} q_{ik} p_{kj}(t)$ is uniformly convergent, it is continuous for $t \geqslant 0$ and by Theorem 2.3.30, $p'_{ij}(t)$ is also continuous for $t > 0$. Therefore (B_{ij}) must hold for all $t > 0$ and since it also holds at $t = 0$ because of $p'_{ij}(0) = q_{ij}$, we see that it holds for all $t \geqslant 0$. The system (F_{ij}) requires more delicate handling than (B_{ij}), since we can no longer assert a priori that $\sum_{k\in X} p_{ik}(t)\, q_{kj}$ is continuous. However, this can be proved [see e.g. CHUNG (1967, p. 248)], so that the assertion is also true for (F_{ij}).

2.3.4.3.3. Now we give a criterion for the validity of the backward equations.

Theorem 2.3.31. *For any given $i \in X$, (B_{ij}) holds for all $j \in X$ iff $q_i < \infty$ and*

$$\sum_{j\in X-\{i\}} q_{ij} = q_i .$$

Proof. a) Necessity. If (B_{ij}) holds then it is, of course, understood that q_i is finite. Thus let $q_i < \infty$. By summing (B_{ij}) for $j \in X$ we get on account of Theorem 2.3.30

$$-\,p'_{i\mathfrak{d}}(t) = \sum_{k\in X} q_{ik}(1 - p_{k\mathfrak{d}}(t)).$$

Letting $t \to 0+$, we obtain, since the series is uniformly convergent

$$-\,q_{i\mathfrak{d}} = \sum_{k\in X} q_{ik} ,$$

which is identical to the equation to be proved.

b) Sufficiency. For all $h > 0$ we have

$$\frac{p_{ij}(t+h) - p_{ij}(t)}{h} = -\frac{1 - p_{ii}(h)}{h} p_{ij}(t) + \sum_{k \neq i} \frac{p_{ik}(h)}{h} p_{kj}(t).$$

For a given $\varepsilon > 0$ choose a finite set $E \subset \overline{X} - \{i\}$ such that

$$q_i - \sum_{k \in E} q_{ik} < \varepsilon.$$

We have

$$0 \leqslant \sum_{k \in \overline{X} - E - \{i\}} \frac{p_{ik}(h)}{h} p_{kj}(t) \leqslant \sum_{k \in \overline{X} - E - \{i\}} \frac{p_{ik}(h)}{h} =$$

$$= \frac{1 - p_{ii}(h)}{h} - \sum_{k \in E} \frac{p_{ik}(h)}{h} \xrightarrow[h \to 0+]{} q_i - \sum_{k \in E} q_{ik} < \varepsilon.$$

Therefore

$$\overline{\lim_{k \to 0+}} \left| \frac{p_{ij}(t+h) - p_{ij}(t)}{h} - \left\{ q_i p_{ij}(t) + \sum_{k \in E} q_{ik} p_{kj}(t) \right\} \right| < \varepsilon.$$

Now use Theorem 2.3.30. ◇

Corollary. (B_{ij}) *holds for all* $j \in X$ *when* $\sum_{j \in X} q_{ij} = 0$.

2.3.4.3.4. For the forward equations there is no counterpart to Theorem 2.3.31. A criterion for the validity of (F_{ij}) for all $i, j \in X$ was given by REUTER (1957, p. 14) in terms of semigroup theory [42]. We indicate here a sufficient condition.

Proposition 2.3.32. *If* $q_j < \infty$ *and the limit*

$$\lim_{h \to 0+} \frac{p_{ij}(h)}{h} = q_{ij}$$

is uniform with respect to $i \neq j$, *then* (F_{ij}) *holds for all* $i \in X$.

---

[42]) REUTER's approach is to define operators $\mathfrak{Q}$ and $\mathfrak{Q}_0$ as follows: the domain $\mathcal{D}(\mathfrak{Q})$ of $\mathfrak{Q}$ is the set of vectors $\mathbf{x} \in l$ such that $\sum_{i \in X} x_i q_{ij}$ converges absolutely for each j and $\sum_{j \in X} \left| \sum_{i \in X} x_i q_{ij} \right| < \infty$. $\mathfrak{Q}$ is given by

$$(\mathfrak{Q}\mathbf{x})_j = \sum_{i \in X} x_i q_{ij}, \qquad \mathbf{x} \in \mathcal{D}(\mathfrak{Q}).$$

The domain $\mathcal{D}(\mathfrak{Q}_0)$ is the set of vectors $\mathbf{x} \in l$ with only finitely many non-zero components; note that $\mathcal{D}(\mathfrak{Q}_0) \subset \mathcal{D}(\mathfrak{Q})$. $\mathfrak{Q}_0$ is the restriction of $\mathfrak{Q}$ to $\mathcal{D}(\mathfrak{Q}_0)$. It can be shown that ($B_{ij}$) holds for all $i, j \in X$ iff the infinitesimal generator $\mathfrak{A}$ (see Subparagraph 2.3.4.1.2) is an extension of $\mathfrak{Q}_0$; similarly (F_{ij}) holds for all $i, j \in X$ iff $\mathfrak{A}$ is a restriction of $\mathfrak{Q}$.

Proof. The assertion follows from the identity

$$\frac{p_{ij}(t+h)-p_{ij}(t)}{h}=p_{ij}(t)\frac{p_{jj}(h)-1}{h}+\sum_{k\neq j}p_{ik}(t)\frac{p_{kj}(h)}{h}$$

on account of Theorem 2.3.30. ◇

If the forward equations (F_{ij}), $i, j \in X$, hold, then we can obtain a system of equations for the absolute probabilities

$$p_i(t) = \mathsf{P}(\xi(t) = i), \quad i \in X, \quad t \geqslant 0.$$

We have

$$p_i(t) = \sum_{j \in X} p_j(0)\, p_{ji}(t),$$

where $p_j(0) = p_j$, $j \in X$, is the initial distribution

$$p_j = \mathsf{P}(\xi(0) = j).$$

The forward equations yield

$$p_{ij}(t) = \delta_{ij} + \sum_{k \in X} \int_0^t q_{kj} p_{ik}(u)\, du, \quad i, j \in X.$$

Multiplying the above equation by p_i and summing over $i \in X$ produce

$$p_j(t) = p_j + \sum_{k \in X} \int_0^t q_{kj} p_k(u)\, du, \quad j \in X.$$

It follows easily (by using also the backward equations) that $(p_j)_{j \in X}$ is a *stationary distribution* (i.e. $p_j(t) = p_j(0) = p_j$, $j \in X$, $t \geqslant 0$) iff

$$\sum_{k \in X} q_{kj} p_k = 0, \quad j \in X.$$

More generally, we may ask for a *stationary measure*, i.e. a family $(u_j)_{j \in X}$ of non-negative numbers such that

$$u_j = \sum_{i \in X} u_i p_{ij}(t), \quad j \in X, \quad t \geqslant 0.$$

Such a measure will be characterized by

$$\sum_{k \in X} q_{kj} u_k = 0, \quad j \in X.$$

2.3.4.3.5. We know from Corollary to Theorem 2.3.8 that there always exists a (perhaps substochastic) transition matrix function $\mathbf{P}_{\min}(\cdot) = (p_{ij}^{\min}(\cdot))_{i,j \in X}$ for which both the backward and forward equations hold [43]. According to the proof of the theorem quoted we have

$$p_{ij}^{\min}(t) = \lim_{n\to\infty} f_{ij}^{(n)}(t),$$

where $f_{ij}^{(n)}$ are defined recursively by

$$f_{ij}^{(0)}(t) = \delta_{ij}\, \mathrm{e}^{-q_i t},$$

$$f_{ij}^{(n+1)}(t) = \mathrm{e}^{-q_i t}\left[\delta_{ij} + \int_0^t \left(\sum_{k\neq i} q_{ik} f_{kj}^{(n)}(u)\right) \mathrm{e}^{q_i u}\, \mathrm{d}u\right], \qquad (2.3.43)$$

or alternatively by

$$f_{ij}^{(n+1)}(t) = \mathrm{e}^{-q_j t}\left[\delta_{ij} + \int_0^t \left(\sum_{k\neq j} f_{ik}^{(n)}(u)\, q_{kj}\right) \mathrm{e}^{q_j u}\, \mathrm{d}u\right].$$

In the present case the conclusion of Proposition 2.3.9 can be strengthened in the sense that

$$p_{ij}(\cdot) \geqslant p_{ij}^{\min}(\cdot)$$

for any substochastic transition matrix function $\mathbf{P}(\cdot) = (p_{ij}(\cdot))_{i,j\in X}$ whose intensity matrix is $\mathbf{Q}$. This can be proved by writing the inequality (2.3.42) as

$$\frac{\mathrm{d}}{\mathrm{d}t}\left(p_{ij}(t)\mathrm{e}^{q_i t}\right) \geqslant \left(\sum_{k\neq i} q_{ik}\, p_{kj}(t)\right)\mathrm{e}^{q_i t},$$

and integrating, which gives

$$p_{ij}(t) \geqslant \mathrm{e}^{-q_i t}\left[\delta_{ij} + \int_0^t \left(\sum_{k\neq i} q_{ik}\, p_{kj}(u)\right)\mathrm{e}^{q_i u}\, \mathrm{d}u\right].$$

On comparing this inequality with (2.3.43), an easy induction argument yields

$$p_{ij}(t) \geqslant f_{ij}^{(n)}(t), \quad n \in N,$$

[43] They obviously hold for almost all $t \geqslant 0$. The considerations of Subparagraph 2.3.4.3.2 show that they must then hold for all $t \geqslant 0$. It should also be noted that for $\mathbf{P}_{\min}(\cdot)$ equation (2.3.39) is valid together with its consequences.

and hence

$$p_{ij}(t) \geqslant p_{ij}^{\min}(t) = \lim_{n\to\infty} f_{ij}^{(n)}(t)$$

as asserted.

2.3.4.3.6. The above considerations together with Theorems 2.3.20 and 2.3.21 lead to the conclusion that if X is finite or, more generally, if $\sum_{j\in X} q_{ij} = 0$, $i \in X$, and $\sup_{i\in X} q_i < \infty$, then the only transition matrix function $\mathbf{P}(\cdot)$ such that $\mathbf{P}'(0) = \mathbf{Q}$ is given by

$$\mathbf{P}(t) = \exp(t\mathbf{Q}) = \mathbf{I} + \sum_{n\in N^*} \frac{t^n\mathbf{Q}^n}{n!}, \quad t \geqslant 0.$$

In practice, if $X = \{1, \ldots, m\}$, we determine the eigenvalues $\lambda_1, \ldots, \lambda_m$ of $\mathbf{Q}$ and a system $(\mathbf{v}_1, \ldots, \mathbf{v}_m)$ of m distinct associated right eigenvectors. Then

$$\mathbf{P}(t) = \mathbf{V}\,\mathbf{\Lambda}(t)\,\mathbf{V}^{-1}$$

where $\mathbf{V}$ is the matrix whose column vectors are, respectively $\mathbf{v}_1, \ldots, \mathbf{v}_m$ and $\mathbf{\Lambda}(t)$ is the diagonal matrix $(\delta_{ij}\, e^{\lambda_i t})_{1\leqslant i,\, j\leqslant m}$.

Finally, we note that for the case where X is finite, problems similar to those considered in Subparagraph 1.1.4.1.6 were investigated by KEMENY and SNELL (1961).

2.3.4.3.7. The question if $\mathbf{P}_{\min}$ is the only solution to $(\mathbf{B}_{ij})$, $i, j \in X$, is answered by

Theorem 2.3.33 (G. E. H. REUTER). *A necessary and sufficient condition that $\mathbf{P}_{\min}$ be the only solution to $(\mathbf{B}_{ij})$, $i, j \in X$, is the following: For some $\lambda > 0$ the system of equations*

$$(\lambda + q_i)x_i = \sum_{k\neq i} q_{ik}\, x_k \tag{2.3.44}$$

has no bounded solution (equivalently, no bounded non-negative solution) other than $x_i = 0$, $i \in X$. If this condition does not hold, then there exist infinitely many such solutions.

Proof. See REUTER (1957, pp. 25—27). ◇

Notice that by Theorems 2.3.13 and 2.3.33 if $\sum_{j\in X} q_{ij} = 0$, and $\mathbf{P}_{\min}$ is substochastic then there will exist infinitely many solutions to $(\mathbf{B}_{ij})$, $i, j \in X$.

To answer the same question for the forward equations (F_{ij}), $i, j \in X$, let us denote by d the maximum number of linearly independent non-negative solutions $(y_i)_{i\in X}$ of the system

$$(\lambda + q_j)y_j = \sum_{k\neq j} y_k\, q_{kj} \tag{2.3.45}$$

with $\sum_{i \in X} y_i < \infty$ (one proves that d can be found by considering the above system for any particular $\lambda > 0$). We then have

Theorem 2.3.34 (G. E. H. REUTER).

i) *If* $\mathbf{P}_{\min}$ *is stochastic, or if* $\mathbf{P}_{\min}$ *is substochastic but* $d = 0$, *then* $\mathbf{P}_{\min}$ *is the only solution to* (F_{ij}), $i, j \in X$.

ii) *If* $\mathbf{P}_{\min}$ *is substochastic and* $d = 1$, *then there are infinitely many solutions to* (F_{ij}), $i, j \in X$, *and exactly one of them is stochastic.*

iii) *If* $\mathbf{P}_{\min}$ *is substochastic and* $d > 1$, *then there are infinitely many solutions to* (F_{ij}), $i, j \in X$, *including infinitely many stochastic ones.*

Proof. See REUTER (1957, p. 32). ◇

Corollary. *If the* q_i *are bounded, then both* (B_{ij}), $i, j, \in X$, *and* (F_{ij}), $i, j \in X$ *have* $\mathbf{P}_{\min}$ *as their only solution*[44].

Proof. Let $\lambda > 0$ and let $(x_i)_{i \in X}$ be a bounded solution to (2.3.44). Then

$$|x_i| \leqslant \frac{1}{\lambda + q_i}\left(\sum_{k \neq i} q_{ik}\right) \sup_{i \in X} |x_i|$$

$$\leqslant \frac{q_i}{\lambda + q_i} \sup_{i \in X} |x_i|,$$

whence the boundedness of the q_i leads to $\sup_{i \in X} |x_i| \leqslant 0$, that is $x_i = 0$ for all $i \in X$.

Now let $(y_i)_{i \in X}$ be a non-negative solution to (2.3.45) with $\sum_{i \in X} y_i < \infty$ Summing over j in (2.3.45) we get

$$\lambda \sum_{j \in X} y_j + \sum_{j \in X} q_j y_j = \sum_{j \in X}\left(\sum_{k \neq j} y_k q_{kj}\right) = \sum_{k \in X}\left(\sum_{j \neq k} q_{kj}\right) y_k \leqslant \sum_{k \in X} q_k y_k,$$

whence $\sum_{j \in X} y_j \leqslant 0$ that is $y_j = 0$, $j \in X$. ◇

For further comments and examples concerning uniqueness or non-uniqueness for the Kolmogorov equations we refer the reader to KENDALL (1956), and REUTER (1957, pp. 35—39).

---

[44] If $\sum_{j \in X} q_{ij} = 0$ for all $i \in X$, this result is already known to hold (Theorem 2.3.13). We note also that the q_i are bounded iff $\lim_{t \to 0+} p_{ij}(t) = \delta_{ij}$ uniformly with respect to $i, j \in X$. Moreover, for a stochastic transition matrix function the boundedness of the q_i implies that $\sum_{j \in X} q_{ij} = 0$, $i \in X$. See CHUNG (1967, p. 260).

A very elegant treatment of these equations by means of Laplace transform is to be found in FELLER [(1954 b), (1966 b, pp. 457 – 465)]. An approach using approximation by finite systems of differential equations was given by REUTER and LEDERMANN (1953) and another by a perturbation method by KATO (1954). For a probabilistic treatment see CHUNG (1967, pp. 248–250).

2.3.4.4. Continuous parameter regenerative phenomena

2.3.4.4.1. As in the discrete parameter case, the successive entrances into a given state of a homogeneous discrete state space Markov process constitute an important example of a so-called (continuous parameter) regenerative phenomenon. This notion is due to KINGMAN (1963 d) who developed it in a series of very remarkable papers [see KINGMAN (1964 a, 1965 a, b)].

Let $(\Omega, \mathcal{K}, \mathsf{P})$ be a probability space and $\mathcal{E} = (E_t)_{t \geqslant 0}$ a family of events (i.e. elements of the σ-algebra $\mathcal{K}$) on it.

Definition 2.3.35. The family $\mathcal{E}$ is said to be a *regenerative phenomenon* iff $E_0 = \Omega$ and

$$\mathsf{P}\left(\bigcap_{r=1}^{s+1} E_{t_r}\right) = \mathsf{P}(E_{t_1}) \prod_{r=1}^{s} \mathsf{P}(E_{t_{r+1}-t_r})$$

for all $s \in N^*$ and all increasing sequences $0 \leqslant t_1 < \ldots < t_{r+1}$.

Clearly, if $(\xi(t))_{t \geqslant 0}$ is a homogeneous Markov process with countable state space X, then for all $i \in X$ the family $\mathcal{E} = ((\xi(t) = i))_{t \geqslant 0}$ is a regenerative phenomenon on $(\Omega, \mathcal{K}, \mathsf{P}_i)$.

2.3.4.4.2. Let us write

$$p(t) = \mathsf{P}(E_t)$$

and call the function p the *p-function* of the regenerative phenomenon $(E_t)_{t \geqslant 0}$. A p-function is said to be *standard* if

$$\lim_{t \to 0+} p(t) = 1.$$

Write $\mathcal{P}$ for the class of all standard p-functions and $\mathcal{PM}$ for the class of all diagonal transition probabilities $p_{ii}(\cdot)$ associated with homogeneous Markov processes with countable state space so that $\mathcal{PM} \subset \mathcal{P}$. KINGMAN (1964 a) showed that the inclusion is proper. Thus in contrast to the discrete parameter case, the concept of a (continuous parameter) regenerative phenomenon is more general than that of a state of a homogeneous Markov process with countable state space [45].

[45] If we are willing to go over to a non-countable state space, then the representation of a regenerative phenomenon by means of a Markov process is possible. See KINGMAN (1964 a, p. 217).

2.3.4.4.3. The fundamental result in Kingman's theory of regenerative phenomena is a representation formula for the Laplace transform

$$p^*(\lambda) = \int_0^\infty p(t)\, e^{-\lambda t} dt, \qquad \mathrm{Re}\, \lambda > 0,$$

of the standard p-function p. This formula can be expressed as

$$p^*(\lambda) = \left[\lambda + \int_{(0,\infty]} (1 - e^{-\lambda x}) \mu\,(dx)\right]^{-1} \tag{2.3.46}$$

where μ is a positive measure on $(0, \infty]$, uniquely determined by p. Conversely, if μ is any positive measure on $(0, \infty]$ with

$$\int_{(0,\infty]} (1 - e^{-x})\, \mu\,(dx) < \infty, \tag{2.3.47}$$

then there exists a unique continuous function with Laplace transform given by (2.3.46), and this is a standard p-function. Therefore, there is a one-to-one correspondence between standard p-functions and positive measures satisfying (2.3.47).

For a proof and consequences of the fundamental formula (2.3.46) we refer the reader to KINGMAN (1964 a). For further investigations of the class $\mathcal{P}$ see DAVIDSON (1968, 1969) and KENDALL (1968). We merely remark that (2.3.46) leads to Fourier integral representations for the diagonal transition probabilities $p_{ii}(\cdot)$, $i \in X$, similar to those arising in the discrete parameter case. Such representations (and similar ones for the non-diagonal transition probabilities $p_{ij}(\cdot)$, $i \neq j$) were first obtained by KENDALL (1959) whose arguments were subsequently greatly simplified by FELLER (1966 a). To make possible a comparison with the behaviour of the non-diagonal transition probabilities, KINGMAN (1965 a) extended his theory by considering the so-called *delayed* regenerative phenomena and then regenerative phenomena *of order n*. [cf. KENDALL (1967 b)].

We also mention in the present context that the difficult problem of characterizing the class $\mathcal{PM}$ or, more generally, the functions which can arise as transition probabilities of a homogeneous Markov process with countable state space has been recently solved by KINGMAN (1971).

2.3.4.5. Properties of trajectories

2.3.4.5.1. In this paragraph we shall assume that the state space X is endowed with the discrete topology. The set X, if infinite, will be compactified by the one-point compactification: we adjoin the fictive state $\mathfrak{d}$ to X and postulate that every infinite subset of $\overline{X} = X \cup \{\mathfrak{d}\}$ is closed iff it contains $\mathfrak{d}$ [46].

Notice that if $\xi = (\xi(t))_{t \geqslant 0}$ is any homogeneous Markov process with transition matrix function $\overline{\mathbf{P}}(\cdot) = (p_{ij}(\cdot))_{i,j \in \overline{X}}$, then the continuity condition (2.3.36) implies that ξ is stochastically continuous [47]. This follows from the proof of Proposition 2.3.14. Furthermore, it can be proved that there exists a separable and strong Markov version of ξ such that for all $\omega \in \Omega$ and $t > 0$, $\xi(r, \omega)$ has at most one limiting value in X as rational r increases to t (this does not exclude the possibility of its having $\mathfrak{d}$ as a limiting value or even its converging to $\mathfrak{d}$); as rational r decreases to t, there are only two possibilities: either $\xi(t, \omega) = \mathfrak{d}$ and $\xi(r, \omega)$ tends to $\mathfrak{d}$ or $\xi(t, \omega) \in X$ and $\xi(r, \omega)$ has precisely one limiting value in X, namely $\xi(t, \omega)$. For details see CHUNG (1967, pp. 160 — 165, 172—178) and DOOB (1968).

2.3.4.5.2. For every separable version of ξ we can write

$$\mathsf{P}(\xi(u) = i, u \in (s, t] \mid \xi(s) = i) = \begin{cases} \exp\{-q_i(t-s)\} & \text{if } q_i < \infty \\ 0 & \text{if } q_i = \infty \end{cases}$$

for arbitrary $i \in X$, $0 \leqslant s < t$. The case $q_i < \infty$ follows directly from Theorem 2.3.15, while the case $q_i = \infty$ requires some changes in the proof [see CHUNG (1967, p. 153)]. The above result leads to the classification of states into *stable* or *instantaneous* according as $q_i < \infty$ or $q_i = \infty$ [48]. In particular, if $q_i = 0$ then i is said to be *absorbing*. This is justified by Theorem 2.3.29 which shows that $q_i = 0$ iff $p_{ii}(t) = 1$, $t \geqslant 0$. If we introduce the sets

$$\mathfrak{S}_i(\omega) = \{t : \xi(t, \omega) = i\}$$

for all $i \in X$, then it is easy to prove that for an instantaneous state i the probability that $\mathfrak{S}_i(\omega)$ contains an open interval equals zero, while for a stable state i, the conditional probability given that $\xi(s) = i$, that $\mathfrak{S}_i(\omega)$ contains an open interval containing s equals one. [see CHUNG (1967, p. 153)].

2.3.4.5.3. Now suppose that all the states are stable and non-absorbing, i.e. $0 < q_i < \infty$, $i \in X$, and $\sum_{j \in X} q_{ij} = 0$, $i \in X$. According to Subparagraph 2.3.2.4.7, the minimal process associated with a given $\mathbf{Q}$ satis-

---

[46] Without any loss of generality X may be thought as identical with N.

[47] The converse is also true. See CHUNG (1967, p. 146).

[48] The first example for which $q_i = \infty$ for some $i \in X$ was given independently by A. N. KOLMOGOROV and P. LÉVY. There exist examples in which all states are instantaneous [see CHUNG (1967, p. 283)]. Instantaneous states are not only of interest to theorists. They arise e.g. in the theory of dams. See KINGMAN (1963 c).

fying the above requirements can be simply described as follows. If the process is in state i it remains in this state an exponentially distributed length of time with mean q_i^{-1} and then jumps to state j with probability q_{ij}/q_i; after hitting j it remains there an exponentially distributed length of time with mean q_j^{-1} and then jumps to state k with probability q_{jk}/q_j and so on [49].

Formally, let us define the random variables τ_n, ξ_n, $n \in N$, as follows: $\tau_0 = 0$, $\xi_0 =$ arbitrary in X, and for $n > 0$ given the preceding choices, we choose τ_{n+1} and ξ_{n+1} so that

$$\mathsf{P}(\tau_{n+1} < t \mid \tau_0, \xi_0, \ldots, \tau_n, \xi_n) = 1 - \exp(-q_{\xi_n} t),$$

$$\mathsf{P}(\xi_{n+1} = j \mid \tau_0, \xi_0, \ldots, \tau_n, \xi_n, \tau_{n+1}) = q_{\xi_n j}/q_{\xi_n}.$$

We have

$$\xi_{\min}(t, \omega) = \xi_n(\omega)$$

for $t \in \left[\sum_{k=0}^{n} \tau_k(\omega), \quad \sum_{k=0}^{n+1} \tau_k(\omega) \right)$ and

$$\xi_{\min}(t, \omega) = \mathfrak{d}$$

for $t \geqslant \tau_\infty = \sum_{k \in N} \tau_k(\omega) \neq \infty$. It can be proved that if $\tau_\infty(\omega) < \infty$, then the trajectory $\xi_{\min}(\cdot, \omega)$ tends to $\mathfrak{d}$ as t approaches $\tau_\infty(\omega)$ from the left. This is the reason why $\tau_\infty(\omega)$ is called *the first infinity* of $\xi_{\min}(\cdot, \omega)$.

All processes with the same $\mathbf{Q}$ agree up to time τ_∞ with the minimal process. To obtain them we must continue the construction with τ_∞ as τ_0. The interested reader is referred for details to CHUNG (1967, pp. 257—271).

2.3.4.6. Discrete skeletons and classification of states

2.3.4.6.1. With every homogeneous discrete state space Markov process $\xi = (\xi(t))_{t \geqslant 0}$ we associate a family of Markov chains $\Sigma_h = (\xi(nh))_{n \in N}$, $h > 0$. Clearly, the transition matrix of Σ_h is $(p_{ij}(h))_{i, j \in \overline{X}}$. The Markov chain Σ_h is called the *discrete skeleton* (with scale parameter h) of ξ [50]. Discrete skeletons can be used to derive theorems about

---

[49] While q_i and q_{ij}/q_i have interesting probabilistic interpretations, the meaning of q_{ij} itself is obscure. An interpretation of q_{ij} which does not depend on q_i, finite or infinite, was given by FREEDMAN (1968 b).

[50] We mention in this context a problem known as the *imbedding problem* of ELFVING; determine when the transition matrix of a denumerable Markov chain can be identified with the transition matrix of a discrete skeleton associated with a Markov process with the same state space. This problem, posed by G. ELFVING in 1937, is excessively difficult. For some partial results see KINGMAN (1962), RUNNENBURG (1962), SCHEFFER (1962).

Markov processes by applying known theorems for Markov chains. Here we consider classification of states. For further considerations concerning discrete skeletons see CONNER (1967) and KINGMAN (1963 b).

2.3.4.6.2. Write $i \to j$ (*i leads to j*) if there is $t > 0$ such that $p_{ij}(t) > 0$. If $i \to j$ and $j \to i$, we say that i and j *communicate* and write $i \leftrightarrow j$. Theorem 2.3.27 implies immediately that $i \to j$ or $i \leftrightarrow j$ iff these relations hold for any Σ_h, $h > 0$. In particular, the classification of all states into classes of those which communicate is identical for ξ and all the Σ_h. Similarly, the notion of essentialness is the same. We note an important simplification, namely, that in any Σ_h, all the states are reflexive and aperiodic. This follows from (2.3.36) which implies that $p_{ii}(h) > 0$ for all $i \in X$, $h > 0$. The continuous parameter analogue of Theorem 1.1.15 may be easily obtained from that theorem itself.

Theorem 2.3.36 (P. LÉVY). *The limit*

$$\lim_{t\to\infty} p_{ij}(t) = \pi_{ij}$$

exists for all $i, j \in X$.

Proof. On account of Theorem 1.1.15,

$$\lim_{n\to\infty} p_{ij}(nh) = \pi_{ij}(h)$$

exists for all $h > 0$, $i, j \in \overline{X}$. Therefore, given $\varepsilon > 0$ there is an $n_0 \in N^*$ such that

$$|p_{ij}(nh) - p_{ij}(n'h)| < \frac{\varepsilon}{3}$$

if $n, n' \geqslant n_0$. Since p_{ij} is uniformly continuous in $[0, \infty)$ by Proposition 2.3.26 we can choose h so that the oscillation of p_{ij} in any interval of length h is less than $\varepsilon/3$. It follows that

$$|p_{ij}(t) - p_{ij}(t')| < \varepsilon$$

if t, $t' > n_0 h$, which proves the theorem. ◇

The π_{ij} are sometimes called *final probabilities*. Paralleling the discrete parameter case they satisfy the relations

$$\pi_{ij} = \sum_{k\in\overline{X}} \pi_{ik} p_{kj}(t) = \sum_{k\in\overline{X}} p_{ik}(t)\pi_{kj} = \sum_{k\in\overline{X}} \pi_{ik}\pi_{kj}, \quad t \geqslant 0,$$

and if $\pi_{ii} > 0$, $i \in \overline{X}$, then the matrix $(\pi_{ij})_{i,j\in\overline{X}}$ is stochastic. For further considerations of this matrix see JENSEN and KENDALL (1971), KENDALL and REUTER (1957). A continuous parameter analogue of the theory of geometric ergodicity developed by D. G. KENDALL and D. VERE JONES for Markov chains was given by KINGMAN (1963 a, b).

2.3.4.6.3. We come now to the notion of recurrence and define i to be *recurrent (nonrecurrent)* iff it is recurrent in some Σ_h and thus by Theorem 1.1.6, iff $\sum_{n\in N} p_{ii}(nh)$ diverges (converges). It can be proved that i is recurrent (nonrecurrent) if it is recurrent (nonrecurrent) in every Σ_h, and a necessary and sufficient condition for recurrence (nonrecurrence) is the divergence (convergence) of the integral $\int_0^\infty p_{ii}(t)\,dt$. Moreover, i is recurrent (nonrecurrent) if the probability $\mathsf{P}(\mathfrak{S}_i(\omega)$ is unbounded $|\,\xi(0)=i)$ equals one (zero). For details see CHUNG (1967, pp. 184—186).

Theorems 1.1.14 and 2.3.36 suggest defining i as *positive* or *null* according as $\pi_{ii} > 0$ or $\pi_{ii} = 0$. Thus i is positive (null) in ξ iff it is positive (null) in any Σ_h.

We conclude by noting that, in case ξ is uniquely defined by $\mathbf{Q}$, a stable non-absorbing state i is recurrent in ξ iff it is recurrent in the *imbedded* Markov chain $(\xi_n)_{n\in N}$ with transition matrix

$$r_{ij} = \begin{cases} q_{ij}/q_i & \text{if } i \neq j \\ 0 & \text{if } i = j. \end{cases}$$

(See Subparagraph 2.3.2.4.6).

However, this equivalence does not extend to positiveness. For examples see R. G. MILLER JR. (1963). In MILLER's paper the imbedded Markov chain is also used to obtain criteria for positiveness in terms of stationary measures (see Theorem 1.1.43).

2.3.4.7. Birth-and-death processes

2.3.4.7.1. In this paragraph we illustrate the general theory above with the important special case of birth-and-death processes introduced by W. FELLER in 1939. These processes have since been used to describe stochastic models for population growth, queues, epidemics, and several other situations. See e.g. BAILEY (1957), KARLIN and MCGREGOR (1958 a, b), LAHRES (1964), and PRABHU (1965 b).

Let b_i, d_i, $i \in N$, $d_0 = 0$, be non-negative numbers. A homogeneous Markov jump process ξ with state space N is said to be a *birth-and-death process* with infinitesimal birth-and-death rates b_i and d_i if its transition intensity matrix is defined by

$$q_{ij} = \begin{cases} -(b_i + d_i) & \text{if } j = i \\ b_i & \text{if } j = i+1, \quad i, j \in N. \\ d_i & \text{if } j = i-1 \\ 0 & \text{otherwise} \end{cases} \tag{2.3.48}$$

The states may be interpreted as the size of a "population". When $b_0 = 0$, the state 0 will be absorbing so that the population cannot recover once it has become extinct; when $b_0 > 0$, the population may "revive" through "immigration" at a rate b_0 [51].

Clearly, if at time t the process is in state n, it can, after a random waiting time, move only to $n+1$ (a "birth") or to $n-1$ (a "death"), and therefore it can be considered as the continuous parameter version of the non-homogeneous random walk studied in Paragraph 1.2.1.5.

2.3.4.7.2. On account of Theorem 2.3.31 the transition matrix function of every birth-and-death process satisfies the backward equations

$$\begin{cases} p'_{ij}(t) = d_i\, p_{i-1,j}(t) - (b_i + d_i)\, p_{ij}(t) + b_i\, p_{i+1,j}(t) \\ p'_{0j}(t) = -\, b_0\, p_{0j}(t) + b_0\, p_{1j}(t),\ i \in N^*, j \in N. \end{cases}$$

The forward equations reduce to

$$p'_{ij}(t) = b_{j-1}\, p_{i,j-1}(t) - (b_j + d_j)\, p_{ij}(t) + d_{j+1}\, p_{i,j+1}(t)$$
$$p'_{i0}(t) = -b_0\, p_{i0}(t) + d_1\, p_{i1}(t),\ \ i \in N, j \in N^*.$$

We shall see that the forward equations do not hold for all birth-and-death processes.

Concerning the above systems, we have

Theorem 2.3.37. (D. G. KENDALL). *Assume* $b_i > 0$, $i \in N$, $d_j > 0$, $j \in N^*$. *The general solution* [52] *to the differential equations*

$$a'_{ij}(t) = \sum_{k \in N} q_{ik}\, a_{kj}(t) = \sum_{k \in N} a_{ik}(t)\, q_{kj},$$

$0 < t < \infty$, $i, j \in N$ *where the* q_{ij} *are given by* (2.3.48), *is*

$$a_{ij}(t) = u_j\, P_i\left(\frac{\mathrm{d}}{\mathrm{dt}}\right) P_j\left(\frac{\mathrm{d}}{\mathrm{dt}}\right) c(t),$$

where $c(\cdot)$ *is an arbitrary function of class* C^∞ *(which is, in fact,* $a_{00}(\cdot)$*), and where the numbers* u_j *and the polynomials* P_j *are determined recurrently by the equations*

$$u_0 = 1,\ \ d_{j+1}\, u_{j+1} = b_j\, u_j,\ \ j \in N.$$

---

[51] Sometimes a transition from zero to some ignored state -1 is allowed by assuming $d_0 > 0$.

[52] This means we do not require the $a_{ij}(\cdot)$ to be entries of a transition matrix function.

and

$$P_0(x) = 1, \; x P_0(x) = -b_0 P_0(x) + b_0 P_1(x),$$

$$x P_k(x) = d_k P_{k-1}(x) - (d_k + b_k) P_k(x) + b_k P_{k+1}(x), \quad k \in N^*.$$

Proof. See KENDALL (1960 a, p. 178). ◇
For further developments see KEMPERMAN (1962).

2.3.4.7.3. Concerning classification of the states of birth-and-death processes we shall briefly describe the results obtained by KARLIN and MCGREGOR (1957 a, b). Assume there exists one and only one birth-and-death process whose transition matrix function satisfies the forward equations. (A criterion for this to hold will be given later in Theorem 2.3.39). Suppose that $d_k > 0$, $k \in N^*$ and set

$$u_0 = 1, \quad u_k = \frac{b_0 \dots b_{k-1}}{d_1 \dots d_k}, \qquad k \in N^*.$$

It is easily verified that $(u_i)_{i \in N}$ is a stationary measure. (See Subparagraph 2.3.4.3.4). Moreover, if $\sum_{i \in N} u_i < \infty$, then by using the backward equations we get

$$\pi_{ij} \left(= \lim_{t \to \infty} p_{ij}(t)\right) = \frac{u_j}{\sum_{k \in N} u_k}$$

for all $i, j \in N$.

If $b_i > 0$, $i \in N$, $d_j > 0$, $j \in N^*$, the states form an essential class, and we have the following criteria [53]:

(a) The states are positive iff

$$\sum_{n \in N} u_n < \infty \text{ and } \sum_{n \in N} \frac{1}{b_n u_n} = \infty;$$

(b) The states are null recurrent iff

$$\sum_{n \in N} u_n = \infty \text{ and } \sum_{n \in N} \frac{1}{b_n u_n} = \infty;$$

(c) The states are nonrecurrent iff

$$\sum_{n \in N} u_n = \infty \text{ and } \sum_{n \in N} \frac{1}{b_n u_n} < \infty.$$

---

[53] Compare with Subparagraph 1.2.1.5.1.

These results are based on an integral representation, namely

$$p_{ij}(t) = \pi_{ij} \int_0^\infty e^{-xt} Q_i(x) Q_j(x) \, d\psi(x),$$

where $(Q_i)_{i \in N}$ aré polynomials defined recursively by

$$Q_0(x) = 1$$

$$-xQ_0(x) = -b_0 Q_0(x) + b_0 Q_1(x)$$

$$-xQ_n(x) = d_n Q_{n-1}(x) - (b_n + d_n) Q_n(x) + b_n Q_{n+1}(x), n \in N^*,$$

and ψ is a distribution function on $[0, \infty)$ with respect to which the Q_i are orthogonal. The above representation reveals a very intimate connection between the theory of birth-and-death processes and the Stieltjes moment problem. It should be noted that this representation could also be deduced from Theorem 2.3.37. For details see KENDALL (1960 a).

2.3.4.7.4. Now, following REUTER (1957, pp. 41—44), we apply Theorems 2.3.33 and 2.3.34 to birth-and-death processes.

The systems (2.3.44) and (2.3.45) reduce respectively to

$$\begin{cases} (\lambda + b_0) x_0 = b_0 x_1 \\ (\lambda + b_n + d_n) x_n = d_n x_{n-1} + b_n x_{n+1}, \quad n \in N^*. \end{cases} \tag{2.3.49}$$

$$\begin{cases} (\lambda + b_0) y_0 = d_1 y_1 \\ (\lambda + b_n + d_n) y_n = b_{n-1} y_{n-1} + d_{n+1} y_{n+1}, \; n \in N^*. \end{cases} \tag{2.3.50}$$

Rewrite (2.3.49) as

$$b_n (x_{n+1} - x_n) = \lambda x_n + d_n (x_n - x_{n-1}), \quad n \in N^*. \tag{2.3.51}$$

When $b_0 = 0$, then $x_0 = 0$, x_1 is arbitrary, and $x_2, x_3, \ldots$ are uniquely determined; when $b_0 > 0$, x_0 is arbitrary, and $x_1, x_2, \ldots$ are then determined. We take $x_1 = 1$ when $b_0 = 0$ and $x_0 = 1$ (hence $x_1 = 1 + \lambda/b_0$) when $b_0 > 0$; in either case, $x_1 > x_0 \geqslant 0$ and an easy induction from (2.3.51) shows that $x_{n+1} \geqslant x_n$, $n \in N^*$. We then have to decide whether x_n is bounded; if it is, then (2.3.44) has a non-trivial solution.

For (2.3.50), we sum the first $n+1$ equations and write $z_m = y_0 + \ldots + y_m$, $m \in N$. This leads to

$$\begin{cases} (\lambda + b_0 + d_1)\, z_0 = d_1 z_1 \\ d_{n+1}(z_{n+1} - z_n) = \lambda z_n + b_n(z_n - z_{n-1}), \quad n \in N^*. \end{cases} \tag{2.3.52}$$

Taking $z_0 = 1$, we see by induction that $z_n > z_{n-1}$. We have to decide whether (2.3.50) has a non-negative solution with $\sum_{m \in N} y_m < \infty$ i.e. whether z_m is bounded; if it is, then (2.3.45) has a non-trivial solution.

It is now clear that (2.3.51) and (2.3.52) can be treated together by means of

Lema 2.3.38. *Suppose that* $\alpha_n > 0$, $\beta_n > 0$, $n \in N^*$, *that* $0 \leqslant u_0 < u_1 < u_2 < \ldots$ *and that*

$$u_{n+1} - u_n = \alpha_n u_n + \beta_n (u_n - u_{n-1}), \quad n \in N^*. \tag{2.3.53}$$

Then u_n *is bounded iff*

$$\sum_{n \in N^*} (\alpha_n + \beta_n \alpha_{n-1} + \ldots + \beta_n \beta_{n-1} \ldots \beta_2 \alpha_1 + \beta_n \ldots \beta_2 \beta_1) < \infty.$$

Proof. Repeated applications of (2.3.53) yield

$$u_{n+1} - u_n = \alpha_n u_n + \beta_n \alpha_{n-1} u_{n-1} + \ldots +$$

$$+ \beta_n \beta_{n-1} \ldots \beta_2 \alpha_1 u_1 + \beta_n \ldots \beta_2 \beta_1 (u_1 - u_0).$$

Hence

$$u_{n+1} - u_n \leqslant A_n u_n,$$

$$u_{n+1} - u_n \geqslant A_n (u_1 - u_0),$$

where

$$A_n = \alpha_n + \beta_n \alpha_{n-1} + \ldots + \beta_n \beta_{n-1} \ldots \beta_2 \alpha_1 + \beta_n \ldots \beta_2 \beta_1, \quad n \in N^*.$$

The above two inequalities give

$$u_1 + (u_1 - u_0) \sum_{k=1}^{n-1} A_k \leqslant u_n \leqslant u_1 \sum_{k=1}^{n-1} (1 + A_k), \qquad n > 1,$$

and this implies that u_n is bounded iff $\sum_{k \in N^*} A_k < \infty$, as asserted. ◇

To apply the lemma to (2.3.51), take $u_n = x_n$, $\alpha_n = \lambda/b_n$, and $\beta_n = d_n/b_n$. Then

$$A_n = \lambda\left(\frac{1}{b_n} + \frac{d_n}{b_n b_{n-1}} + \dots + \frac{d_n \dots d_2}{b_n \dots b_2 b_1}\right) + \frac{d_n \dots d_1}{b_n \dots b_1}. \quad (2.3.54)$$

The convergence of $\sum_{n \in N^*} A_n$ does not, of course, depend on the choice of $\lambda > 0$, nor is it affected by omitting the last term $(d_n \dots d_1)/(b_n \dots b_1)$ in (2.3.54). Thus x_n in (2.3.51) is bounded iff $A < \infty$, where

$$A = \sum_{n \in N^*}\left(\frac{1}{b_n} + \frac{d_n}{b_n b_{n-1}} + \dots + \frac{d_n \dots d_2}{b_n \dots b_2 b_1}\right). \quad (2.3.55)$$

Similarly we apply the lemma to (2.3.52) by taking $u_n = z_n$, $\alpha_n = \lambda/d_{n+1}$, and $\beta_n = b_n/d_{n+1}$. Then

$$A_n = \lambda\left(\frac{1}{d_{n+1}} + \frac{b_n}{d_{n+1}d_n} + \dots + \frac{b_n \dots b_2}{d_{n+1} \dots d_2}\right) + \frac{b_n \dots b_1}{d_{n+1} \dots d_2}; \quad (2.3.56)$$

again the convergence of $\sum_{n \in N^*} A_n$ does not depend on the choice of λ, and we may also change the last term in (2.3.56) to $(b_n \dots b_1)/(d_{n+1} \dots d_1)$. Thus z_n in (2.3.52) is bounded (equivalently, y_n in (2.3.50) satisfies $\sum_{n \in N^*} y_n < \infty$) iff $B < \infty$, where

$$B = \sum_{n \in N^*}\left(\frac{1}{d_{n+1}} + \frac{b_n}{d_{n+1}d_n} + \dots + \frac{b_n \dots b_1}{d_{n+1} \dots d_2 d_1}\right). \quad (2.3.57)$$

Finally, we show that A and B are both finite iff C is finite, where

$$C = \sum_{n \in N^*}\left(\frac{d_n \dots d_2}{b_n \dots b_2 b_1} + \frac{b_n \dots b_1}{d_{n+1} \dots d_2 d_1}\right) \quad (2.3.58)$$

Certainly, C is finite if A and B are, because $C \leqslant A + B$. On the other hand, we may write

$$A = \sum_{n \in N^*} \frac{d_n \dots d_2}{b_n \dots b_2 b_1}\left(1 + \frac{b_1}{d_2} + \dots + \frac{b_1 \dots b_{n-1}}{d_2 \dots d_n}\right),$$

$$B = \sum_{n \in N^*} \frac{b_n \dots b_1}{d_{n+1} \dots d_2 d_1}\left(1 + \frac{d_1}{b_1} + \dots + \frac{d_1 \dots d_n}{b_1 \dots b_n}\right).$$

If C is finite, then $\sum_{n \in N^*} (d_n \dots d_2)/(b_n \dots b_1)$ and $\sum_{n \in N^*} (b_n \dots b_1)/(d_{n+1} \dots d_1)$ both converge, and the factors

$$1 + \frac{b_1}{d_2} + \dots + \frac{b_1 \dots b_{n-1}}{d_2 \dots d_n}, \quad 1 + \frac{d_1}{b_1} + \dots + \frac{d_1 \dots d_n}{b_1 \dots b_n}$$

are both bounded; hence A and B are finite.

Now we can state

Theorem 2.3.39. *Let* A, B, C *be defined by* (2.3.55), (2.3.57), (2.3.58),

(i) *If* $A = \infty$, *there is exactly one corresponding birth-and-death process; its transition matrix function is stochastic and satisfies the forward equations* (R. L. DOBRUŠIN).

(ii) *If* $A < \infty$ *and* $B = \infty$, *there are infinitely many corresponding birth-and-death processes. The transition matrix function of only one of these satisfies the forward equations, but it is substochastic* (G. E. H. REUTER).

(iii) *If* $C < \infty$, *equivalently if* $A < \infty$ *and* $B < \infty$, *there are infinitely many corresponding birth-and-death processes whose transition matrix functions satisfy the forward equations. Exactly one of these is stochastic* (S. KARLIN and J. MCGREGOR).

Proof. Use Theorem 2.3.34 and the above considerations. ◇

We notice that the criteria have been stated so as not to involve b_0. They can (as should be clear from their derivation) be modified so as not to involve any pre-assigned finite subset of the b_n and d_n.

For criteria in case infinitely many d_n are null see REUTER and LEDERMANN (1953).

2.3.4.7.5. An interesting special case of the birth-and-death processes are the *linear* birth-and-death processes *(Feller-Arley processes)* for which:

$$b_i = bi, \; d_i = di, \; i \in N, \; b, \; d > 0.$$

It follows from Theorem 2.3.39 that for every given pair of positive constants b and d there exists exactly one linear birth-and-death process.

Clearly, a linear birth-and-death process is a homogeneous Markov branching process with

$$\mathfrak{b} = b + d, \; \mathfrak{p}_0 = \frac{d}{b + d}, \; \mathfrak{p}_2 = \frac{b}{b + d}.$$

Therefore, if we consider the generating functions

$$F_i(z, t) = \sum_{k \in N} p_{ik}(t) \, z^k, \; i \in N,$$

then on account of (2.3.32) we can write

$$\frac{\partial F_1}{\partial t} = [bz^2 - (b+d)z + d]\frac{\partial F_1}{\partial z} \tag{2.3.59}$$

If $b \neq d$ the general solution of (2.3.59) is

$$F_1(z, t) = f\left[\frac{d - bz}{1 - z} \exp\{(d-b)t\}\right],$$

where f is an arbitrary differentiable function. Using the initial condition $F_1(z, 0) = z$ we obtain

$$z = f\left(\frac{d - bz}{1 - z}\right),$$

whence

$$f(w) = \frac{d - w}{b - w}.$$

Therefore

$$F_i(z, t) = [F_1(z, t)]^i = \left[\frac{u(t) + [1 - u(t) - v(t)]\, z}{1 - v(t)\, z}\right]^i, \quad i \in N,$$

where

$$u(t) = \frac{d[\exp\{(b-d)t\} - 1]}{b\exp\{(b-d)t\} - d},$$

$$v(t) = \frac{b[\exp\{(b-d)t\} - 1]}{b\exp\{(b-d)t\} - d}.$$

Straightforward computations yield

$$p_{ij}(t) = \sum_{0 \leqslant l \leqslant \min(i,\, j)} \binom{i}{l}\binom{i+j+l-1}{j-l} [u(t)]^{i-l}[v(t)]^{j-l}[1-u(t)-v(t)]^l$$

for all $i \in N^*$, $j \in N$, $t \geqslant 0$. Thus

$$p_{10}(t) = u(t)$$

and hence the extinction probability is

$$q = \lim_{t\to\infty} p_{10}(t) = \begin{cases} 1 & \text{if } b < d \\ d/b & \text{if } b > d. \end{cases}$$

In case $b = d$, the general solution of (2.3.59) is

$$F_1(z, t) = f\left(t + \frac{1}{b(1-z)}\right).$$

We get

$$f(w) = 1 - \frac{1}{bw},$$

then

$$F_1(z, t) = \frac{bt + (1 - bt)\,z}{1 + bt - btz,},$$

$$p_{ij}(t) = \sum_{0 \leqslant l \leqslant \min(i,\, j)} \binom{i}{l} \binom{i+j-l-1}{j-l} (1 - bt)^l (1 + bt)^{l-i-j} (bt)^{i+j-2l}$$

for all $i \in N^*$, $j \in N$, $t \geqslant 0$. In this case

$$p_{10}(t) = \frac{bt}{1 + bt},$$

hence

$$q = \lim_{t\to\infty} p_{10}(t) = 1.$$

A linear birth-and-death process will be said to be *subcritical*, *critical*, or *supercritical* according as $b < d$, $b = d$, or $b > d$.

By differentiating the generating function we easily deduce the mean and the variance of the Feller-Arley process. We have (assuming $\xi(0) = 1$)

$$\mathsf{E}\xi(t) = \begin{cases} \exp\{(b-d)t\} & \text{if } b \neq d \\ 1 & \text{if } b = d. \end{cases}$$

$$\mathsf{D}\,\xi(t) = \begin{cases} \dfrac{b+d}{b-d} \exp\{(b-d)t\}\left(\exp\{(b-d)t\} - 1\right) & \text{if } b \neq d \\ 2\,bt & \text{if } b = d. \end{cases}$$

2.3.4.7.6. Another important special case of the birth-and-death processes are the *pure birth processes* for which

$$d_i = 0, \quad i \in N.$$

For a pure birth process the system (2.3.44) reduces to

$$(\lambda + b_n)\, x_n = b_n\, x_{n+1},\ n \in N. \tag{2.3.60}$$

Suppose $b_n > 0$, $n \in N^*$. When $b_0 = 0$, then $x_0 = 0$, x_1 is arbitrary, and $x_2, x_3 \ldots$ are then uniquely determined; when $b_0 > 0$, then x_0 is arbitrary and $x_1, x_2 \ldots$ are then determined. Taking $x_1 = 1$ in both cases, we deduce

$$x_n = \prod_{k=1}^{n-1} \left(1 + \frac{\lambda}{b_k}\right),\ n > 1.$$

Thus x_n is positive and is bounded iff $\sum_{n \in N^*} b_n^{-1} < \infty$. We conclude that there are one or infinitely many birth processes according as $\sum_{n \in N^*} b_n^{-1}$ diverges or converges [54]. This result was first obtained in 1940 by W. FELLER and O. LUNDBERG. On the other hand, (2.3.45) becomes

$$(\lambda + b_0)\, y_0 = 0$$

$$(\lambda + b_n)\, y_n = b_{n-1}\, y_{n-1}, \quad n \in N^*,$$

so that $y_i = 0$, $i \in N$, and the forward equations have a unique solution (the minimal one). It is easy to prove that in case the b_i are all distinct this solution is given by

$$p_{ij}(t) = \begin{cases} \left(\prod_{k=i}^{j-1} b_k\right) \sum_{k=i}^{j} \dfrac{e^{-b_k t}}{\prod_{r=i,\ r \neq k}^{j} (b_r - b_k)} & \text{if } j > i \\ e^{-b_i t} & \text{if } j = i \\ 0 & \text{if } j < i, \end{cases}$$

$i, j \in N$, (see e.g. PRABHU (1965 b, p. 135)).

---

[54] It is also easy to see that, in case $b_n > 0$ for $n \geqslant m$, then there are one or infinitely many pure birth processes according as $\sum_{n \geqslant m} b_n^{-1}$ diverges or converges On the other hand, if $b_n = 0$ for infinitely many n, then there is only one corresponding pure birth process.

On account of Subparagraph 2.3.4.5.3 we can assert that the sojourn times $\tau_1, \tau_2, \ldots$ in the successive states of the pure birth process are independent random variables with

$$\mathsf{P}(\tau_n \leqslant t \mid \xi(0) = 1) = 1 - e^{-b_n t}, \quad t \geqslant 0, \quad n \in N^*.$$

This result generalizes (ii) Theorem 2.2.10; the homogeneous Poisson process $P(\lambda)$ is a pure birth process with constant birth rates $b_i = \lambda$, $i \in N$.

A well-known pure birth process is the *linear* birth process (the *Yule-Furry process*) for which

$$b_i = bi, i \in N, \quad b > 0.$$

Clearly, for a given $b > 0$ there is only one corresponding Yule-Furry process with transition probabilities

$$p_{ij}(t) = \begin{cases} \binom{j-1}{j-i} e^{-ibt}(1 - e^{-bt})^{j-i} & \text{if } j \geqslant i \\ 0 & \text{if } j < i, \quad i, \quad j \in N. \end{cases}$$

On account of Subparagraph 2.3.4.7.5 the mean of the pure birth process $\xi(t)$ equals e^{bt}. It is easily verified that

$$W_t = \xi(t)\, e^{-bt}, \quad t \geqslant 0$$

is a continuous parameter non-negative martingale [55]. If we consider a separable version of ξ the Doob theorem tells us that W_t converges almost surely to a positive finite limit W. KENDALL (1966, p. 396) showed that

$$\mathsf{P}(W < w \mid \xi(0) = r) = \frac{1}{(r-1)!} \int_0^w e^{-u} u^{r-1}\, du, \quad w \geqslant 0.$$

Further, ξ possesses a conditional distribution with W as conditioning random variable; this may be described informally by saying that

$$(\xi(b^{-1} \log [1 + (t/W)]))_{t \geqslant 0}$$

---

[55] Compare with Subparagraph 1.2.2.6.1.

is a Poisson process $P(1)$ when W is given. It can be deduced that

$$\log \xi(t) = bt + \log W + \zeta(t)\, e^{-bt/2} \sqrt{\frac{2 \log bt}{W}},$$

where the cluster-set of limit points of $\zeta(t)$ as $t \to \infty$ consists exactly of the segment $[-1, 1]$.

2.3.4.7.7. In a similar manner we can consider *pure death processes* i.e. birth-and-death processes for which

$$b_i = 0, \quad i \in N.$$

Using Theorem 2.3.33 it is easily seen that for arbitrarily given death rates d_i, $i \in N$, $(d_0 = 0)$, there is only one pure death process and its transition matrix function is stochastic.

In the *linear* case $d_i = di$, $i \in N$, $d > 0$, the transition probabilities are

$$p_{ij}(t) = \begin{cases} \binom{i}{j} e^{-jdt} (1 - e^{-dt})^{i-j} & \text{if } j \leqslant i \\ 0 \quad \text{if} \quad j > i, & i, \quad j \in N, \end{cases}$$

and the time to extinction

$$\nu = \inf(t \colon \xi(t) = 0)$$

has the distribution

$$\mathsf{P}(\nu < t \mid \xi(0) = i) = p_{i0}(t) = (1 - e^{-dt})^i, \quad t > 0, \quad i \in N.$$

An interesting relation between linear birth and linear death processes was noticed by KERRIDGE (1964).

2.3.5. Markov diffusion processes

In this subsection we shall consider Markov processes whose states (points in R^n) vary continuously in such a way that only small changes of state occur during small intervals of time. We would expect the trajectories of such processes to be continuous functions. Strictly speaking, a diffusion process is a continuous strong Markov process.

A full mathematical discussion of diffusion processes is possible[56] but beyond the aims of this book. To pursue it, a good knowledge of semigroup theory and stochastic differential equations is needed. The interested reader may consult DYNKIN (1965), GIHMAN and SKOROHOD (1968), ITÔ (1963), ITÔ and MCKEAN (1965), MANDL (1968) and UENO (1967). We shall be concerned here with the so-called classical Markov diffusion processes first considered by A. N. KOLMOGOROV in 1931.

2.3.5.1. Classical diffusion processes

2.3.5.1.1. Let us first consider the one-dimensional case $n = 1$. A Markov process ξ with transition function P on $(R, \mathcal{B})$ is said to be a *classical diffusion process* if the following *Hinčin conditions* are satisfied: there exist two real-valued functions a and b defined on $T \times R$ such that:

$$\int_{|y-x|>\varepsilon} P(s, x; t, \mathrm{d}y) = o(t-s), \tag{2.3.61}$$

$$\int_{|y-x|\leqslant\varepsilon} (y-x)^2 P(s, x; t, \mathrm{d}y) = a(s, x)(t-s) + o(t-s), \tag{2.3.62}$$

$$\int_{|y-x|\leqslant\varepsilon} (y-x) P(s, x; t, \mathrm{d}y) = b(s,x)(t-s) + o(t-s), \tag{2.3.63}$$

for any $x \in R$, $\varepsilon > 0$ uniformly with respect to $s, t \in T_1$, $s < t$, on every compact interval $T_1 \subset T$.

The first limit condition (2.3.61) amounts to the fact that in a small time interval the state of the process is almost sure to remain in the immediate neighborhood of its initial state[57]. The relations (2.3.62) and (2.3.63) are conditions on the truncated variance and mean displacement of the process in a small time interval[58]. The functions a and b are called, respectively, the *diffusion* and the *drift* coefficients.

2.3.5.1.2. Now let us consider the general case $n \geqslant 1$. A (vector valued) Markov process ξ with transition function P on $(R, \mathcal{B}^n)$ is said

---

[56] The same is true for the most general type of Markov processes in which both continuous and jump transitions may occur.

[57] If the discussed condition holds uniformly with respect to x in every compact interval of R, then by Theorem 2.1.14 we can assert that ξ is indeed continuous.

[58] It is convenient to regard $\xi(t)$ as the position of a "particle" at time t.

to be a *classical (multidimensional) diffusion process* if the following *Hinčin conditions* are satisfied: there exist real-valued functions a_{ij} and b_i, $1 \leqslant i, j \leqslant n$, defined on $T \times R^n$ such that [59]

$$\int_{|\mathbf{y}-\mathbf{x}|>\varepsilon} P(s, \mathbf{x}; t, \mathrm{d}\mathbf{y}) = o(t-s), \tag{2.3.64}$$

$$\int_{|\mathbf{y}-\mathbf{x}|\leqslant\varepsilon} (y_i - x_i)(y_j - x_j) P(s, \mathbf{x}; t, \mathrm{d}y) = a_{ij}(s, \mathbf{x})(t-s) + o(t-s), \tag{2.3.65}$$

$$\int_{|\mathbf{y}-\mathbf{x}|\leqslant\varepsilon} (y_i - x_i) P(s, \mathbf{x}; t, \mathrm{d}y) = b_i(s, \mathbf{x})(t-s) + o(t-s), \tag{2.3.66}$$

for any $\mathbf{x} = (x_1, \dots, x_n) \in R^n$, $\varepsilon > 0$, uniformly with respect to $s, t \in T_1$, $s < t$ on every compact interval $T_1 \subset T$.

We note that an easily verified condition implying Hinčin conditions is that of Kolmogorov: that there exists a $\delta > 0$ such that

$$\int_{R^n} |\mathbf{y} - \mathbf{x}|^{2+\delta} P(s, \mathbf{x}; t, \mathrm{d}\mathbf{y}) = o(t-s)$$

for any $\mathbf{x} \in R^n$ uniformly with respect to $s, t \in T_1$, $s < t$, on every compact interval $T_1 \subset T$, and (2.3.65) and (2.3.66) are fulfilled only when $\varepsilon = \infty$. Indeed, for $k = 0, 1, 2$ we have

$$\int_{|\mathbf{y}-\mathbf{x}|>\varepsilon} |\mathbf{y} - \mathbf{x}|^k P(s, \mathbf{x}; t, \mathrm{d}\mathbf{y}) \leqslant$$

$$\leqslant \frac{1}{\varepsilon^{2+\delta-k}} \int_{R^n} |\mathbf{y} - \mathbf{x}|^{2+\delta} P(s, \mathbf{x}; t, \mathrm{d}\mathbf{y}) = o(t-s).$$

for all $\varepsilon > 0$.

2.3.5.1.3. In the homogeneous case $P(s, \mathbf{x}; t, A)$ depends only on $t - s$. We shall set $P(t; \mathbf{x}, A) = P(0, \mathbf{x}; t, A)$, thus $P(s, \mathbf{x}; t, A) = P(t-s; \mathbf{x}, A)$.

---

[59] In what follows $|\mathbf{y} - \mathbf{x}|$ denotes the Euclidean distance between $\mathbf{x}, \mathbf{y} \in R^n$.

A homogeneous (vector valued) Markov process ξ is said to be a classical *homogeneous* (multidimensional) diffusion process if the following Hinčin conditions are satisfied: there exist real-valued function a_{ij} and b_i, $1 \leqslant i, j \leqslant n$, defined on R^n such that

$$\int_{|\mathbf{y}-\mathbf{x}|>\varepsilon} P(t; \mathbf{x}, d\mathbf{y}) = o(t), \tag{2.3.64'}$$

$$\int_{|\mathbf{y}-\mathbf{x}|\leqslant\varepsilon} (y_i - x_i)(y_j - x_j) P(t; \mathbf{x}, d\mathbf{y}) = a_{ij}(\mathbf{x})t + o(t), \tag{2.3.65'}$$

$$\int_{|\mathbf{y}-\mathbf{x}|\leqslant\varepsilon} (y_i - x_i) P(t; \mathbf{x}, d\mathbf{y}) = b_i(\mathbf{x})t + o(t), \tag{2.3.66'}$$

for any $\mathbf{x} = (x_1, \ldots, x_n) \in R^n$, $\varepsilon > 0$.

Correspondingly, Kolmogorov's condition amounts to the following: there exists a $\delta > 0$ such that

$$\int_{R^n} |\mathbf{y} - \mathbf{x}|^{2+\delta} P(t; \mathbf{x}, d\mathbf{y}) = o(t)$$

for any $\mathbf{x} \in R^n$ and (2.3.65') and (2.3.66') are fulfilled only when $\varepsilon = \infty$.

2.3.5.2. The Kolmogorov equations

2.3.5.2.1. We shall show that certain differential equations can be derived from (2.3.64), (2.3.65), and (2.3.66). Consider first the one-dimensional case $n = 1$.

Theorem 2.3.40 (A.N. Kolmogorov). *Let f be a real-valued, continuous and bounded function defined on R and suppose that the partial derivatives $\partial u/\partial x$ and $\partial^2 u/\partial x^2$ of the function*

$$u(s, x) = \int_R f(y) P(s, x; t, dy)$$

exist and are continuous on $T \times R$. If $a(., x)$ and $b(., x)$ are continuous for any $x \in R$, then for $0 \leqslant s < t$, $x \in R$ the derivative $\partial u/\partial s$ exists and satisfies the equation

$$-\frac{\partial u}{\partial s} = b(s, x)\frac{\partial u}{\partial x} + \frac{1}{2} a(s, x)\frac{\partial^2 u}{\partial x^2} \tag{2.3.67}$$

with the initial condition

$$\lim_{s \to t-} u(s, x) = f(x).$$

Proof. The initial condition follows from the equations

$$u(s, x) = \int_R f(y) P(s, x; t, \mathrm{d}y) = f(x) + \int_R [f(y) - f(x)] P(s, x; t, \mathrm{d}y) =$$

$$= f(x) + \int_{|y-x| \leqslant \varepsilon} [f(y) - f(x)] P(s, x; t, \mathrm{d}y) + o(t - s)$$

by taking into account the fact that f is a continuous bounded function.

To deduce (2.3.67) we first note that the Chapman-Kolmogorov equation implies that

$$u(s_1, x) = \int_R P(s_1, x; s_2, \mathrm{d}z)\, u(s_2, z), \qquad 0 \leqslant s_1 < s_2 \leqslant t.$$

Then, on account of Taylor's formula,

$$u(s, z) - u(s, x) = \frac{\partial u(s, x)}{\partial x}(z - x) +$$

$$+ \frac{1}{2}\left[\frac{\partial^2 u(s, x)}{\partial x^2} + \eta_\varepsilon\right](z - x)^2, \qquad |z - x| \leqslant \varepsilon,$$

where

$$|\eta_\varepsilon| \leqslant \sup_{\substack{|z-x| \leqslant \varepsilon \\ 0 \leqslant s \leqslant t}} \left| \frac{\partial^2 u(s, x)}{\partial x^2} - \left(\frac{\partial^2 u(s, x)}{\partial x^2}\right)_{x=z} \right|.$$

Therefore we can write

$$u(s_1, x) - u(s_2, x) = \int_R [u(s_2, z) - u(s_2, x)] P(s_1, x; s_2, \mathrm{d}z) =$$

$$= \int_{|z-x| \leqslant \varepsilon} [u(s_2, z) - u(s_2, x)] P(s_1, x; s_2, dz) + o(s_2 - s_1) =$$

$$= \frac{\partial u(s_2, x)}{\partial x} \int_{|z-x| \leqslant \varepsilon} (z - x) P(s_1, x; s_2, \mathrm{d}z) +$$

$$+ \frac{1}{2}\left(\frac{\partial^2 u(s_2, x)}{\partial x^2} + \eta_\varepsilon\right) \int_{|z-x| \leqslant \varepsilon} (z - x)^2 P(s_1, x; s_2, \mathrm{d}z) + o(s_2 - s_1).$$

Thus

$$\frac{u(s_1, x) - u(s_2, x)}{s_2 - s_1} = b(s_1, x)\frac{\partial u(s_2, x)}{\partial x} +$$

$$+ \frac{1}{2} a(s_1, x)\left(\frac{\partial^2 u(s_2, x)}{\partial x^2} + \eta_\varepsilon\right) + \frac{o(s_2 - s_1)}{s_2 - s_1},$$

whence (2.3.67) follows on letting $s_2 - s_1 \to 0$ and $\varepsilon \to 0$. ◇

Theorem 2.3.40 shows that the problem of finding the transition function P given a and b reduces to a Cauchy problem for the equation (2.3.67) with f chosen conveniently (for example $f(y) = e^{i\theta y}$). However, a more precise result is available.

Proposition 2.3.41. *Suppose that the derivatives*

$$\frac{\partial}{\partial x} P(s, x; t, A) \quad \textit{and} \quad \frac{\partial^2}{\partial x^2} P(s, x; t, A)$$

exist and are continuous with respect to $s \in T$, $s < t$, $x \in R$ *and that* $a(\cdot, x)$ *and* $b(\cdot, x)$ *are continuous for any* $x \in R$. *Then for* $0 \leqslant s < t$ *the derivative* $\partial P/\partial s$ *exists and satisfies the equation*

$$-\frac{\partial P(s, x; t, A)}{\partial s} = b(s, x)\frac{\partial P(s, x; t, A)}{\partial x} +$$

$$+ \frac{1}{2} a(s, x)\frac{\partial^2 P(s, x; t, A)}{\partial x^2}.$$

Proof Use Theorem 2.3.40 with $f(y) = P(t, y; u, A)$ where $u > t$ and $A \in \mathcal{B}$ are fixed. ◇

Now suppose that P has a *transition density function* p, that is,

$$P(s, x; t, A) = \int_A p(s, x; t, y)\,dy,$$

for any $A \in \mathcal{B}$. We then have

Proposition 2.3.42. *Suppose that the derivatives*

$$\frac{\partial}{\partial x} p(s, x; t, y) \quad \textit{and} \quad \frac{\partial^2}{\partial x^2} p(s, x; t, y)$$

exist and are continuous with respect to $s \in T$, $s < t$, $x \in R$, *and that* $a(\cdot, x)$ *and* $b(\cdot, x)$ *are continuous for any* $x \in R$. *Then for* $0 \leqslant s < t$ *and* $x, y \in R$, *the derivative* $\partial p/\partial s$ *exists and satisfies the so-called backward Kolmogorov equation*:

$$-\frac{\partial p}{\partial s} = b(s, x)\frac{\partial p}{\partial x} + \frac{1}{2} a(s, x)\frac{\partial^2 p}{\partial x^2}. \tag{2.3.68}$$

Proof. Notice that

$$p(s, x; t, y) = \int_R p(s, x; u, z)\, p(u, z; t, y)\, \mathrm{d}z, \qquad s < u < t.$$

It remains to apply Theorem 2.3.40 with $f(y) = p(t, y; u, z)$[60] where $u > t$ and $z \in R$ are fixed. ◇

Theorem 2.3.43. *Let* (2.3.61—63) *be fulfilled uniformly with respect to* $x \in R$. *Suppose that the partial derivatives*

$$\frac{\partial p(s, x; t, y)}{\partial t}, \qquad \frac{\partial}{\partial y}(b(t, y)p(s, x; t, y)),$$

$$\frac{\partial^2}{\partial y^2}(a(t, y)p(s, x; t, y))$$

exist, the first one being continuous with respect to $(t, y) \in T \times R$, $t > s$ *and the other two with respect to* $y \in R$. *Then* p *satisfies the so-called forward Kolmogorov (or Fokker-Planck) equation*

$$\frac{\partial p(s, x; t, y)}{\partial t} = -\frac{\partial}{\partial y}(b(t, y)\, p(s, x; t, y)) + \frac{1}{2}\frac{\partial^2}{\partial y^2}(a(t, y)\, p(s, x; t, y)). \tag{2.3.69}$$

Proof. Let g be a real-valued function on R identically null outside a finite interval and with a continuous second derivative. As in the

[60] In the proof of Theorem 2.3.40, the boundedness of f is used only to verify the initial condition.

proof of Theorem 2.3.40, it can be shown that, uniformly with respect to $x \in R$

$$\lim_{h \to 0+} \frac{1}{h} \left[\int_R g(y) p(s, x; s+h, y)\, \mathrm{d}y - g(x) \right] = \\ = b(s, x) g'(x) + \frac{1}{2} a(s, x) g''(x). \tag{2.3.70}$$

Using the Chapman-Kolmogorov equation and (2.3.70) we get

$$\frac{\partial}{\partial t} \int_R p(s, x; t, y) g(y)\, \mathrm{d}y = \int_R \frac{\partial}{\partial t} p(s, x; t, y)\, g(y)\, \mathrm{d}y =$$

$$= \lim_{h \to 0+} \frac{1}{h} \left[\int_R p(s, x; t+h, y) g(y)\, \mathrm{d}y - \int_R p(s, x; t, z) g(z)\, \mathrm{d}z \right] =$$

$$= \lim_{h \to 0+} \frac{1}{h} \int_R p(s, x; t, z) \left[\int_R p(t, z; t+h, y) g(y)\, \mathrm{d}y - g(z) \right] \mathrm{d}z =$$

$$= \int_R p(s, x; t, z) \left[b(t, z) g'(z) + \frac{1}{2} a(t, z) g''(z) \right] \mathrm{d}z.$$

Integration by parts yields

$$\int_R \frac{\partial}{\partial t} p(s, x; t, y) g(y)\, \mathrm{d}y =$$

$$= \int_R \left[-\frac{\partial}{\partial z} (b(t, z) p(s, x; t, z)) + \frac{1}{2} \frac{\partial^2}{\partial z^2} (a(t, z) p(s, x; t, z)) \right] g(z) \mathrm{d}z.$$

Hence, g being arbitrary, we obtain (2.3.69). ◇

2.3.5.2.2. Similar results can be obtained for the general case $n \geqslant 1$. We shall merely state them, the proofs being left to the reader.

Theorem 2.3.44. *Let f be a real-valued, continuous and bounded function defined on R^n and suppose that the partial derivatives $\partial u / \partial x_i$ and $\partial^2 u / \partial x_i \partial x_j$, $1 \leqslant i, j \leqslant n$, of the function*

$$u(s, \mathbf{x}) = \int_{R^n} f(y)\, P(s, \mathbf{x}; t, \mathrm{d}y)$$

exist and are continuous on $T \times R^n$. *If* $a_{ij}(\cdot, \mathbf{x})$ *and* $b_i(\cdot, \mathbf{x})$, $1 \leqslant i, j \leqslant n$, *are continuous for any* $\mathbf{x} \in R^n$, *then for* $0 \leqslant s < t$, $\mathbf{x} \in R^n$, *the derivative* $\partial u/\partial s$ *exists and satisfies the equation*

$$-\frac{\partial u}{\partial s} = \sum_{i=1}^{n} b_i(s, \mathbf{x}) \frac{\partial u}{\partial x_i} + \frac{1}{2} \sum_{i,j=1}^{n} a_{ij}(s, \mathbf{x}) \frac{\partial^2 u}{\partial x_i \partial x_j} \qquad (2.3.71)$$

with the initial condition

$$\lim_{s \to t-} u(s, \mathbf{x}) = f(\mathbf{x}).$$

Proposition 2.3.45. *Suppose that the derivatives*

$$\frac{\partial}{\partial x_i} P(s, \mathbf{x}; t, A), \qquad \frac{\partial^2}{\partial x_i \partial x_j} P(s, \mathbf{x}; t, A), \qquad 1 \leqslant i, j \leqslant n,$$

exist and are continuous with respect to $s \in T$, $s < t$, $\mathbf{x} \in R^n$, *and that* $a_{ij}(\cdot, \mathbf{x})$ *and* $b_i(\cdot, \mathbf{x})$, $1 \leqslant i, j \leqslant n$, *are continuous for any* $\mathbf{x} \in R^n$. *Then for* $0 \leqslant s < t$ *the derivative* $\partial P/\partial s$ *exist and satisfies the equation*

$$-\frac{\partial P(s, \mathbf{x}; t, A)}{\partial s} = \sum_{i=1}^{n} b_i(s, \mathbf{x}) \frac{\partial P(s, \mathbf{x}; t, A)}{\partial x_i} + $$

$$+ \frac{1}{2} \sum_{i,j=1}^{n} a_{ij}(s, \mathbf{x}) \frac{\partial^2 P(s, \mathbf{x}; t, A)}{\partial x_i \partial x_j}.$$

Now suppose that P has a *transition density function* p, that is,

$$P(s, \mathbf{x}; t, A) = \int_A p(s, \mathbf{x}; t, \mathbf{y}) \, d\mathbf{y}$$

for any $A \in \mathcal{B}^n$.

Proposition 2.3.46. *Suppose that the derivatives*

$$\frac{\partial}{\partial x_i} p(s, \mathbf{x}; t, \mathbf{y}), \quad \frac{\partial^2}{\partial x_i \partial x_j} p(s, \mathbf{x}; t, \mathbf{y}), \quad 1 \leqslant i, j \leqslant n,$$

exist and are continuous with respect to $s \in T$, $s < t$, $\mathbf{x} \in R^n$, *and that* $a_{ij}(\cdot, \mathbf{x})$ *and* $b_i(\cdot, \mathbf{x})$, $1 \leqslant i, j \leqslant n$, *are continuous for any* $\mathbf{x} \in R^n$. *Then for* $0 \leqslant s < t$, $\mathbf{x}, \mathbf{y} \in R^n$, *the derivative* $\partial p(s, \mathbf{x}; t, \mathbf{y})/\partial s$ *exists and satisfies the so-called backward Kolmogorov equation*

$$-\frac{\partial p}{\partial s} = \sum_{i=1}^{n} b_i(s, \mathbf{x}) \frac{\partial p}{\partial x_i} + \frac{1}{2} \sum_{i,j=1}^{n} a_{ij}(s, \mathbf{x}) \frac{\partial^2 p}{\partial x_i \partial x_j}. \qquad (2.3.72)$$

Theorem 2.3.47. *Let* (2.3.64—66) *be fulfilled uniformly with respect to* $\mathbf{x} \in R^n$. *Suppose that the partial derivatives*

$$\frac{\partial p(s, \mathbf{x}; t, \mathbf{y})}{\partial t}, \quad \frac{\partial}{\partial y_i}(b_i(t, \mathbf{y})\, p(s, \mathbf{x}; t, \mathbf{y})),$$

$$\frac{\partial^2}{\partial y_i\, \partial y_j}(a_{ij}(t, \mathbf{y})\, p(s, \mathbf{x}; t, \mathbf{y})), \quad 1 \leqslant i, j \leqslant n$$

exist, the first one being continuous with respect to $(t, \mathbf{y}) \in T \times R^n$, $t > s$, *and the other ones with respect to* $\mathbf{y} \in R^n$. *Under these conditions p satisfies the so-called forward Kolmogorov (or Fokker-Planck) equation*

$$\frac{\partial p(s, \mathbf{x}; t, \mathbf{y})}{\partial t} = -\sum_{i=1}^{n} \frac{\partial}{\partial y_i}[b_i(t, \mathbf{y})\, p(s, \mathbf{x}; t, \mathbf{y})] +$$

$$+ \frac{1}{2} \sum_{i,j=1}^{n} \frac{\partial^2}{\partial y_i\, \partial y_j}[a_{ij}(t, \mathbf{y})\, p(s, \mathbf{x}; t, \mathbf{y})] \tag{2.3.73}$$

2.3.5.2.3. In the homogeneous case, equations (2.3.68), (2.3.69), (2.3.72), and (2.3.73) become, respectively, (with $p(t; \mathbf{x}, \mathbf{y}) = p(0, \mathbf{x}; t, \mathbf{y})$)

$$\frac{\partial p}{\partial t} = b(x)\frac{\partial p}{\partial x} + \frac{1}{2} a(x) \frac{\partial^2 p}{\partial x^2}, \tag{2.3.68'}$$

$$\frac{\partial p}{\partial t} = -\frac{\partial}{\partial y}(b(y)\, p) + \frac{1}{2} \frac{\partial^2}{\partial y^2}(a(y)\, p), \tag{2.3.69'}$$

$$\frac{\partial p}{\partial t} = \sum_{i=1}^{n} b_i(\mathbf{x}) \frac{\partial p}{\partial x_i} + \frac{1}{2} \sum_{i,j=1}^{n} a_{ij}(\mathbf{x}) \frac{\partial^2 p}{\partial x_i\, \partial x_j}, \tag{2.3.72'}$$

$$\frac{\partial p}{\partial t} = -\sum_{i=1}^{n} \frac{\partial}{\partial y_i}[b_i(\mathbf{y})\, p(t; \mathbf{x}, \mathbf{y})] + \frac{1}{2} \sum_{i,j=1}^{n} \frac{\partial^2}{\partial y_i\, \partial y_j}[a_{ij}(\mathbf{y})\, p(t; \mathbf{x}, \mathbf{y})]. \tag{2.3.73'}$$

The conditions under which they hold are easily deduced from the preceding subparagraphs.

2.3.5.2.4. Existence and uniqueness of the solutions of (2.3.72) and (2.3.73)[61] can be proved for the so-called fundamental solutions.

---

[61] The first approach to this problem is due to FELLER (1936). FELLER's treatment parallels that for the backward and forward equations of jump processes.

A function $p(s, \mathbf{x}; t, \mathbf{y})$ $(0 \leqslant s < t < \infty, \mathbf{x}, \mathbf{y} \in R^n)$ is said to be a *fundamental solution* in $T \times R^n$ to (2.3.72) if

i) p and the derivatives $\partial p/\partial x_i$, $\partial^2 p/\partial x_i\, \partial x_j$, $1 \leqslant i, j \leqslant n$, $\partial p/\partial s$ are continuous with respect to all the variables and p as function of s and $\mathbf{x}$ satisfies (2.3.72).

(ii) for every real-valued bounded continuous function f defined on R^n

$$\lim_{s \to t-} \int_{R^n} p(s, \mathbf{x}; t, \mathbf{y}) f(\mathbf{y})\, d\mathbf{y} = f(\mathbf{x}), \quad (t, \mathbf{x}) \in T \times R^n,$$

the convergence being uniform with respect to $\mathbf{x}$ on every bounded set in R^n.

(iii) p is bounded on the set $t - s + |\mathbf{x} - \mathbf{y})| \geqslant \delta$ for any $\delta > 0$.

This definition was suggested by a special case, namely that where $a_{ij}(s, \mathbf{x})$=const., $b_i(s, \mathbf{x}) \equiv 0$, $1 \leqslant i, j \leqslant n$, with $\mathbf{A} = (a_{ij})_{1 \leqslant i,j \leqslant n}$ positive definite.

In this case (2.3.72) is satisfied by

$$p(s, \mathbf{x}; t, \mathbf{y}) = [2\pi(t - s)]^{-\frac{n}{2}} [\det \mathbf{A})]^{-\frac{1}{2}} \exp\left(-\frac{(\mathbf{y} - \mathbf{x})\mathbf{A}^{-1}(\mathbf{y} - \mathbf{x})'}{2(t - s)}\right)$$

which fulfills conditions i)—iii) above.

The study of fundamental solutions to the parabolic equations has been carried out by M. GÉVREY, F. G. DRESSEL, W. POGORZELSKI, and S. D. EIDELMAN.

Theorem 2.3.48. *Suppose that*

j) *the functions* a_{ij} *and* b_i, $1 \leqslant i, j \leqslant n$, *are bounded and continuous on* $T \times R^n$ *and satisfy a Hölder condition with exponent* $\lambda > 0$ *with respect to* $\mathbf{x} \in R^n$ *and* $s \in R$; *that is there is a constant* $K > 0$ *such that*

$$| a_{ij}(s, \mathbf{x}) - a_{ij}(s, \mathbf{y})| \leqslant K |\mathbf{x} - \mathbf{y}|^\lambda,$$

$$| a_{ij}(s, \mathbf{x}) - a_{ij}(t, \mathbf{x}) | \leqslant K |t - s|^\lambda,$$

$$| b_i(s, \mathbf{x}) - b_i(s, \mathbf{y})| \leqslant K |\mathbf{x} - \mathbf{y}|^\lambda,$$

for all $\mathbf{x}, \mathbf{y} \in R^n$, $s, t \in T$, $1 \leqslant i, j \leqslant n$.

jj) *there is a constant* $\gamma > 0$ *such that*

$$\sum_{i,j=.}^{n} a_{ij}(s, \mathbf{x})\, \lambda_i \lambda_j \geqslant \gamma \sum_{i=1}^{n} \lambda_i^2$$

for all $(s, \mathbf{x}) \in T \times R^n$, $(\lambda_1, \ldots, \lambda_n) \in R^n$.

Under these conditions there is a unique fundamental solution p to (2.3.72)[62] *with the following properties*

a) *p is a transition density function and* $p(s, \mathbf{x}; t, \mathbf{y}) > 0$ *for all* $0 \leqslant s < t < \infty$, $\mathbf{x}, \mathbf{y} \in R^n$;

b) *there are positive constants M and* α *such that*

$$p(s, \mathbf{x}; t, \mathbf{y}) \leqslant M(t-s)^{-\frac{n}{2}} \exp\left(-\frac{\alpha |\mathbf{y}-\mathbf{x}|^2}{t-s}\right),$$

$$\left|\frac{\partial p(s, \mathbf{x}; t, \mathbf{y})}{\partial x_i}\right| \leqslant M(t-s)^{-\frac{n+1}{2}} \exp\left(-\frac{\alpha |\mathbf{y}-\mathbf{x}|^2}{t-s}\right),$$

$$\left|\frac{\partial^2 p(s, \mathbf{x}; t, \mathbf{y})}{\partial x_i \partial x_j}\right|, \quad \left|\frac{\partial p(s, \mathbf{x}; t, \mathbf{y})}{\partial t}\right| \leqslant M(t-s)^{-\frac{n+2}{2}} \exp\left(-\frac{\alpha |\mathbf{y}-\mathbf{x}|^2}{t-s}\right)$$

for all $0 \leqslant s < t < \infty$, $\mathbf{x}, \mathbf{y} \in R^n$, $1 \leqslant i, j \leqslant n$.

c) *if the derivatives* $\partial a_{ij}/\partial x_j$, $\partial^2 a_{ij}/\partial x_i \partial x_j$, $\partial b_i/\partial x_i$, $1 \leqslant i, j \leqslant n$, *exist, are bounded and continuous on* $T \times R^n$, *and satisfy a Hölder condition with exponent* $\lambda > 0$ *with respect to* $\mathbf{x} \in R^n$, *then p as function of t and* $\mathbf{y}$ *satisfies equation* (2.3.73).

d) *for every bounded function f belonging to the class* $C^{(2,\lambda)}$ *on* R^n *(i.e. its partial derivatives of order two satisfy a Hölder condition with exponent* $\lambda > 0$*) the function*

$$u(s, \mathbf{x}) = \int_{R^n} p(s, \mathbf{x}; t, \mathbf{y}) f(\mathbf{y}) \, d\mathbf{y}$$

satisfies

$$\lim_{s \to t-} \frac{\partial u(s, \mathbf{x})}{\partial x_i} = \frac{\partial f(\mathbf{x})}{\partial x_i}, \quad \lim_{s \to t-} \frac{\partial^2 u(t, \mathbf{x})}{\partial x_i \partial x_j} = \frac{\partial^2 f(\mathbf{x})}{\partial x_i \partial x_j}$$

for all $(t, \mathbf{x}) \in T \times R^n$, $1 \leqslant i, j \leqslant n$.

Proof. See IL'IN, KALAŠNIKOV and OLEĬNIK (1962) (for (a)—(c)) and FRIEDMAN (1958) (for (d)). ◇

Theorem 2.3.49. *Suppose that conditions* (j) *and* (jj) *in the previous theorem are fulfilled. Then to the fundamental solution p in* $T \times R^n$ *to*

---

[62] All the conclusions of Theorem 2.3.48 are also valid for liner parabolic equations containing a term of the form $c(s, \mathbf{x})p$, the function c satisfying the same conditions as the b_i. Such a term may appear if P is a substochastic transition function.

(2.3.72) *there corresponds the (perhaps substochastic) transition function*

$$P(s, \mathbf{x}; t, A) = \int_A p(s, \mathbf{x}; t, \mathbf{y})\, d\mathbf{y}$$

satisfying (2.3.64—66).

Proof. See DYNKIN [(1961, Ch. 6), (1965, Ch. 5, § 6)]. ◇

In the homogeneous case the above considerations simplify as follows:

A function $p(t; \mathbf{x}, \mathbf{y})$ $(t > 0, \mathbf{x}, \mathbf{y} \in R^n)$ is said to be a *fundamental solution* in $T \times R^n$ to (2.3.72′) if

i′) p and the derivatives $\partial p/\partial x_i$, $\partial^2 p/\partial x_i \partial x_j$, $1 \leqslant i, j \leqslant n$, $\partial p/\partial t$ are continuous on $T \times R^n$, and p as function of t and $\mathbf{x}$ satisfies (2.3.72′);

ii′) for every real-valued bounded continuous function f defined on R^n

$$\lim_{t \to 0+} \int_{R^n} p(t; \mathbf{x}, \mathbf{y}) f(\mathbf{y})\, d\mathbf{y} = f(\mathbf{x}), \ \mathbf{x} \in R^n,$$

the convergence being uniform with respect to $\mathbf{x}$ on every bounded set in R^n.

iii′) p is bounded on the set $t + |\mathbf{x} - \mathbf{y}| \geqslant \delta$ for any $\delta > 0$.

Theorem 2.3.48′. *Suppose that*

j′) *the functions* a_{ij} *and* b_i, $1 \leqslant i, j \leqslant n$, *are bounded and continuous on* R^n *and satisfy a Hölder condition with exponent* $\lambda > 0$.

jj′) *there is a constant* $\gamma > 0$ *such that*

$$\sum_{i,j=1}^{n} a_{ij}(\mathbf{x}) \lambda_i \lambda_j \geqslant \gamma \sum_{i=1}^{n} \lambda_i^2$$

for all $\mathbf{x} \in R^n$, $(\lambda_1, \ldots, \lambda_n) \in R^n$.

Under these conditions, there is a unique fundamental solution p *to* (2.3.72′) *with the following properties*

a′) p *is a transition density function and* $p(t; \mathbf{x}, \mathbf{y}) > 0$ *for all* $t > 0$, $\mathbf{x}, \mathbf{y} \in R^n$.

b′) *there are positive constants* M *and* α *such that*

$$p(t; \mathbf{x}, \mathbf{y}) \leqslant M t^{-\frac{n}{2}} \exp\left(-\frac{\alpha |\mathbf{y} - \mathbf{x}|^2}{t}\right),$$

$$\left|\frac{\partial p(t; \mathbf{x}, \mathbf{y})}{\partial x_i}\right| \leqslant M t^{-\frac{n+1}{2}} \exp\left(-\frac{\alpha |\mathbf{y} - \mathbf{x}|^2}{t}\right),$$

$$\left|\frac{\partial^2 p(t; \mathbf{x}, \mathbf{y})}{\partial x_i\, \partial x_j}\right|, \left|\frac{\partial p(t; \mathbf{x}, \mathbf{y})}{\partial t}\right| \leqslant M t^{-\frac{n+2}{2}} \exp\left(-\frac{\alpha |\mathbf{y} - \mathbf{x}|^2}{t}\right)$$

for all $t > 0$, $\mathbf{x}, \mathbf{y} \in R^n$, $1 \leqslant i, j \leqslant n$.

c′) *if the derivatives* $\partial a_{ij}/\partial x_j$, $\partial^2 a_{ij}/\partial x_i \partial x_j$, $\partial b_i/\partial x_i$, $1 \leqslant i, j \leqslant n$, *exist, are bounded and satisfy a Hölder condition with exponent* $\lambda > 0$, *then* $p(t; \mathbf{x}, \mathbf{y})$ *as function of* t *and* $\mathbf{y}$ *satisfies equation* (2.3.73′).

d′) *for every bounded function f belonging to the class $C^{(2,\lambda)}$ on R^n (i.e. its partial derivatives of order two satisfy a Hölder condition with exponent $\lambda > 0$) the function*

$$u(t, \mathbf{x}) = \int_{R^n} p(t; \mathbf{x}, \mathbf{y}) f(\mathbf{y})\, d\mathbf{y}$$

satisfies

$$\lim_{t\to 0+} \frac{\partial u(t, \mathbf{x})}{\partial x_i} = \frac{\partial f(\mathbf{x})}{\partial x_i}, \quad \lim_{t\to 0+} \frac{\partial^2 u(t, \mathbf{x})}{\partial x_i\, \partial x_j} = \frac{\partial^2 f(\mathbf{x})}{\partial x_i\, \partial x_j}$$

for all $\mathbf{x} \in R^n$, $1 \leqslant i, j \leqslant n$.

Proof. Theorem 2.3.48. ◇

Theorem 2.3.49′. *Suppose that conditions* (j′) *and* (jj′) *in the previous theorem are fulfilled. Then to the fundamental solution p in $T \times R^n$ to* (2.3.72′) *there corresponds a Feller (perhaps substochastic) transition function*

$$P(t; \mathbf{x}, A) = \int_A p(t; \mathbf{x}, \mathbf{y})\, d\mathbf{y}$$

satisfying (2.3.64′–66′).

Proof. See Dynkin (1965, Ch. 5, § 6). ◇

2.3.5.3. Approximations

2.3.5.3.1. From now on we shall consider only the homogeneous one-dimensional case $n = 1$. First we shall show that the backward and forward equations can be obtained by means of a limiting procedure applied to a suitable Markov chain in which changes of states and the time intervals in which they occur are small.

Consider a particle taking small steps of amounts $-\Delta x$ or Δx in small time intervals of length Δt.

Suppose that if the particle is at x at time t then the probabilities that it will be at $x \pm \Delta x$ at time $t + \Delta t$ are $\frac{1}{2}\left(1 \pm \frac{b(x)}{a(x)}\Delta x\right)$. Now denoting by $p(t; x, y)$ the probability that the particle is at y at time t given that it starts at x at time $t = 0$, we can write

$$p(t + \Delta t; x, y) =$$

$$= \frac{1}{2}\left(1 + \frac{b(x)}{a(x)}\Delta x\right) p(t; x + \Delta x, y) +$$

$$+ \frac{1}{2}\left(1 - \frac{b(x)}{a(x)}\Delta x\right) p(t; x - \Delta x, y),$$

whence

$$\frac{p(t+\Delta t; x, y) - p(t; x, y)}{\Delta t} =$$

$$= \frac{p(t; x+\Delta x, y) - 2p(t; x, y) + p(t; x-\Delta x, y)}{2(\Delta x)^2} \cdot \frac{(\Delta x)^2}{\Delta t} +$$

$$+ \frac{b(x)}{2a(x)} \frac{(\Delta x)^2}{\Delta t} \left[\frac{p(t; x+\Delta x . y) - p(t; x, y)}{\Delta x} + \right.$$

$$\left. + \frac{p(t; x, y) - p(t; x-\Delta x, y)}{\Delta x} \right].$$

Imposing the condition that $\Delta x \sim \sqrt{a(x)\,\Delta t}$ and letting $\Delta t \to 0$ we get formally

$$\frac{\partial p(t; x, y)}{\partial t} = b(x) \frac{\partial p(t; x, y)}{\partial x} + \frac{1}{2} a(x) \frac{\partial^2 p(t; x, y)}{\partial x^2},$$

i.e. the backward equation. Similarly, by writing

$$p(t; x, y) =$$

$$= \frac{1}{2} p(t-\Delta t; x, y-\Delta y) \left(1 + \frac{b(y)}{a(y)} \Delta y\right) +$$

$$+ \frac{1}{2} p(t-\Delta t, x, y+\Delta y) \left(1 - \frac{b(y)}{a(y)} \Delta y\right),$$

one can obtain the forward equation (2.3.69′).

Moreover, it is possible to show as in Theorem 2.2.16 that a suitable sequence of random polygonal functions associated with the Markov chain considered converges (weakly) to a diffusion process governed by the Kolmogorov equations we have obtained. For details see GIHMAN and SKOROHOD (1965, Ch. 9, § 4).

2.3.5.3.2. Another useful procedure is the reverse one of using a diffusion process to study a discrete (-state or-parameter) process (Markovian or otherwise).

The usefulness of this procedure is due to the fact that mathematical methods associated with the continuum (differential equations, integration, etc.) are easier to work with than those associated with discrete variations (difference equations, summation, etc.).

Diffusion approximations of discrete parameter Markov processes were first considered by A. JA. HINČIN in 1933. Similar approximations have proved useful in population genetics [KIMURA (1964), WATTERSON (1962)], mathematical learning theory [NORMAN (1972)], and other areas [IGLEHART (1968 b)].

If we wish to use a diffusion approximation for a given process we must write down the equivalents of the diffusion and drift coefficients, namely the mean and variance per unit time of the increments of the process.

As an example [cf. FELLER (1951)] let us take a Galton-Watson chain $(\xi_n)_{n\in N}$. The mean and variance of the increment $\xi_{n+1}-\xi_n$, conditional on $\xi_n = k$, are (see Subparagraph 1.2.2.1.3)

$$\mathsf{E}(\xi_{n+1} - \xi_n | \xi_n = k) = km - k = k(m-1),$$

$$\mathsf{D}(\xi_{n+1} - \xi_n | \xi_n = k) = k\sigma^2.$$

Since these quantities are proportional to the value of ξ_n, we are led to take as diffusion approximation a diffusion process ξ with drift coefficient βy and diffusion coefficient αy, that is, both proportional to y, the value of ξ at time t. The forward equation will be

$$\frac{1}{2}\alpha\frac{\partial^2}{\partial y^2}(yp) - \beta\frac{\partial}{\partial y}(yp) = \frac{\partial p}{\partial t}.$$

Introducing the moment generating function

$$\psi(x, t, \theta) = \int_0^\infty e^{-\theta y} p(t; x, y)\, dy,$$

we find that

$$\frac{\partial\psi}{\partial t} = \left(\beta\theta - \frac{1}{2}\alpha\theta^2\right)\frac{\partial\psi}{\partial\theta},$$

that is, a Lagrange equation whose solution is

$$\psi(x, t, \theta) = \exp\left\{\frac{-xe^{\beta t}\theta}{1+\left(\frac{1}{2}\alpha/\beta\right)(e^{\beta t}-1)\theta}\right\}.$$

Hence we find

$$\mathsf{E}(\xi(t)|\xi(0) = x) = xe^{\beta t},$$

$$\mathsf{D}(\xi(t)|\xi(0) = x) = \frac{\alpha x}{\beta} e^{\beta t}(e^{\beta t} - 1).$$

The similarity of these formulas to the corresponding ones for Galton-Watson chains is apparent.

2.3.5.4. Boundaries

2.3.5.4.1. We know that under sufficiently general conditions, two given functions a and b determine a Markov diffusion process. It is possible to obtain from such a process other ones by restricting the state space R to a finite or semi-infinite interval (r_1, r_2) where $-\infty \leqslant r_1 < r_2 \leqslant \infty$ and imposing boundary conditions at r_1 and / or r_2. It is also possible that a and b may be defined only on an interval $(r_1, r_2) \subset R$, that is the state space is restricted by definition. In the latter case boundary conditions at r_1 and / or r_2 are in general also required in order to determine the process uniquely. The important thing is that boundary conditions can be imposed iff a boundary is actually reached. In what follows we briefly describe some results by FELLER (1952), obtained from the theory of semigroups.

2.3.5.4.2. Suppose $a(x) > 0$ for $x \in (r_1, r_2)$ and introduce E. Hille's function

$$h(x) = \exp\left\{-\int_{x'}^{x} \frac{2b(u)}{a(u)}\, du\right\},$$

where $x' \in (r_1, r_2)$ is fixed. Put

$$g(x) = \frac{2}{a(x)\, h(x)},$$

and consider the quantities

$$\sigma_1 = \iint\limits_{r_1<y<x<x'} g(x)\, h(y)\, dx\, dy, \quad \mu_1 = \iint\limits_{r_1<y<x<x'} h(x) g(y) dx\, dy,$$

$$\sigma_2 = \iint\limits_{r_2>y>x>x'} g(x)\, h(y)\, dx\, dy, \quad \mu_2 = \iint\limits_{r_2>y>x>x'} h(x)\, g(y)\, dx\, dy.$$

Feller's classification criteria are as follows [cf. ITÔ (1963, pp. 103—106)]. The *boundary* r_i $(i = 1,2)$ is said to be

— *regular* if $\sigma_i < \infty$, $\mu_i < \infty$;
— an *exit* boundary if $\sigma_i < \infty$, $\mu_i = \infty$;
— an *entrance* boundary if $\sigma_i = \infty$, $\mu_i < \infty$;
— *natural* if $\sigma_i = \infty$, $\mu_i = \infty$;

(this classification does not depend on the particular choice of x').

Boundaries which are regular or exit are called *accessible*, the others *inaccessible*. The probabilistic meaning of Feller's classification amounts to the fact that the probability of reaching an accessible boundary in finite time is positive, while the same probability for an inaccessible boundary in zero.

As concerns the solutions to (2.3.68′) and (2.3.69′) for $x, y \in (r_1, r_2)$, the situation is as follows. When none of the boundaries is regular, then there exists exactly one solution common to (2.3.68′) and (2.3.69′). When the boundary r_i, or both boundaries, are regular, there exist infinitely many common solutions. If $p(t; x, y)$ is a solution to (2.3.68′) then

$$\alpha_i \lim_{x \to r_i} p(t; x, y) + (1 - \alpha_i)(-1)^i \lim_{x \to r_i} h^{-1}(x) \frac{\partial p(t; x, y)}{\partial x} = 0$$

for some constant $0 \leqslant \alpha_i \leqslant 1$. Conversely, to an arbitrarily prescribed constant α_i there corresponds uniquely a solution common to (2.3.68′) and (2.3.69′). The corresponding solutions to (2.3.69′) satisfy the boundary condition

$$\alpha_i \lim_{y \to r_i} h(y)\, a(y) p(t; x, y) +$$

$$+ (1 - \alpha_i)(-1)^i \left\{ \frac{\partial}{\partial y} (a(y) p(t; x, y)) - 2b(y) p(t; x, y) \right\} = 0$$

and this condition determines the solution to (2.3.69′).

By analogy with random walk, for $\alpha_i = 0$ the boundary r_i is said to be a *reflecting* barrier, and for $\alpha_i = 1$ an *absorbing* one.

2.3.5.5. Brownian motion as diffusion process

2.3.5.5.1. It follows from Proposition 2.2.3 that one-dimensional Brownian motion w is a homogeneous Markov process with transition density function

$$p(t; x, y) = \frac{1}{\sqrt{2\pi t}} \exp\left\{ -\frac{(y - x)^2}{2t} \right\}.$$

It is easily verified that w is a classical diffusion process with diffusion and drift coefficients $a(x) \equiv 1$, $b(x) \equiv 0$, $x \in R$, so that (2.3.68′) and (2.3.69′) become

$$\frac{\partial p}{\partial t} = \frac{1}{2} \frac{\partial^2 p}{\partial x^2}, \tag{2.3.74}$$

$$\frac{\partial p}{\partial t} = \frac{1}{2} \frac{\partial^2 p}{\partial y^2}. \tag{2.3.75}$$

Such equations are well known in the theories of diffusion (of a substance) and heat conduction. This is one reason for the nomenclature "diffusion process".

It is now known [see Itô and McKean (1965, Ch. 5)] that Brownian motion can be transformed into any given diffusion (i.e. continuous strong Markov process), by a suitable random time change[63].

2.3.5.5.2. By imposing various boundary conditions on one of equations (2.3.74) or (2.3.75) we obtain density transition functions corresponding to modifications of the Brownian motion w. (As is easily seen, every $x \in R$ acts as a regular boundary and $-\infty$ and ∞ are natural boundaries). Let us begin with the case of an absorbing barrier at $a > 0$. We shall solve the equation

$$\frac{\partial q(t; y)}{\partial t} = \frac{1}{2} \frac{\partial^2 q(t; y)}{\partial y^2} \tag{2.3.76}$$

for $y < a$, $t > 0$ subject to the condition

$$q(t; a) = 0, \; t > 0. \tag{2.3.77}$$

The method of attack is Lord Kelvin's method of images [see e.g. Sommerfeld (1949)]. Imagine the barrier as a mirror and place an "image source" at $x = 2a$, the image of the origin in the mirror. This leads to a solution

$$q(t; y) = p(t; 0, y) + Ap) \; t; 2a, y),$$

and the question now is whether a suitable value of A can be found to satisfy (2.3.77). Clearly $A = -1$, and therefore

$$q(t; y) = \frac{1}{\sqrt{2\pi t}} \left[\exp\left(-\frac{y^2}{2t}\right) - \exp\left\{-\frac{(y-2a)^2}{2t}\right\}\right].$$

We have for $u \leqslant a$

$$\int_{-\infty}^{u} q(t, y) \, dy = \mathsf{P}(w^a_{\text{abs}}(t) < u),$$

where

$$w^a_{\text{abs}}(t) = \begin{cases} w(t) & \text{if } t \leqslant \tau_a \\ a & \text{if } t > \tau_a, \end{cases}$$

$$\tau_a = \inf\{t : w(t) = a\}.$$

[63] Proposition 2.3.5 shows that Brownian motion itself is a strong Markov process.

The process $w^{a}_{ab,\xi}$ is said to be a *Brownian motion with absorbing barrier* at the point a.

Observe that $\int_{-\infty}^{a} q(t; y)\,dy$ is the probability that the Brownian free trajectory does not reach the point a before time t. Therefore

$$-\frac{d}{dt}\int_{-\infty}^{a} q(t; y)\,dy = \frac{a}{\sqrt{2\pi t^3}} \exp\left(-\frac{a^2}{2t}\right), \quad t > 0,$$

is the probability density of the first passage time τ_a from 0 to a of a free Brownian motion. For a detailed treatment of first passage times for homogeneous diffusion processes we refer the reader to Cox and MILLER (1965, p. 230).

Similar methods apply to the case of two absorbing barriers, one at $-b\,(b > 0)$ and the other at $a > 0$[64]. The solution will satisfy (2.3.76) for $-b < y < a$ subject to the conditions

$$q(t; -b) = q(t; a), \ t > 0.$$

It is given by

$$q(t; y) =$$

$$= \frac{1}{\sqrt{2\pi t}} \sum_{k \in Z} \left[\exp\left\{-\frac{(y - 2k(a + b))^2}{2t}\right\}\right.$$

$$\left.- \exp\left\{-\frac{(y + (2k - 2)a + 2kb)^2}{2t}\right\}\right].$$

In the case of a reflecting barrier at $a > 0$ the solution will satisfy (2.3.76) for $y < a$ subject to the condition

$$\left.\frac{\partial q(t; y)}{\partial y}\right|_{y=a} = 0, \ t > 0.$$

It is given by

$$q(t; y) = \frac{1}{\sqrt{2\pi t}} \left[\exp\left(-\frac{y^2}{2t}\right) + \exp\left\{-\frac{(y - 2a)^2}{2t}\right\}\right].$$

---

[64] Another approach is to use the method of separation of variables which leads to a Fourier series solution.

We have for $u \leqslant a$

$$\int_{-\infty}^{u} q(t; y)\, \mathrm{d}y = \mathsf{P}(w_{\mathrm{ref}}^{a}(t) < u)$$

where

$$w_{\mathrm{ref}}^{a}(t) = \begin{cases} w(t) & \text{if } w(t) \leqslant a, \\ 2a - w(t) & \text{if } w(t) > a. \end{cases}$$

The process w_{ref}^{a} is said to be a *Brownian motion with reflecting barrier* at the point a.

For details see Cox and Miller (1965, pp. 219—225).

2.3.5.5.3. Finally we consider a transformation of the Brownian motion, namely the process

$$u(t) = \mathrm{e}^{-\rho t}\, w\left(\frac{\alpha}{2\rho}\, \mathrm{e}^{2\rho t}\right),$$

where α and ρ are positive constants. This transformation is known as the *Ornstein-Uhlenbeck process* and was introduced as a model for an actual Brownian motion in which a particle has a well-defined velocity $u(t)$ at time t. It is clear that u is still a Markov process with transition density function

$$p(t; x, y) = \frac{1}{\sqrt{2\pi\, D(t)}} \exp\left\{-\frac{(y - x\mathrm{e}^{-\rho t})^2}{2D(t)}\right\},$$

where

$$D(t) = \frac{\alpha}{2\rho}(1 - \mathrm{e}^{-2\rho t}).$$

Straightforward computations show that the Ornstein-Uhlenbeck process is a classical diffusion process with diffusion and drift coefficients $a(x) \equiv \alpha$ and $b(x) = -\rho x$, $x \in R$. Details can be found in Cox and Miller (1965, p. 225) and Rosenblatt (1962, p. 137).

2.3.6. Extensions of Markov processes

There exist non-Markov processes which can be analysed by considering Markov processes associated with them in a suitable manner. The most direct method of dealing with a process that is not Markovian is to add a number of supplementary variables sufficient to create a process with the Markov property. In this subsection we shall consider such a case.

For a detailed discussion of methods for dealing with non-Markov processes we refer the reader to COX and MILLER (1965, Ch. 6).

2.3.6.1. Semi-Markov processes

2.3.6.1.1. Semi-Markov processes are generalizations of both Markov chains and Markov processes with countable state spaces. They were introduced independently by P. LÉVY, W. L. SMITH and L. TAKÁCS in 1954. Roughly speaking, a semi-Markov process is a stochastic process which moves from one to another of at most a countable number of states with the successive states visited forming a Markov chain which stays in a given state a random length of time, the distribution function of which may depend on the present state as well as on the one to be visited next. (Thus we have a situation similar to that in which Markov branching processes were generalized to obtain age-dependent branching processes).

2.3.6.1.2. Let X be at most a countable set. Let $\mathbf{F} = (F_{ij}(\cdot))_{i,j \in X}$ be a matrix-valued function whose entries are non-negative, non-decreasing functions defined on $[0, \infty)$ and satisfying $F_{ij}(0) = 0$, $\sum_{j \in X} F_{ij}(+\infty) = 1$, $i \in X$. Finally, let $\mathbf{p} = (p_i)_{i \in X}$ be a probability distribution on X.

Consider a Markov chain $(\xi_n, \tau_n)_{n \in N}$, $\tau_0 \equiv 0$, with state space $X \times [0, \infty)$, initial distribution

$$\mathsf{P}(\xi_0 = i, \quad \tau_0 \equiv 0) = p_i, \, i \in X,$$

and transition function

$$\mathsf{P}(\xi_n = j, \tau_n \in [0, t] \mid \xi_{n-1}, \tau_{n-1}) = F_{\xi_{n-1} j}(t), \quad t > 0, \quad n \in N^*.$$

Definition 2.3.50. The process $(\xi(t))_{t \geqslant 0}$ defined by

$$\xi(t) = \xi_{n-1}$$

if

$$\sum_{i=0}^{n-1} \tau_i \leqslant t < \sum_{i=0}^{n} \tau_i, \quad n \in N^*,$$

is said to be a *semi-Markov process* determined by $(X, \mathbf{p}, \mathbf{F})$.

The above definition makes precise the description of a semi-Markov process given in the previous subparagraph if we interpret $\tau_1, \tau_2, \ldots$

as the sojourn times in the successively visited states and $F_{ij}(t)$ as the probability that a sojourn at i has duration $\leqslant t$ and ends by a jump into j.

Clearly, in the case of homogeneous denumerable state space Markov jump processes with all states stable one has

$$F_{ij}(t) = \frac{q_{ij}}{q_i}(1 - e^{-q_i t})(1 - \delta_{ij}), \quad i, j \in X.$$

Note also that if we define

$$\eta(t) = \begin{cases} t & \text{if } \xi(s) = \xi(t), \quad 0 \leqslant s \leqslant t, \\ t - \sup\,\{s : 0 \leqslant s \leqslant t,\ \xi(s) \neq \xi(t)\} & \text{otherwise,} \end{cases}$$

then $(\xi(t), \eta(t))_{t \geqslant 0}$ will be a Markov process. Thus a semi-Markov process may be considered as "half" of a Markov one.

Notice that $\xi(t, \omega)$ is defined only for $t \leqslant \tau(\omega) = \sum_{i \in N} \tau_i$. PYKE (1961 a) showed that if X is a finite set, then $\tau < \infty$ with probability zero, but for an infinite X it can happen that τ is finite with positive probability.

As in Subparagraph 2.3.4.5.3 we shall call τ the "first infinity" of the semi-Markov process. A general definition of a semi-Markov process which allows the possibility of infinities was given by PYKE and SCHAUFELE (1964). See also ÇINLAR (1969), and YACKEL (1966, 1968).

2.3.6.1.3. We now note that Definition 2.3.50 implies the following easily verified relations

$$\begin{aligned} &\mathsf{P}(\tau_n \leqslant t \mid \xi_0, \ldots, \xi_{n-1}) = \mathcal{S}_{\xi_{n-1}}(t), \\ &\mathsf{P}(\xi_n = j \mid \xi_0, \ldots, \xi_{n-1}) = r_{\xi_{n-1} j}, \\ &\mathsf{P}(\tau_n \leqslant t \mid \xi_0, \ldots, \xi_n) = f_{\xi_{n-1} \xi_n}(t), \quad n \in N^*, t \geqslant 0, \\ &\mathsf{P}(\tau_{n_\nu} \leqslant t_\nu, \quad 1 \leqslant \nu \leqslant k \mid \xi_n, \quad n \in N) = \\ &= \mathsf{P}(\tau_{n_\nu} \leqslant t_\nu, \ 1 \leqslant \nu \leqslant k \mid \xi_m, 0 \leqslant m \leqslant n_k) = \\ &= \prod_{\nu=1}^{k} f_{\xi_{n_\nu - 1} \xi_{n_\nu}}(t_\nu), \quad 0 < n_1 < \ldots < n_k, \end{aligned} \tag{2.3.78}$$

all equalities holding with probability one. Here

$$\mathcal{S}_i = \sum_{j \in X} F_{ij}, \quad i \in X,$$

$$r_{ij} = F_{ij}(+\infty), \; i, j \in X,$$

$$f_{ij}(t) = \begin{cases} r_{ij}^{-1} F_{ij}(t) & \text{if } r_{ij} \neq 0, \\ \delta_t([1, \infty)) & \text{if } r_{ij} = 0. \end{cases}$$

Consequently, $\mathcal{S}_i$ is the absolute distribution of the sojourn time at $i \in X$ and r_{ij} is the probability that a jump into i is followed by a jump to j. It follows from the last two relations (2.3.78) that $\tau_{n_1}, \ldots, \tau_{n_k}$ are conditionally independent given $\xi_{n_1-1}, \xi_{n_1}, \ldots, \xi_{n_k-1}, \xi_{n_k}$.

2.3.6.1.4. Set

$$p_{ij}(t) = \mathsf{P}\,(\xi(s+t) = j \mid \xi(s) = i, \; \xi(s-) \neq i), \quad i, \; j \in X,$$

i.e. $p_{ij}(t)$ is the conditional probability of the event $(\xi(s+t) = j)$ given that a jump to i occurs at s. Considering the first jump following s we see that

$$p_{ij}(t) = \delta_{ij}(1 - \mathcal{S}_i(t)) + \sum_{k \in X} \int_0^t p_{kj}(t-u)\,\mathrm{d}F_{ik}(u), \quad i, j \in X. \tag{2.3.79}$$

In the Markovian case this system reduces to the Kolmogorov backward system. Consider the Laplace transforms

$$F_{ij}^*(\lambda) = \int_0^\infty \mathrm{e}^{-\lambda t}\,\mathrm{d}F_{ij}(t),$$

$$p_{ij}^*(\lambda) = \int_0^\infty \mathrm{e}^{-\lambda t}\,p_{ij}(t)\,\mathrm{d}t,$$

where $\lambda > 0$, and denote

$$\mathbf{F}^* = (F_{ij}^*)_{i,j \in X}, \quad \mathbf{P}^* = (p_{ij}^*)_{i,j \in X},$$

$$\Sigma = \Big(\sum_{k \in X} F_{ik}^*\,\delta_{ij}\Big)_{i,j \in X}.$$

The system (2.3.79) leads to the equation

$$\mathbf{P}^* = \lambda^{-1}[\mathbf{I} - \Sigma] + \mathbf{F}^*\mathbf{P}^*.$$

The properties of the system (2.3.79) are similar to those of the Kolmogorov backward system for Markov processes.

Theorem 2.3.51 (W. FELLER). *The system* (2.3.79) *possesses a minimal solution whose Laplace transform is*

$$\mathbf{P}^* = (\mathbf{I} + \mathbf{F}^* + (\mathbf{F}^*)^2 + \dots)(\mathbf{I} - \Sigma)\lambda^{-1}. \qquad (2.3.80)$$

This solution is unique iff it is stochastic. A necessary and sufficient condition for uniqueness is the existence of a $\lambda > 0$ *such that* $\mathbf{F}^*\mathbf{x} = \mathbf{x}$, $\mathbf{0} \leqslant \mathbf{x} \leqslant \mathbf{1}$ *implies* $\mathbf{x} = \mathbf{0}$. *(Here* $\mathbf{x}$ *and* $\mathbf{1}$ *are, respectively, the column-vectors* $(x_i)_{i \in X}$ *and* $(\delta_{ii})_{i \in X}$*).*

The minimal solution also satisfies the equation [65].

$$\mathbf{P}^* = \lambda^{-1}(\mathbf{I} - \Sigma) + \mathbf{P}^*(\mathbf{I} - \Sigma)^{-1}\mathbf{F}^*(\mathbf{I} - \Sigma).$$

Proof. See FELLER (1964). ◇

2.3.6.1.5. The asymptotic behaviour of the $p_{ij}(t)$ as $t \to \infty$ is closely related to the structure of the Markov chain $(\xi_n)_{n \in N}$ with transition matrix $(r_{ij})_{i,j \in X}$.

Theorem 2.3.52 (W. FELLER)[66]. *If all states of the Markov chain* $(\xi_n)_{n \in N}$ *are recurrent, then the minimal solution is stochastic. If this Markov chain is irreducible and recurrent, then for the (unique) solution of the system* (2.3.79) *we have*

$$\lim_{t \to \infty} p_{ij}(t) = a_j,$$

with

$$a_j = \begin{cases} \dfrac{u_j m_j}{\sum_{i \in X} u_i m_i} & \text{if } \sum_{i \in X} u_i m_i < \infty, \\ 0 & \text{if } \sum_{i \in X} u_i m_i = \infty, \end{cases} \qquad j \in X,$$

where

$$m_j = \int_0^\infty x \, d\mathcal{S}_j(x), \quad j \in X,$$

and $(u_j)_{j \in X}$ *is the unique (positive) solution of the system*[67].

$$u_j = \sum_{i \in X} u_i r_{ij}, \quad j \in X, \quad u_0 = 1.$$

Proof. See FELLER (1964). ◇

---

[65] This equation plays the role of the Kolmogorov forward system, but its structure and probabilistic meaning are far from obvious.

[66] For earlier less definitive results see FABENS (1961) and SMITH (1955).

[67] See Theorem 1.1.43.

For further results (concerning asymptotic properties and classification of states) we refer the reader to PYKE (1961 a, b), PYKE and SCHAUFELE (1964), and YACKEL (1966).

2.3.6.1.6. We close by mentioning a class of stochastic processes related to the semi-Markov processes. Set

$$\nu(t) = \sup\left\{n \geqslant 0: \sum_{i=0}^{n} \tau_i \leqslant t\right\},$$

and for $i \in X$ denote by $N_i(t)$ the number of indices k for which $\xi_k = i$, $0 < k \leqslant \nu(t)$. The process $(N_i)_{i \in X}$ is called a *Markov renewal process* determined by $(X, \mathbf{p}, \mathbf{F})$. Clearly

$$\nu(t) = \sum_{i \in X} N_i(t), \quad \xi(t) = \xi_{\nu(t)}.$$

A Markov renewal process is thus a process which records at each time t the number of times the process has been in each of the possible states up to t, if the movement from state to state is described by a Markov chain and if the time required for each successive move is a random variable whose distribution function may depend on the two states between which the move is being made.

Markov renewal processes were first investigated by PYKE (1961 a, b).

2.3.6.2. Renewal processes

2.3.6.2.1. Renewal processes are a mathematical model of the following real situation. Consider a population of individuals (industrial machinery, electric bulbs, etc.) such that when any individual fails, it is instantaneously replaced by a new one of age zero. The items live and fail independently of each other. Suppose that at time $t = 0$ the population consists of a single item whose residual life time $\tau_1 \geqslant 0$ has a distribution function K and that the items which successively replace the initial one have life times $\tau_2, \tau_3, \ldots \geqslant 0$, the random variables $\tau_1, \tau_2, \tau_3, \ldots$, being independent with

$$\mathsf{P}(\tau_k \leqslant t) = F(t), \quad t \geqslant 0, \ k \geqslant 2.$$

If we call a replacement a "renewal", then the renewals occur at times $\eta_1, \eta_2, \ldots$ where

$$\eta_n = \sum_{i=1}^{n} \tau_i, \quad n \in N^*.$$

and the random variable

$$N(t) = \max\{n: \eta_n \leqslant t\}, \quad t \geqslant 0,$$

is the number of renewals during the time interval $(0, t]$.

The process $(N(t))_{t\geqslant 0}$, $N(0) = 0$, is said to be a *renewal process* determined by (K, F).

It is easily seen that $(N(t))_{t\geqslant 0}$ is a semi-Markov process determined by $(X, \mathbf{p}, \mathbf{F})$ with

$$X = N,$$

$$p = (\delta_{i0})_{i\in N}$$

$$F_{ij}(t) = \begin{cases} K(t) & \text{if } i = 0,\ j = 1, \\ F(t) & \text{if } j - i = 1,\ i \neq 0, \\ 0 & \text{otherwise.} \end{cases}$$

It follows that in the case where

$$K(t) = F(t) = 1 - \mathrm{e}^{-\lambda t},$$

$(N(t))_{t\geqslant 0}$ is the Poisson process $P(\lambda)$.

Note also that in case $K = F$, the process $(N(t))_{t\geqslant 0}$ may also be thought of as the Markov renewal process determined by $(X, \mathbf{p}, \mathbf{F})$ where X reduces to a single point and $\mathbf{F} = (F)$. Thus, we now have a justification for the nomenclature Markov renewal process.

2.3.6.2.2. We shall ignore the case $\mathsf{P}(\tau_n = 0) = 1$, $n \geqslant 2$, corresponding to the situation in which items almost surely fail instantaneously. The random variable $N(t)$ is almost surely finite and has finite moments of all orders for all $t>0$. Indeed there is an $a > 0$, such that $\mathsf{P}(\tau_m \geqslant a) > 0$, $m \geqslant 2$, and without any loss of generality we can suppose $a = 1$; then $\mathsf{P}(\tau_m \geqslant 1) = p > 0$, $m \geqslant 2$. Now define

$$\bar{\tau}_m = \begin{cases} 0 \text{ if } \tau_m < 1, \\ 1 \text{ if } \tau_m \geqslant 1, \end{cases}$$

and set $\bar{\eta}_0 = \bar{\eta}_1 = 0$, $\bar{\eta}_m = \sum_{i=2}^{m} \bar{\tau}_i$, $m \geqslant 2$, $\overline{N}(t) = \max\{n \in N: \bar{\eta}_n \leqslant t\}$. Since $\bar{\eta}_n \leqslant \eta_n$ we have $N(t) \leqslant \overline{N}(t)$. Clearly, $\overline{N}(t) = \nu_0 + \ldots + \nu_{[t]}$, where the ν_i are integer-valued, independent, and identically distributed with

$$\mathsf{P}(\nu_0 = k) = p(1-p)^{k-1}, \quad k \in N^*.$$

Thus, the moment generating function of $\overline{N}(t)$ equals

$$\mathsf{E}(e^{\theta \overline{N}(t)}) = \Big(\sum_{k \in N^*} e^{k\theta} p(1-p)^{k-1} \Big)^{[t]+1} =$$

$$= \left(\frac{pe^{\theta}}{1-(1-p)\,e^{\theta}} \right)^{[t]+1}, \quad \operatorname{Re}\theta < \log \frac{1}{1-p}.$$

It follows that $\mathsf{P}(\overline{N}(t) < \infty) = 1$ and $\mathsf{E}[\overline{N}(t)]^k < \infty$, $k \in N^*$. Therefore $N(t)$ will have the same properties.

2.3.6.2.3. Set

$$F_n(t) = \mathsf{P}(\gamma_n \leqslant t), \; n \in N^*,$$

$$F_0(t) \equiv 1, \; t \geqslant 0.$$

Clearly

$$F_1 = K,$$

$$F_{n+1}(t) = \int_0 F_n(t-u)\, \mathrm{d}F(u), \; n \in N^*.$$

A straightforward computation yields

$$\mathsf{P}(N(t) = n) = F_n(t) - F_{n+1}(t), \quad n \in N,$$

$$U(t) = \mathsf{E}(N(t)) = \sum_{n \in N^*} F_n(t).$$

The function U, which has been shown to be finite, is called the *renewal function*. A huge literature is devoted to this function and to renewal theory and its applications. The reader is referred to Cox (1962), Feller (1966 b, Ch. 11), Smith (1958), and Prabhu (1965 a, Chs. 5,6). Here we state some of the best known results.

Theorem 2.3.53. *The renewal function U satisfies the integral equation*

$$U(t) = K(t) + \int_0^t U(t-u)\, \mathrm{d}F(u). \tag{2.3.81}$$

Moreover, U is the unique solution of (2.3.81) *which is bounded on any finite interval.*

Proof. See e.g. Prabhu (1965 a, p. 157). ◇

Denote

$$\int_0^{\infty} t \, \mathrm{d}F(t) = \mu,$$

if the integral is convergent, and $\mu = \infty$ otherwise.

Theorem 2.3.54 (W. FELLER). *We have*

$$\lim_{t \to \infty} \frac{U(t)}{t} = \frac{1}{\mu},$$

the limit being interpreted as zero when $\mu = \infty$.

Proof. See e.g. PRABHU (1965 a, p. 160). ◇

References

ACZÉL, J.

(1966) Lectures on functional equations and their applications. New York: Academic Press.

ADKE, S. R.

(1964) The generalized birth and death process and Gaussian diffusion. J. Math. Anal. Appl. **9**, 336–340.

ARBIB, M. A.

(1965) Hitting and martingale characterizations of one-dimensional diffusions. Z. Wahrscheinlichkeitstheorie **4**, 232–247.

ARLEY, N.

(1943) On the theory of stochastic processes and their applications to the theory of cosmic radiation. Copenhagen: G. F. C. Gads.

(1967) On the general birth-and-death-with-immigration stochastic process. Skand. Aktuarietidskr. 175–182.

ATHREYA, K. B.

(1968) Some results on multitype continuous time Markov branching processes. Ann. Math. Statist. **39**, 347–357.

(1969 a) On the equivalence of conditions on a branching process in continuous time and on its offspring distribution. J. Math. Kyoto Univ. **9**, 41–53.

(1969 b) Limit theorems for multitype continuous time Markov branching processes. I, II Z. Wahrscheinlichkeitstheorie. **12**, 320–332; **13**, 204–214.

ATHREYA, K. B., and S. KARLIN

(1967) Limit theorems for the split times of branching processes, J. Math. Mech. **17**, 257–277.

(1968) Embedding of urn schemes into continuous time Markov branching processes and related limit theorems. Ann. Math. Statist. **39**, 1801–1817.

AUSTIN, D. G.

(1959) The generalized backward Kolmogorov equation in abstract space. Illinois J. Math. **3**, 532–537.

BACHELIER, L.

(1900) Théorie de la spéculation. Ann. Sci. École Norm. Sup. **17**, 21–86.

BAILEY, N. T. J.

(1957) The mathematical theory of epidemics. London: Griffin.

BARTLETT, M. S.
(1966) An introduction to stochastic processes, with special reference to methods and applications. 2nd Edition. London—New York: Cambridge Univ. Press.

BENEŠ, V. E.
(1967) Existence of finite invariant measures for Markov processes. Proc. Amer. Math. Soc. **18**, 1058—1061.

BEUTLER, F. J., and O. A. Z. LENEMAN
(1966) The theory of stationary point processes. Acta Math. **116**, 159—197.

BHARUCHA-REID, A. T.
(1960) Elements of the theory of Markov processes and their applications. New York—Toronto—London: McGraw-Hill.

BILLINGSLEY, P.
(1968) Convergence of probability measures. New York—London—Sydney: Wiley.

BLACKWELL, D. and D. FREEDMAN
(1968) On the local behaviour of Markov transition probabilities. Ann. Math. Statist. **39**, 2123—2127.

BLUMENTHAL, R. M., and R. K. GETOOR
(1968) Markov processes and potential theory. New York—London: Academic Press.

BORGES, R.
(1966) Zur Existenz von separablen stochastischen Prozessen. Z. Wahrscheinlichkeitstheorie **6**, 125—128.

BOROVKOV, A. A.
(1970) Theorems on the convergence to Markov diffusion processes. Z. Wahrscheinlichkeitstheorie **16**, 47—76.

BREIMAN, L.
(1963) The Poisson tendency in traffic distribution. Ann. Math. Statist. **34**, 308—311.

BRODI, S. M.
(1965) Investigation of queueing systems by means of semi-Markov processes. Kibernetika (Kiev), no. 6, 55—58. (Russian)

ČENCOV, N. N.
(1956) Weak convergence of stochastic processes whose trajectories have no discontinuities of the second kind. Teor. Verojatnost. i Primenen. **1**, 154—161. (Russian)

ČERKASOV, I. D.
(1957) On the equations of Kolmogorov. Uspehi Mat. Nauk (N.S.) **12**, no. 5 (77), 237—244. (Russian)
(1962) On a means of finding transition probabilities for Brownian motion. Rev. Math. Pures Appl. **7**, 369—382.

CHEONG, C. K.
(1967) Geometric convergence of semi-Markov transition probabilities. Z. Wahrscheinlichkeitstheorie **7**, 122—130.

(1970) Quasi-stationary distributions in semi-Markov processes. J. Appl. Probability **7**, 388—399 and 788.

CHUNG, K. L.

(1963) On the boundary theory for Markov chains. Acta Math. **110**, 19—77.

(1966) On the boundary theory for Markov chains. II. Acta Math. **115**, 111—163.

(1967) Markov chains with stationary transition probabilities. 2nd Edition. Berlin — Heidelberg — New York : Springer.

ÇINLAR, E.

(1969 a) On semi-Markov processes on arbitrary spaces. Proc. Cambridge Philos. Soc. **66**, 381—392.

(1969 b) Markov renewal theory. Adv. Appl. Probability **1**, 123—187.

COHEN, J. W.

(1962) Derived Markov chains. I—III. Indag. Math. **24**, 55—82.

(1962) A note on skeleton chains. Nieuw Arch. Wisk. (3) **10**, 180—186.

CONNER, H.

(1967) A note on limit theorems for Markov branching processes. Proc. Amer. Math. Soc. **18**, 76—86.

COX, D. R.

(1955 a) A use of complex probabilities in the theory of stochastic processes. Proc. Cambridge Philos. Soc. **51**, 313—319.

(1955 b) The analysis of non-Markovian stochastic processes by the inclusion of supplementary variables. Proc. Cambridge Philos. Soc. **51**, 433—441.

(1962) Renewal theory. London: Methuen.

COX, D. R., and P. A. W. LEWIS

(1966) Statistical analysis of series of events. London: Methuen.

COX, D. R., and H. D. MILLER

(1965) The theory of stochastic processes. London: Methuen.

CRAMÉR, H.

(1951) A contribution to the theory of stochastic processes. Proc. 2nd Berkeley Symp. Math. Statist. Prob., pp. 329—340. Berkeley: Univ. California Press.

(1966) On stochastic processes whose trajectories have no discontinuities of the second kind. Ann. mat. pura appl. **71**, 85—92.

CRUMP, K. S., and C. J. MODE

(1968—69) A general age-dependent branching process. I, II. J. Math. Anal. Appl. **24**, 494—508; **25**, 8—17.

CUCULESCU, I.

(1968) Markov processes and excessive functions. Bucharest: Publishing House of the Academy. (Romanian)

DANIELS, H. E.

(1963) The Poisson process with a curved absorbing boundary. Bull. Inst. Internat. Statist. **40**, Book 2, 994—1008.

DARLING, D. A., and A. J. F. SIEGERT

(1953) The first passage problem for a continuous Markov process. Ann. Math. Statist. **24**, 624—639.

DARROCH, J. N., and E. SENETA
(1967) On quasi-stationary distributions in absorbing continuous-time finite Markov chains. J. Appl. Probability **4**, 192–196.

DAVIDSON, R.
(1968) Arithmetic and other properties of certain Delphic semi-groups. I, II. *Z.* Wahrscheinlichkeitstheorie **10**, 120–172.
(1969) More Delphic theory and practice. Z. Wahrscheinlichkeitstheorie **13**, 191–203.

DAVIS, A. W.
(1964) On the probability generating functional for the cumulative population in a simple birth-and-death process. Biometrika **51**, 245–249.
(1965) On the theory of birth, death and diffusion processes. J. Appl. Probability **2**, 293–322.

DAWSON, D. A.
(1966) Potential theory and non-markovian chains. Z. Wahrscheinlichkeitstheorie **5**, 118–138.

DOOB, J. L.
(1953) Stochastic processes. New York: Wiley.
(1968) Compactification of the discrete state space of a Markov process. Z. Wahrscheinlichkeitstheorie **10**, 236–251.
(1969) An application of stochastic process separability. Enseignement Math. (2), **15**, 101–105.

DUDLEY, R. M.
(1965) Gaussian processes on several parameters. Ann. Math. Statist. **36**, 771–788.

DUGUÉ, D.
(1967) Fonctions caractéristiques d'intégrales browniennes. Rev. Roumaine Math. Pures Appl. **12**, 1207–1215.

DVORETZKY, A.
(1963) On the oscillation of the Brownian motion process. Israel J. Math. **1**, 212–214.

DVORETZKY, A., P. ERDÖS, and S. KAKUTANI
(1950) Double points of Brownian motion in *n*-space. Acta Sc. Math. (Szeged) **12**, 75–81.
(1954) Multiple points of paths of Brownian motion in the plane. Bul'. Res. Council Israel **3**, 364–371.
(1958) Points of multiplicity c of plane Brownian motion paths. Bull. Res. Council Israel **7 F**, 175–180.
(1961) Nonincrease everywhere of the Brownian motion process. Proc. 4th Berkeley Symp. Math. Statist. Prob. Vol. II, pp. 103–116. Berkeley: Univ. California Press.

DVORETZKY, A., P. ERDÖS, S. KAKUTANI, and S. J. TAYLOR
(1957) Triple points of Brownian paths in 3-space. Proc. Cambridge Philos. Soc. **53**, 856–862.

DYNKIN, E. B.
(1961) Foundations of the theory of Markov processes. Oxford—London—Paris: Pergamon Press.
(1965) Markov processes. Vols. I. II. New York—Berlin—Göttingen—Heidelberg: Academic Press & Springer.
(1967) General boundary conditions for Markov processes with a countable set of states. Teor. Verojatnost. Primenen. **12**, 222—257. (Russian)

DYNKIN, E. B., and A. A. JUŠKEVIČ
(1967) Theorems and problems in Markov processes. Moscow: Nauka. (Russian)

ERDÖS, P., and S. J. TAYLOR
(1961) On the Hausdorff measure of Brownian paths in the plane. Proc. Cambridge Philos. Soc. **57**, 209—222.

FABENS, A. J.
(1961) The solution of queueing and inventory models by semi-Markov processes. J. Roy. Statist. Soc. Ser. B. **23**, 113—127.

FELLER, W.
(1936) Zur Theorie der stochastischen Prozesse (Existenz und Eindeutigkeitssätze.) Math. Ann. **113**, 113—160.
(1940) On the integro-differential equations of purely discontinuous Markov processes. Trans. Amer. Math. Soc. **48**, 488—515. Errata ibid (1945), **58**, 474.
(1951) Diffusion processes in genetics. Proc. 2nd Berkeley Symp. Math. Statist. Prob. pp. 227—246. Berkeley: Univ. California Press.
(1952) The parabolic differential equation and the associated semi-groups of transformations. Ann. Math. **55**, 468—519.
(1954 a) Diffusion processes in one dimension. Trans. Amer. Math. Soc. **77**, 1—31.
(1954 b) On boundaries and lateral conditions for the Kolmogorov differential equations. Ann. Math. **65**, 527—570.
(1959) The birth and death processes as diffusion processes. J. Math. Pures Appl. (9), **38**, 301—345.
(1964) On semi-Markov processes. Proc. Nat. Acad. Sci. U.S.A. **51**, 635—659.
(1966 a) On the Fourier representation for Markov chain and the strong ratio theorem. J. Math. Mech. **15**, 273—283.
(1966 b) An introduction to probability theory and its applications. Vol. 2, New York: Wiley.
(1968) An introduction to probability theory and its applications. Vol. 1, 3rd Edition. New York: Wiley.

FRÉCHET, M.
(1938) Recherches théoriques modernes sur le calcul des probabilités. Vol. II, Méthode des fonctions arbitraires. Théorie des événements en chaîne dans le cas d'un nombre fini d'états possibles. Paris: Hermann.

FREEDMAN, D. A.
(1967 a) An oscillating semigroup. Ann. Math. Statist. **38**, 924—926.
(1967 b) A theorem of Lévy and a peculiar semigroup. Ann. Math. Statist. **38**, 1552—1557.

(1968 a) On the convergence of approximants to a Markov chain. I, II. Proc. Nat. Acad. Sci. U.S.A. **60**, 66–72; 446–449.

(1968 b) On the derivative of a semigroup. Bull. Amer. Math. Soc. **74**, 749–751.

FREĬDLIN, M. I.

(1967) Markov processes and differential equations. Theory of Probability. Mathematical Statistics. Theoretical Cybernetics. 1966, pp. 7–58. Moscow: Akad Nauk SSSR. (Russian).

FRIEDMAN, A.

(1958) Boundary estimates for second order parabolic equations and their applications. J. Math. Mech. **7**, 771–791.

FRISTEDT, B. E.

(1967 a) Sample function behaviour of increasing processes with stationary independent increments. Pacific J. Math. **21**, 21–33.

(1967b) An extension of a theorem of S. J. Taylor concerning the multiple points of the symmetric stable process. Z. Wahrscheinlichkeitstheorie **9**, 62–64.

FUKUSHIMA, M.

(1966) On spectral functions related to birth-and-death processes. J. Math. Kyoto Univ. **5**, 151–161.

GANGOLLI, R.

(1967) Positive definite kernels on homogeneous spaces and certain stochastic processes related to Lévy's Brownian motion of several parameters. Ann. Inst. H. Poincaré Sect. B (N.S.) **3**, 121–226.

GIHMAN, I. I.

(1969) Convergence to Markov processes. Ukrain. Mat. Ž. **21**, 316–324. (Russian)

GIHMAN, I. I., and A. V. SKOROHOD

(1965) Introduction to the theory of random processes. Moscow: Nauka. (Russian).

(1968) Stochastic differential equations. Kiev: Naukova Dumka. (Russian)

GOLDMAN, J. R.

(1967 a) Stochastic point processes; limit theorems. Ann. Math. Statist. **38**, 771–779.

(1967 b) Infinitely divisible point processes in R^n. J. Math. Anal. Appl. **17**, 133–146.

GOODMAN, G. S.

(1970) An intrinsic time for non-stationary finite Markov chains. Z. Wahrscheinlichkeitstheorie **16**, 165–180.

GOODMAN, L. A.

(1967) The probabilities of extinction for birth-and-death processes that are age-dependent or phase-dependent. Biometrika **54**, 579–596.

HARLAMOV, B. P.

(1968) Some properties of branching processes with an arbitrary set of particle types. Teor. Verojatnost. i Primenen. **13**, 82–95. (Russian)

HARRIS, T. E.

(1963) The theory of branching processes. Berlin – Göttingen – Heidelberg: Springer.

HAS'MINSKI, R. Z.

(1962) Probabilistic representation of the solutions to some differential equations. Proc. 6th All-Union Conf. Theory Prob. and Math. Statist. (Vilnius 1960) pp. 177–182, Vilnius; Gos. Izdat. Polit. Naučn. Lit. Litovsk. SSR. (Russian)

HASOFER, A. M.
(1970) On the representation of ignorance in Poisson processes. J. Roy. Statist. Soc. Ser. B, **32**, 268–271.

HEATHCOTE, C. R., and J. E. MOYAL
(1959) The random walk (in continuous time) and its application to the theory of queues. Biometrika **46**, 400–411.

HELMS, L. L.
(1967) Biharmonic functions and Brownian motion. J. Appl. Probability **4**, 130–136.
(1970) Markov processes with creation of mass. II. Z. Wahrscheinlichkeitstheorie **15**, 208–218.

HUNT, G. A.
(1957–58) Markov processes and potentials. I–III. Illinois. J. Math. **1**, 44–93, 316–369; **2**, 151–213.
(1960) Markov chains and Martin boundaries. Illinois J. Math. **4**, 313–340.
(1966) Martingales et processus de Markov. Paris: Dunod.

IGLEHART, D. L.
(1968 a) Limit theorems for multi-urn Ehrenfest model. Ann. Math. Statist. **39**, 864–876.
(1968 b) Diffusion approximations in applied probability, pp. 235–254 in *Mathematics of the decision sciences*, Part 2. Providence: Amer. Math. Society.

IKEDA, N., M. NAGASAWA, and S. WATANABE
(1968–69) Branching Markov processes. I–III. J. Math. Kyoto Univ. **8**, 233–278; 365–410; **9**, 95–160.

IL'IN, A. M., A. S. KALAŠNIKOV, and O. A. OLEĬNIK
(1962) Second-order linear equations of parabolic type. Uspehi Mat. Nauk **17**, no. 3 (105), 3–146. (Russian)

ITÔ, K.
(1942) On stochastic processes. Jap. J. Math. **18**, 261–301.
(1960) Stochastic processes. Vol. 1. Moscow: Izdatel'stvo Inostrannoĭ Literatury. (Russian)
(1963) Stochastic processes. Vol. 2. Moscow: Izdatel'stvo Inostrannoĭ Literatury. (Russian)
(1968) The canonical modification of stochastic processes. J. Math. Soc. Japan **20**, 130–150.

ITÔ, K., and H. P. MCKEAN JR.
(1965) Diffusion processes and their sample paths. Berlin–Heidelberg–New York: Springer.

JACOBSEN, M.
(1972) A characterization of minimal Markov jump processes. Z. Wahrscheinlichkeitstheorie **23**, 32–46.

JAGERS, P.
(1967) Renewal theory and the almost sure convergence of branching processes. Ark. Math. **7**, 495–504.
(1968) Age-dependent branching processes allowing immigration. Teor. Verojatnost. i Primenen. **13**, 230–242.

JANSSEN, J.
(1964–65) Processus de renouvellements Markoviens et processus semimarkoviens. Cahiers Centre Etudes Recherche Opér. **6**, 81–105; **7**, 126–141.
JENSEN, A., and D. KENDALL
(1971) Denumerable Markov processes with bounded generators: a routine for calculating $p_{ij}(\infty)$. J. Appl. Probability **8**, 423–427.
JIŘINA, M.
(1967) General branching processes with continuous time parameter. Proc. 5th Berkeley Symp. Math. Statist. Prob. Vol. II, Part 1, pp. 389–399. Berkeley: Univ. California Press.
JOHN, P. W. M.
(1961) A note on the quadratic birth process. J. London Math. Soc. **36**, 159–160.
JUŠKEVIČ, A. A.
(1957) On strong Markov processes. Teor. Verojatnost. i Primenen. **2**, 187–213. (Russian)
(1959) On differentiability of transition probabilities of homogeneous Markov processes with a countable number of states. Učenye Zapiski MGU **186**, Mat. 9, 141–160. (Russian)
(1967) Some remarks on boundary conditions for birth and death processes. Trans. 4th Prague Conf. Information Theory, Statist. Decision Functions, Random Processes, pp. 381–387. Prague: Academia. (Russian)
KAC, M.
(1947) Random walk and the theory of Brownian motion. Amer. Math. Monthly **54**, 369–391.
(1951) On some connections between probability theory and differential and integral equations. Proc. 2nd Berkeley Symp. Math. Statist. Prob. pp. 189–215. Berkeley: Univ. California Press.
(1952) An application of probability theory to the study of Laplace equation. Roczn. Polsk. towarz. mat. **25**, 122–130.
(1959) Probability and related topics in physical sciences. London–New York: Interscience.
KARLIN, S.
(1966) A first course in stochastic processes. New York–London: Academic Press.
KARLIN, S., and J. MCGREGOR
(1957 a) The differential equations of birth and death processes and the Stieltjes moment problem. Trans. Amer. Math. Soc. **85**, 489–546.
(1957 b) The classification of birth and death processes. Trans. Amer. Math. Soc. **86**, 366–400.
(1958 a) Many server queueing processes with Poisson input and exponential service times. Pacific J. Math. **8**, 87–118.
(1958 b) Linear growth, birth and death processes. J. Math. Mech. **7**, 643–662.
(1959 a) A characterization of birth and death processes. Proc. Nat. Acad. Sci. U.S.A. **445**, 375–379.
(1959 b) Coincidence properties of birth and death processes. Pacific J. Math. **9**, 1109–1140.
(1959 c) Coincidence probabilities. Pacific J. Math. **9**, 1141–1164.

(1960) Classical diffusion processes and total positivity. J. Math. Anal. Appl. **1**, 163–183.

(1961) Occupation time laws for birth and death processes. Proc. 4th Berkeley Symp. Math. Statist. Prob. Vol. II, pp. 249–272. Berkeley: Univ. California Press.

(1968) Embeddability of discrete time simple branching processes into continuous time branching processes. Trans. Amer. Math. Soc. **132**, 115–136.

KATO, T.

(1954) On the semigroups generated by Kolmogoroff's differential equations. J. Math. Soc. Japan **6**, 1–15.

KEILSON, J.

(1964–65) A review of transient behaviour in regular diffusion and birth-death processes. I, II J. Appl. Probability **1**, 247–266; **2**, 405–428.

KEMENY, J. G., and L. J. SNELL

(1961) Finite continuous time Markov chains. Teor. Verojatnost. i Primenen. **6**, 110–115.

KEMPERMAN, J. H. B.

(1962) An analytical approach to the differential equations of the birth-and-death process. Michigan Math. J. **9**, 321–361.

(1963) The semigroup property for the birth and death process. J. Reine Angew. Math. **212**, 80–103.

KENDALL, D. G.

(1948) On the generalized birth and death process. Ann. Math. Statist. **19**, 1–15.

(1955) Some analytical properties of continuous stationary Markov transition functions. Trans. Amer. Math. Soc. **78**, 529–540.

(1956) Some further pathological examples in the theory of denumerable Markov processes. Quart. J. Math. Oxford (Ser. 2), **7**, 39–56.

(1959) Unitary dilations of one-parameter semigroups of Markov transition operators, and the corresponding integral representations for Markov processes with a countable infinity of states. Proc. London Math. Soc. (3) **9**, 417–431.

(1960 a) Geometric ergodicity and the theory of queues, pp. 176–195 in K. J. Arrow, S. Karlin and P. Suppes (Eds.) *Mathematical methods in the social sciences, 1959*. Stanford: Stanford University Press.

(1960 b) Hyperstonian spaces associated with Markov chains. Proc. London Math. Soc. (3) **10**, 67–87.

(1963) Hyperstonian spaces associated with Markov chains (II). J. London Math. Soc. **38**, 89–90.

(1965) On the behaviour of a standard Markov transition function near $t = 0$. Z. Wahrscheinlichkeitstheorie **3**, 276–278.

(1966) Branching processes since 1873. J. London Math. Soc. **41**, 385–406.

(1967 a) On Markov groups. Proc. 5th Berkeley Symp. Math. Statist. Prob. Vol. II, Part 2, pp. 165–173. Berkeley: Univ. California Press.

(1967 b) Some recent developments in the theory of denumerable Markov processes. Trans. 4th Prague Conf. Information Theory, Statist. Decision Functions, Random Processes, pp. 11–27. Prague: Academia.

(1968) Delphic semi-groups, infinitely divisible regenerative phenomena, and the arithmetic of p-functions. Z. Wahrscheinlichkeitstheorie **9**, 163–195.

KENDALL, D. G., and G. E. H. REUTER
(1956) Some pathological Markov processes with a denumerable infinity of states and the associated semigroups of operators in *l*. Proc. International Congress of Mathematicians, 1954, Vol. III, pp. 377–415. Amsterdam: North-Holland Publishing Co.
(1957) The calculation of the ergodic projection for Markov chains and processes with a countable infinity of states. Acta Math. **97**, 103–144.
KERRIDGE, D.
(1964) Probabilistic solution of the simple birth process. Biometrika **51**, 258–259.
KIMURA, M.
(1964) Diffusion models in population genetics. J. Appl. Probability **1**, 177–232.
KINGMAN, J. F. C.
(1962) The imbedding problem for finite Markov chains. Z. Wahrscheinlikeitstheorie **1**, 14–24.
(1963 a) The exponential decay of Markov transition probabilities. Proc. London Math. Soc. **13**, 337–358.
(1963 b) Ergodic properties of continuous time Markov processes and their discrete skeletons, Proc. London Math. Soc. **13**, 593–604.
(1963 c) On continuous time models in the theory of dams. J. Australian Math. Soc. **3**, 480–487.
(1963 d) A continuous time analogue of the theory of recurrent events. Bull. Amer. Math. Soc. **69**, 268–272.
(1964 a) The stochastic theory of regenerative events. Z. Wahrscheinlichkeitstheorie **2**, 180–224.
(1964 b) On doubly stochastic Poisson processes. Proc. Cambridge Philos. Soc. **60**, 923–930.
(1965 a) Linked systems of regenerative events. Proc. London Math. Soc. **15**, 125–150.
(1965 b) Some further analytical results in the theory of regenerative events. J. Math. Anal. Appl. **11**, 422–433.
(1967 a) Completely random measures. Pacific J. Math. **21**, 59–78.
(1967 b) Markov transition probabilities I. Z. Wahrscheinlichkeitstheorie **7**, 248–270.
(1967 c) Markov transition probabilities II. Completely monotonic functions. Z. Wahrscheinlichkeitstheorie **9**, 1–9.
(1968 a) Markov transition probabilities III. General state spaces. Z. Wahrscheinlichkeitstheorie **10**, 87–101.
(1968 b) Markov transition probabilities IV. Recurrence time distributions. Z. Wahrscheinlichkeitstheorie **11**, 9–17.
(1968 c) Measurable *p*-functions. Z. Wahrscheinlichkeitstheorie **11**, 1–8.
(1970) Stationary regenerative phenomena. Z. Wahrscheinlichkeitstheorie **15**, 1–18.
(1971) Markov transition probabilities. V. Z. Wahrscheinlichkeitstheorie **17**, 89–103.
KINNEY, J. R.
(1966) Convex hull of Brownian motion in *d*-dimensions. Israel J. Math. **4**, 139–143.

KNAPP, A. W.

(1965) Connection between Brownian motion and potential theory. J. Math. Anal. Appl. **12**, 328–349.

KNIGHT, F. B.

(1967) Equivalence of probabilistic solutions of heat equation. Rev. Roumaine Math. Pures Appl. **12**, 1305–1309.

KOLMOGOROV, A. N.

(1931) Über die analytischen Methoden in der Wahrscheinlichkeitsrechnung. Math. Ann. **104**, 415–458.

(1951) On the differentiability of the transition probabilities of stationary Markov processes with a denumerable number of states. Učenye Zapiski MGU **148**, 53–59. (Russian)

KOLMOGOROV, A. N., and JU. V. PROHOROV

(1956) Zufällige Funktionen und Grenzverteilungssätze. Bericht über die Tagung Wahrscheinlichkeitsrechnung und mathematische Statistik in Berlin, Oktober 1954, pp. 113–126. Berlin; Deutscher Verlag der Wissenschaften.

KRYLOV, N. V.

(1966) On quasi-diffusion processes. Teor. Verojatnost. i Primenen. **11**, 424–443. (Russian)

KURTZ, T. G.

(1969) A note on sequences of continuous parameter Markov chains. Ann. Math. Statist. **40**, 1078–1082.

(1971) Comparison of semi-Markov and Markov processes. Ann. Math. Statist. **42**, 991–1002.

LADOHIN, V. I.

(1963) The distribution of the instant of first attainment of the maximum for a trajectory of a Wiener process. Kazan. Gos. Univ. Učen. Zap. **123**, kn. 6, 43–55. (Russian)

LAHRES, H.

(1964) Einführung in die diskreten Markoff-Prozesse und ihre Anwendungen. Braunschweig: Fr. Vieweg & Sohn.

LAMBOTTE, J. P.

(1968) Processus semi-Markoviens et files d'attente. Cahiers Centre Etudes Recherche Opér. **10**, 21–31.

LEADBETTER, M. R.

(1968) On three basic results in the theory of stationary point processes. Proc. Amer. Math. Soc. **19**, 115–117.

LEDERMANN, W., and G. E. H. REUTER

(1954) Spectral theory for the differential equations of simple birth and death processes. Philos. Trans. Roy. Soc. London Ser. A, **246**, 321–369.

Lévy P.
(1965) Processus stochastiques et mouvement Brownien. 2e Edition. Paris: Gauthier-Villars.
(1967) Remarques sur les états instantanés des processus markoviens et stationnaires, à une infinité dénombrable d'états possibles. C. R. Acad. Sci. Paris Sér. A – B **264**, A844 – A848.
(1969) Conjectures relatives aux points multiples de certaines variétés. Rev. Roum. Math. Pures Appl. **14**, 819 – 827.

Leysieffer, F. V.
(1967) Functions of finite Markov chains. Ann. Math. Statist. **38**, 206 – 212.

Loève, M.
(1963) Probability theory. 3rd Edition, Princeton – Toronto – New York – London: Van Nostrand.

Mandl, P.
(1964) An elementary proof of the ergodic property of birth and death processes. Časopis Pěst. Mat. **89**, 354 – 358. (Czech)
(1968) Analytical treatment of one-dimensional Markov processes. Prague – Berlin – Heidelberg – New York: Academia & Springer.

Marczewski, E.
(1953) Remarks on the Poisson stochastic process II. Studia Math. **13**, 130 – 136.

McFadden, J. A.
(1965) The mixed Poisson process. Sankhyā Ser. A. **27**, 83 – 92.

McKean Jr., H. P.
(1963) Brownian motion with a several-dimensional time. Teor. Verojatnost. Primenen. **8**, 357 – 378.
(1969) Stochastic integrals. New York – London: Academic Press.

Meyer, P.-A.,
(1963) Brelot's axiomatic theory of the Dirichlet problem and Hunt's theory. Ann. Inst. Fourier (Grenoble) **13**, 357 – 372.
(1966) Probabilités et potentiel. Paris: Hermann.
(1967) Processus de Markov. Lecture Notes in Mathematics No. 26. Berlin – New York: Springer.
(1968) Guide détaillé de la théorie «générale» des processus. Séminaire de Probabilités (Univ. Strasbourg. 1967), Vol. II, pp. 140 – 165. Berlin: Springer.

Milch, P. R.
(1968) A multi-dimensional linear growth birth and death process. Ann. Math. Statist. **39**, 727 – 754.

Miller Jr., R. G.
(1963) Stationarity equations in continuous time Markov chains. Trans. Amer. Math. Soc. **109**, 35 – 44.

Mode, C. J.
(1971) Multitype branching processes. Theory and applications. New York: American Elsevier.

Moyal, J. E.
(1957) Discontinuous Markoff processes. Acta Math. **98**, 221 – 264.

NELSON, E.
(1967) Dynamical theories of Brownian motion. Princeton: Princeton Univ. Press.
NEUTS, M. F.
(1964) Generating functions for Markov renewal processes. Ann. Math. Statist. **35**, 431–434.
NEVEU, J.
(1961) Lattice methods and submarkovian processes. Proc. 4th Berkeley Symp. Math. Statist. Prob. Vol. II, pp. 347–391. Berkeley: Univ. California Press.
(1968) Sur la structure des processus ponctuels stationnaires. C.R. Acad. Sci. Paris Sér. A–B **267**, A 561–A564.
NORMAN, M. F.
(1972) Markov processes and learning models. New York – London: Academic Press.
ONICESCU, O.
(1960) Sur les équations de la théorie des probabilités. Bull. Math. Soc. Sci. Math. Phys. R. P. Roumaine (N.S) **4 (52)**, 61–86.
ONICESCU, O., G. MIHOC, and C. T. IONESCU TULCEA
(1956) Probability theory and applications. Bucharest: Publishing House of the Academy. (Romanian)
OREY, S.
(1962) Non-differentiability of absolute probabilities of Markov chains. Quart. J. Math. Oxford (Ser. 2), **13**, 252–254.
PARTHASARATHY, K. R.
(1967) Probability measures on metric spaces. New York–London: Academic Press.
PARZEN, E.
(1962) Stochastic processes. San Francisco: Holden-Day.
PRABHU, N. U.
(1965 a) Queues and inventories. New York: Wiley.
(1965 b) Stochastic processes. New York–London: Macmillan.
PROHOROV, JU. V.
(1956) Convergence of stochastic processes and limit theorems in the theory of probability. Teor. Verojatnost. i Primenen. **1**, 177–238. (Russian)
(1961) The method of characteristic functionals. Proc. 4th Berkeley Symp. Math. Statist. Prob. Vol. II, pp. 403–419. Berkeley: Univ. California Press.
PRUITT, W. E.
(1963) Bilateral birth-and-death processes. Trans. Amer. Math. Soc. **107**, 508–525.
PURI, P. S.
(1968) Some further results on the birth-and-death process and its integral. Proc. Cambridge Philos. Soc. **64**, 141–154.
PYKE, R.
(1959) The supremum and infimum of the Poisson process. Ann. Math. Statist. **30**, 568–576.
(1961 a) Markov renewal processes: definitions and preliminary properties. Ann. Math. Statist. **32**, 1231–1242.

(1961 b) Markov renewal processes with finitely many states. Ann. Math. Statist. **32**, 1243–1259.

PYKE, R., and R. SCHAUFELE

(1964) Limit theorems for Markov renewal processes. Ann. Math. Statist. **35**, 1746–1764

(1966) The existence and uniqueness of stationary measures for Markov renewal processes. Ann. Math. Statist. **37**, 1439–1462.

RAO, M., and H. WEDEL

(1969) Poisson processes as renewal processes invariant under translations. Ark. Math. **7**, 539–541.

RÉNYI, A

(1962) Wahrscheinlichkeitsrechnung mit einem Anhang über Informationstheorie. Berlin: Deutscher Verlag der Wissenschaften.

(1964) On an extremal property of the Poisson process. Ann. Inst. Statist. Math. **16**, 129–133.

(1967 a) Remarks on the Poisson process. Symposium on Probability Methods in Analysis (Loutraki, 1966), pp. 280–286. Berlin–Heidelberg–New York: Springer.

(1967 b) Probabilistic methods in analysis. I, II. Mat. Lapok **18**, 5–35: 175–194. (Hungarian)

REUTER, G. E. H.

(1957) Denumerable Markov processes and the associated contraction semigroups on *l*. Acta Math. **97**, 1–46.

(1959) Denumerable Markov processes. II. J. London Math. Soc. **34**, 81–91.

(1962) Denumerable Markov processes. III. J. London Math. Soc. **37**, 63–73.

(1967) Nul solution of the Kolmogorov differential equations. Mathematika **14**, 56–61.

(1969) Remarks on a Markov chain example of Kolmogorov. Z. Wahrscheinlichkeitstheorie **13**, 315–321.

REUTER, G. E. H., and W. LEDERMANN

(1953) On the differential equations for the transition probabilities of Markov processes with enumerably many states. Proc. Cambridge Philos. Soc. **49**, 247–262.

ROSENBLATT, M.

(1962) Random processes. New York: Oxford Univ. Press.

(1965) Functions of Markov processes. Z. Wahrscheinlichkeitstheorie **5**, 232–243.

RUNNENBURG, J. T.

(1962) On Elfving's problem of imbedding a time-discrete Markov chain in a time-continuous one for finitely many states. I. Indag. Math. **24**, 536–541.

RYLL-NARDZEWSKI, C.

(1953) On the non-homogeneous Poisson process (I). Studia Math. **14**, 124–128.

(1954) Remarks on the Poisson stochastic process (III). Studia Math. **14**, 314–318.

SACK, R. A.

(1959) Equivalence of two absorption problems with Markovian transitions and continuous or discrete time parameters. Proc. Cambridge Philos. Soc. **55**, 177–180.

SAKS, S.
(1937) Theory of the integral. New York: Haffner.
SATO, K., and T. UENO
(1964—65) Multi-dimensional diffusion and the Markov process on the boundary. J. Math. Kyoto Univ. 4, 529—605.
SCHEFFER, C. L.
(1962) On Elfving's problem of imbedding a time-discrete Markov chain in a time-continuous one for finitely many states. II. Indag. Math. 24, 542—548.
SENČENKO, D. V.
(1966) Unique determination of Markov processes with a finite number of states. Mat. Sb. (N.S.) 71 (113), 30—42. (Russian)
(1968) The characteristics of inhomogeneous Markov processes with a finite number of states. Teor. Verojatnost. i Primenen. 13, 548—555. (Russian)
SENETA, E.
(1967) On imbedding discrete chains in continuous time. Austral. J. Statist. 9, 1—7.
SEVAST'JANOV, B. A.
(1951) The theory of random branching processes. Uspehi Mat. Nauk (N.S.) 6, no. 6 (46), 47—99. (Russian)
(1968 a) The theory of branching processes. Theory of Probability. Mathematical Statistics. Theoretical Cybernetics. 1967, pp. 5—46. Moskow Akad. Nauk SSSR. (Russian)
(1968 b) The mathematical expectations of functions of sums with random numbers of terms. Mat. Zametki 3, 387—394. (Russian)
(1968 c) Branching processes. Mat. Zametki 4, 239—251. (Russian)
(1971) Branching processes. Moscow: Nauka.
SILVERSTEIN, M. L.
(1968) Markov processes with creation of particles. Z. Wahrscheinlichkeitstheorie. 9, 235—257.
SKOROHOD, A. V.
(1964) Random processes with independent increments. Moscow: Nauka. (Russian)
(1965 a) Studies in the theory of random processes, Reading, Mass.: Addison-Wesley.
(1965 b) Constructive methods of specifying stochastic processes. Uspehi Mat. Nauk 20, 3 (123), 67—87. (Russian)
(1967) Homogeneous Markov processes without discontinuities of the second kind. Teor. Verojatnost. i Primenen. 12, 258—278. (Russian)
SKOROHOD A. V., and N. P. SLOBODENJUK
(1970) Limit theorems for random walks. Kiev: Naukova Dumka. (Russian
SMITH, G.,
(1964) Instantaneous states of Markov processes. Trans. Amer. Math. Soc. 110, 185—195.
SMITH, W. L.
(1955) Regenerative stochastic processes. Proc. Roy. Soc. Ser. A 232, 6—31.
(1958) Renewal theory and its ramifications. J. Roy. Statist. Soc. Ser. B 20, 243—302.

(1960) Remarks on the paper "Regenerative stochastic processes". Proc. Roy. Soc. Ser. A **256**, 495–501.

SOMMERFELD, A.

(1949) Partial differential equations in physics. New York: Academic Press.

SPEAKMAN, J. M. O.

(1967) Some problems relating to Markov groups. Proc. 5th Berkeley Symp. Math. Statist. Prob. Vol. II, Part 2, pp. 175–186. Berkeley: Univ. California Press.

STONE, C.

(1963) Limit theorems for random walk, birth and death processes, and diffusion processes. Illinois J. Math. **7**, 638–660.

(1965) On characteristic functions and renewal theory. Trans. Amer. Math. Soc. **120**, 327–342.

STÖRMER, H.

(1970) Semi-Markoff-Prozesse mit endlich vielen Zuständen. Theorie und Anwendungen. Berlin–Heidelberg–New York: Springer.

STROOK, D. W., and S. R. S VARADHAN

(1969) Diffusion processes with continuous coefficients, I, II. Comm. Pure Appl. Math. **22**, 345–400; 479–530.

ŠUR, M. G.

(1967) Ergodic theorems for a class of Markov processes. Teor. Verojatnost. i Primenen. **12**, 493–505. (Russian)

SWEET, A. L., and J. C. HARDIN

(1970) Solutions for some diffusion processes with two barriers. J. Appl. Probability **7**, 423–431.

SZÁSZ, D.

(1967) On the general branching process with continuous time parameter. Studia Sci. Math. Hungar. **2**, 227–247.

TAKÁCS, L.

(1960) Stochastic processes: problems and solutions. London: Methuen.

(1967) Combinatorial methods in the theory of stochastic processes. New York–London–Sydney: Wiley.

TAYLOR, S. J.

(1953) The Hausdorff α-dimensional measure of Brownian paths in n-space. Proc. Cambridge Philos. Soc. **49**, 31–39.

(1964) The exact Hausdorff measure of the sample path for planar Brownian motion. Proc. Cambridge Philos. Soc. **60**, 253–258.

(1966) Multiple points for the sample paths of the symmetric stable process. Z. Wahrscheinlichkeitstheorie **5**, 247–264.

THEDÉEN, T.

(1964) A note on the Poisson tendency in traffic distribution. Ann. Math. Statist. **35**, 1823–1824.

UENO, T.

(1967) A survey on the Markov process on the boundary of multidimensional diffusion. Proc. 5th Berkeley Symp. Math. Statist. Prob. Vol. II, Part 2, pp. 111–130. Berkeley: Univ. California Press.

(1969) A class of purely discontinuous Markov processes with interactions. I, II. Proc. Japan Acad. **45**, 348–353; 437–440.

VRANCEANU, G. G.

(1964) Geometric interpretations in the theory of Markov processes. Stud. Cerc. Mat. **15**, 15–43. (Romanian)

(1969) Interprétation géométrique des processus probabilistiques continus. Paris: Gauthier-Villars.

WATANABE, S.

(1969) On two-dimensional Markov processes with branching property. Trans. Amer. Math. Soc. **136**, 447–466.

WATTERSON, G. A.

(1962) Some theoretical aspects of diffusion theory in population genetics. Ann. Math. Statist. **33**, 939–957.

WAUGH, W. A. O'N.

(1970) Transformation of a birth process into a Poisson process. J. Roy. Statist. Soc. Ser. B. **32**, 418–431.

(1972 a) Uses of the sojourn time series for the Markovian birth process. Proc. 6th Berkeley Symp. Math. Statist. Prob. (to appear).

(1972 b) Taboo extinction, sojourn times, and asymptotic growth for the Markovian birth and death process. Ann. Math. Statist. **43** (to appear).

WEINER, H.

(1966) On age-dependent branching processes. J. Appl. Probability **3**, 383–402.

WHITTLE, P.

(1963) Stochastic processes in several dimensions. Bull. Inst. Internat. Statist. **40**, Book 2, 974–994.

WIENER, N.

(1923) Differential space. J. Math. Phys. **2**, 131–174.

WILLIAMS, D.

(1964 a) The process extended to the boundary. Z. Wahrscheinlichkeitstheorie **2**, 332–339.

(1964 b) On the construction problem for Markov chains. Z. Wahrscheinlichkeitstheorie **3**, 227–246.

(1966 a) On the construction problem for Markov chains. II. Z. Wahrscheinlichkeitstheorie **5**, 296–299.

(1966 b) A new method of approximation in Markov-chain theory and its application to some problems in the theory of random time substitution. Proc. London Math. Soc. (3) **16**, 213–240.

(1967 a) A note on the Q-matrices of Markov chains. Z. Wahrscheinlichkeitstheorie **7**, 116–121.

(1967 b) Uniform ergodicity in Markov chains. Proc. 5th Berkeley Symp. Math. Statist. Prob. Vol. II, Part 2, pp. 187–191. Berkeley: Univ. California Press.

(1969 a) Fictitious states, coupled laws and local time. Z. Wahrscheinlichkeitstheorie **11**, 288–310.

(1969 b) Markov properties of Brownian local time. Bull. Amer. Math. Soc. **75**, 1035–1036.

(1969 c) On operator semigroups and Markov groups. Z. Wahrscheinlichkeitstheorie **13**, 280–285.

(1970) Decomposing the Brownian path. Bull. Amer. Math. Soc. **76**, 871–873.

WOODROOFE, M.

(1968) On the weak convergence of stochastic processes without discontinuities of the second kind. Z. Wahrscheinlichkeitstheorie **11**, 18–25.

YACKEL, J.

(1966) Limit theorems for semi-Markov processes. Trans. Amer. Math. Soc. **123**, 402–424.

(1968) A random time change relating semi-Markov and Markov processes. Ann. Math. Statist. **39**, 358–364.

YANG, CHAO-QUN.

(1965) A class of birth and death processes. Chinese Math.—Acta **6**, 305–329.

YOSIDA, K.

(1965) Functional analysis. Berlin—Göttingen—Heidelberg: Springer.

Notation index

N^* $= \{1, 2, \ldots, n, \ldots\}$
N $= \{0, 1, 2, \ldots, n, \ldots\}$
$-N = \{\ldots, -n, \ldots, -2, -1, 0\}$
Z $= (-N) \cup N^*$
R = the set of all real numbers
T = $[0, a)$ with either $a > 0$ or $a = \infty$
$\mathbf{A}'$ = the transpose of the matrix $\mathbf{A}$
E = expectation
D = variance
$\circ$ = composition
$x \to y+$ = x approaches y from the right
$x \to y-$ = x approaches y from the left
Re z = real part of the complex number z
i = the imaginary unit $\sqrt{-1}$
a.s. = almost surely
i.o. = infinitely often
iff = if and only if
$\diamond$ = end of a proof

Subject index

The contents should be also consulted for subject matter

Author index